INSTRUMENT TRANSFORMERS

Standards, Practices, Testing Procedures and Data on Electrical Systems

INSTRUMENT TRANSFORMERS

Standards, Practices, Testing Procedures and Data on Electrical Systems

P. K. PATTANAIK

Deputy Manager (Electrical)
Equipment and Meter Relay Division
Orissa Power Transmission Corporation Ltd.
Burla, Sambalpur
Orissa

MJP Publishers

MJP PUBLISHERS
New No. 5, Muthu Kalathy Street
Triplicane
Chennai 600 005

This book has been published in good faith that the work of the author is original. All efforts have been taken to make the material error-free. However, the author and publisher disclaim responsibility for any inadvertent errors.

PREFACE

In practice, voltage and current in the order of "kilo or mega" range cannot be measured directly to study their performance or to analyse their quality. The use of converters plays a vital role in bringing down the values to the allowable working limit for smooth, safe and easy control and operations. These converter are called instrument transformers. The unit that is used for current conversion is called the current transformer and that used for voltage transformation is called voltage transformer or potential transformer (PT). These transformers are widely used for power system network, but engineering students seldom study "Transformer" as a subject.

In this book an attempt has been made to discuss the details of instrument transformers and the critical issues like its application, improved technology, etc. The purpose of this book is to serve as a guide to solve the technical problems on Instrument Transformers, which electrical engineers face during their field works. This book discusses the important aspects of Instrument Transformers (current and potential transformers and capacitive voltage transformers with technical data of the electrical systems), which include the following.

1. Basic construction of the equipment
2. Basic working principle
3. Detailed installation procedure
4. Data on guranteed technical particulars
5. Testing procedures along with monitoring practices
6. Practical maintenance practices
7. Special technical case studies
8. Comparison of the testing and monitoring values with the threshold values.

I gratefully acknowledge the suggestions from all my friends, colleagues and relatives, who have extended their helping hands for preparing this book.

Errors might have crept in despite utmost care taken during the preparation of this book. The author shall be grateful if these are pointed out along with other suggestions.

P. K. Pattanaik

CONTENTS

1

INTRODUCTION

Power is the critical resource for economic development in order to spur economic growth in a liberised environment for any developing country like India. It has to be accepted that the electrical power is the prime mover for the utilization of other powers for the uptrend growth in industrialization, urbanization, economic status, etc.

Though the availability of electrical power is the basic need, only a reasonable blend of quality and reliability can guide the nation or the individual to think in the direction of the road to progress. The poor and erratic quality of power supply adds to the cost of goods and services, thereby affecting the international cost-competitiveness. Poor power quality affects reliable operation of electrical equipment. So reliability also depends upon the power quality.

Power quality, although non-measurable in conventional way, means deviation of power supply from steady, normal, 50 Hz voltage of sinusoidal waveform. The key terms like harmonics, flickers, notch, voltage sag (dip) and voltage swell (surges), distortion, imbalance, etc. are the critical factors to decide the quality of power. The impact of its deviation on electrical installation and equipment may vary from case to case, but is harmful in nature. One may wish or not "quality and reliability" is bound to be the catchword of the future.

To understand the term "Electrical Power", one has to think about the fundamental parameters on which electrical power is defined. It is defined as the product of voltage and current. So quality power is well understood that the parameters like voltage and current must be qualitative to decide its effect on actual practical field. To quantify the electrical power and to study the subjective nature of this important term, the parameters like voltage and current are to be analysed properly. This issue of power quality is one of the important terms for the field of electrical technology. Besides, the other factors like control, operation and measurement of electrical power are also equally important to this technology.

In practice, the voltage and current to the tune of "Kilo or Mega" range cannot be measured directly to study its performance or to analyse its quality. The use of converters plays the vital role to bring down the values to the allowable working limit for smooth, safety and easy control and operations. These converters are called instrument transformers. The unit that is used for current conversion is called current transformer and that used for voltage transformation is called voltage transformer or potential transformer (PT).

These transformers are widely used for power system network, but the study of the same are rarely discussed among engineers. In this book, an attempt has been made to discuss the details of instrument transformers and the critical issues like its application, improved technology, etc.

CONCEPTS IN ELECTRICAL ENGINEERING

Progress in science is a never-ending process and the electrical science in particular is always involved with the parameters like voltage, current, charge, etc., that do not have any visible nature. So, the concepts based upon these parameters are imaginative in nature. However rules, laws, hypothecations are always to be referred to for development of any new ideas related to electrical technology. Some of the fundamental laws/rules like Ohm's law, Faraday's laws of electromagnetic induction, Gauss' law, Maxwell's theorem, Norton's theorem, Thevenin's theorem, etc. are always referred for the fundamental discussion related to any advanced topics on electrical technology. The electrical engineering and the concepts related to the electrical system can be understood by a special chart called skeleton chart which is described as follows.

Skeleton Chart

Skeleton is the assembly of bones and gives the framework of the physical structure and the appearance of any creature. From the skeleton, the idea regarding the strength, vigour and gesture of the creature can be guessed. The skeleton chart described here is the combination of the basic electrical terms that can guide an electrical engineer to enter into the ocean of electrical technology and electrical engineering. So the engineers in the field of electrical technology should know the basic terms associated with this skeleton chart (Figure 1.1).

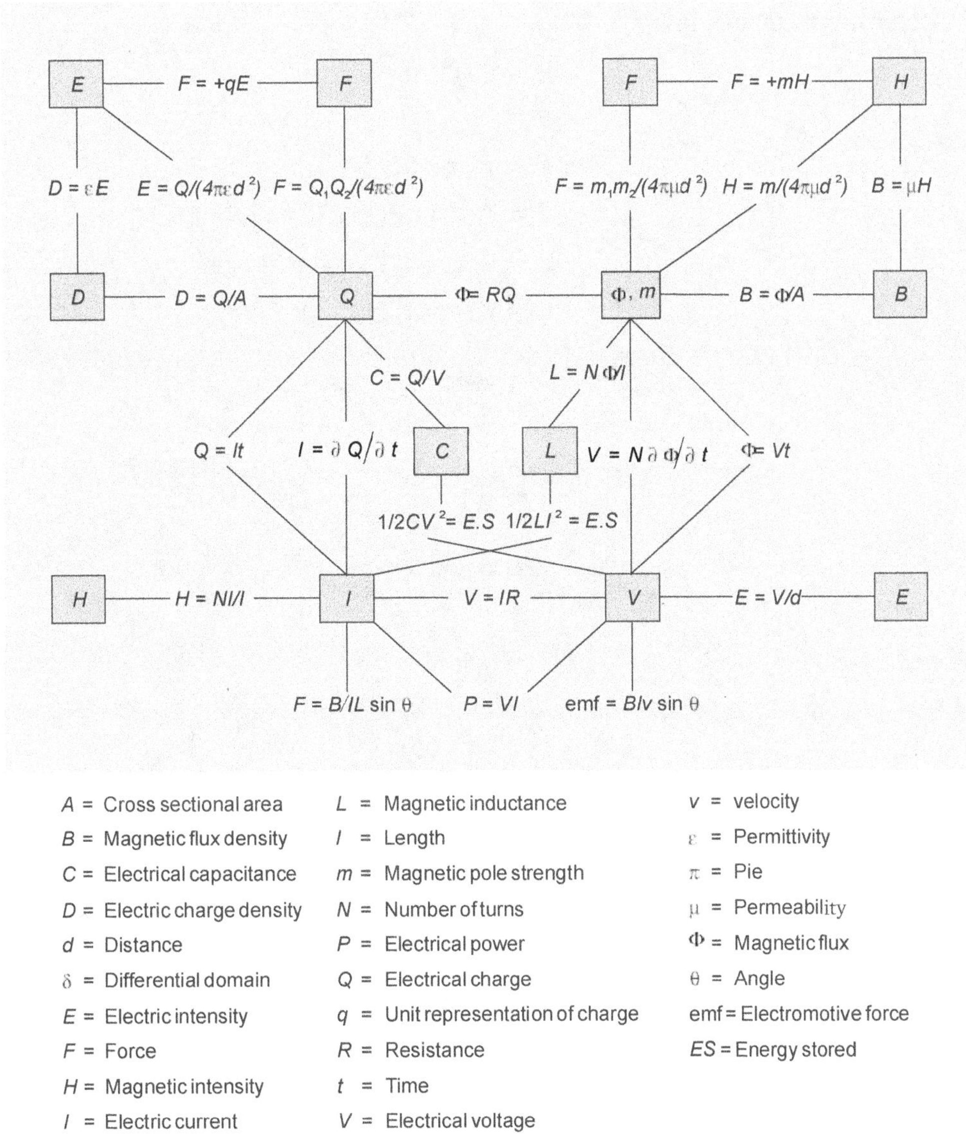

A = Cross sectional area	*L* = Magnetic inductance	*v* = velocity
B = Magnetic flux density	*l* = Length	ε = Permittivity
C = Electrical capacitance	*m* = Magnetic pole strength	π = Pie
D = Electric charge density	*N* = Number of turns	μ = Permeability
d = Distance	*P* = Electrical power	Φ = Magnetic flux
δ = Differential domain	*Q* = Electrical charge	θ = Angle
E = Electric intensity	*q* = Unit representation of charge	emf = Electromotive force
F = Force	*R* = Resistance	*ES* = Energy stored
H = Magnetic intensity	*t* = Time	
I = Electric current	*V* = Electrical voltage	

Figure 1.1   Skeleton chart

Applications of skeleton chart   The formulae that are described in the skeleton chart can be utilized for obtaining the other critical formulae and rules. The basics of the electrical science and the related terms in it can guide the engineer to derive the objectives of any critical theories. Some of the derived applications have been described here for ready references.

Derivation of electrical power

According to Ohm's law

$$R = V/I$$
$$P = VI$$

So,
$$P = VI = (IR)\,I = I^2 R = V^2/R$$
$$I = P/V = \sqrt{(P/R)} = V/R$$
$$V = P/I = \sqrt{(P/R)} = IR$$
$$R = V/I = P/I^2 = V^2/P$$

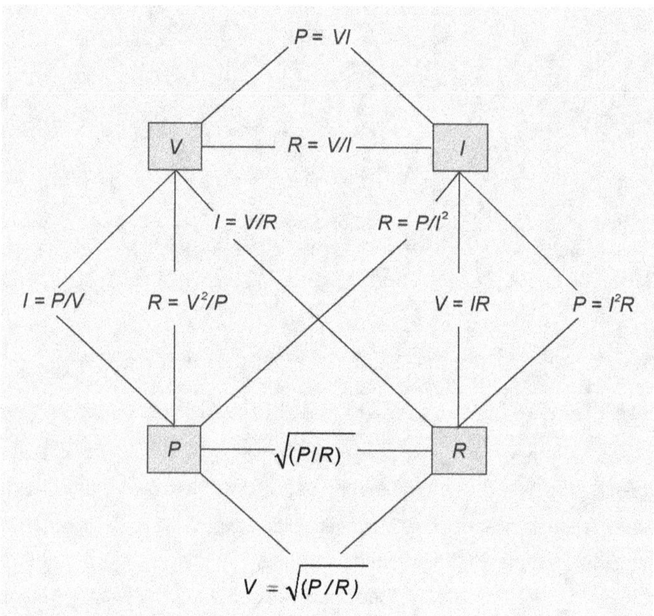

Derivation for capacitance for parallel plates

According to the capacitance fundamental

$$C = Q/V = DA/V = e\,EA/Ed = e\,A/d$$

So capacitance for parallel capacitor $= e\,A/d$

Similarly other fundamental applications can be derived from this skeleton chart.

POWER SCENARIO IN INDIA

Power system, generation, transmission and distribution are the important terms for the electrical world and integration of these systems is the vital link for the growth of the electrical system. A developing country like India, at present, is undergoing a thorough critical change and approaching the stage of the slogan "Power for all by 2012". The journey of electrical power system from its inception till date could be the path-finder index for further growth.

1. First power plant of 130 kW (Hydel) at Sidropong in the year 1897 (10.11.1897).

2. Total installed capacity by 1900 was 1.2 MW (Thermal 1 MW, Hydel 0.2 MW).

3. By independence in 1947, the generation capacity grew up to 1363 MW. In these initial 50 years (1897–1947), the generation and distribution of electricity was achieved by the efforts of the private utilities. Along with hydel plants, coal-based thermal plants were also set-up gradually.

4. Indian Electricity Act, 1910 was then enacted and was meant for the license holders, particularly for the private companies.

5. Electricity Supply Act, 1948, aimed at the rationalization of generation and distribution of electricity in India and formed the different SEBs (State Electricity Boards) to regulate the distribution system in proper manner.

6. Industrial policy resolution, 1956, encouraged different utilities and emphasized the use of electricity for the major activities. Even the SEBs were empowered with the setting up of power generating stations particularly hydel and thermal stations.

7. During the year 1964, the country was demarcated into five regional boards, REB (Regional Electricity Boards). This formation of REBs was to monitor and regulate the available resources in proper manner.

8. During 1975, to meet the increasing demand of electricity, Central Sector generating agencies were like NHPC, NTPC, NPC were formed.

9. After the formation of the regenerating agencies, the practical problem of electrical generation was partially solved. But the transmission area was in neglected condition. So, considering the issues of this transmission problem, during the year 1989, the Central Agency named PGCIL (Power Grid Corporation of India Limited) was framed. This agency was hived off from NTPC and exclusively assigned for the transmission activity and to develop the inter-grid network across the country.

10. The other agencies like REC (Rural Electricity Corporation), PFC (Power Finance Corporation) were also formed. REC was set up to assist the SEBs for the development of electrification work in rural areas. PFC was set up to assist the funding of different state projects.

11. During the year 1991, the power sector reforms were initiated in India. Different SEBs were coupled together to share the individual responsibilities like separate generation, transmission and distribution activities.

12. Even during the year 1991, the Economic Act was enacted with the provision of extending the funds to the private players for different electrical activities.

13. During the year 1998, Electricity Regulatory Commission Act was framed with origination of State Electricity Regulatory Commissions and Central Electricity Regulatory Commissions. The functions of commissions are to control

the electricity tariff, to develop transparent policies regarding subsidies and to promote efficient practices in interested states.

14.   Further during the year 2001, the concept like ABT (Availability-Based Tariff) was introduced. This method helped to develop the grid discipline and a healthy commercial mechanism. Simultaneously, the utilities became conscious for the generation of quality power.

15.   In the year 2001, the Electricity Conservation Act was enacted for the development of awareness to conservation of energy.

16.   Indian Electricity Act, 2003, was the next milestone. The objective of this act was to consolidate the laws relating to all electrical activities like generation, transmission, distribution, trading and utilization of electricity, etc.

17.   During February 2005, the National Electricity Policy was enacted. The prime initiative of this act is to unite all the laws and rules to form the central policy.

18.   During January 2006, the next concept on the basis of uniformity in electricity tariff, National Tariff Policy was initiated.

19.   During August 2006, the Integrated Energy Policy came into force to bring uniformity of different energy resources.

20.   During August 2007, National Electricity Plan was enacted to bring the updated schemes into action.

21.   Finally 12th plan is about to be circulated to bring the electrical system rather in a disciplined manner for all the stakeholders, utilities and private players.

The track record of the power sector and the steps yielded out of it are remarkable. The generation of this electrical power is lower than the rate of consumption of electrical energy due to which there is an ever increasing demand of electrical energy. In India, the population, the problems of political interference and some other factors as discussed below are the obstacles of the correct development of the electrical sector.

Factors of obstacles

1.   Power is the critical infrastructure for the economical growth. The economic acceleration would greatly depend upon the commercially viable power sector, the field in which investment could be done freely for power development. But the rules, methodology and the principles do not allow the investors for the comfortable investment in power sector.

2.   As per the Electricity Act of the year 1948, the tariff policy selection was on the basis of minimum return of 3–4% of the total investment. The state electrical utilities suffered with the shortfall and the basic reasons are

   i.   The greatest threat of power sector was the T and D losses, which was in the order of 30–40% (10–20% technical loss and 20–30% of commercial loss).

   ii.   The interference of political issues, particularly provisions of subsidies to many social sectors was one of the factors for incurring loss to the sector. The calculation reveals that the subsidies were about 30–40% of the generated capacity.

iii. The stringent policy of not proving the flexibility for the power sector was also the prime reason for the loss. The control was with the government and implication of any methodology was based on bureaucracy.

All India Transformation, T&D and AT&C losses

Table 1.1 shows the transmission loss.

Table 1.1 Transmission loss

Year	T & D loss	AT & C loss[*]
2002–03	32.54	32.54
2003–04	32.53	34.78
2004–05	31.25	34.33
2005–06	30.42	34.54
2006–07[#]	28.61	32.07
2007–08[$]	26.91	NA

Source CEA

*—PFC Figure, #—Provisional, $—Estimated

As per the IEP (Integrated Energy Policy) (Table 1.2), the GDP should be at the rate of 9% and the installed capacity should be of 215,000 MW with additional capacity required of 75,000 MW, calling the investment of about 500,000 crores for the 11th plan (2011–12). Similarly for the transmission and other infrastructure, the amount would be to the range of 10 lakhs crores by the end of 2012. Such huge amount could only be attended if the private investors would be called for the active participation to the power sector reform.

Table 1.2 Integrated energy policy

Plan	8% GDP			9% GDP		
	Capacity addition (MW)	Installed capacity by the end of plan (MW)	Billion unit energy consumption	Capacity addition (MW)	Installed capacity by the end of plan (MW)	Billion unit energy consumption
11th (2011–12)	66,000	2,06,000	1029	75,000	2,15,000	1077
12th (2012–17)	97,000	3,03,000	1511	1,16,000	3,31,000	1657
13th (2019–22)	1,42,000	4,45,000	2221	1,79,000	5,10,000	2550

3. The PLF of the power station was also another factor, which resulted in the losses to the sector. The Figures are not encouraging as shown in Table 1.3.

Table 1.3 Thermal PF as per CEA report

2001–02	02–03	03–04	04–05	05–06	06–07	07–08	08–09 Up to October 08
69.9	72.2	72.7	74.8	73.6	76.8	78.6	74.31

4. The per capita consumption is also the factor for deciding the necessity of reforms.

Table 1.4 Per capita consumption

Year	Per capita energy consumption (kWh)
1980–81	176
1990–91	348
2000–01	559
2002–03	567
2003–04	592
2004–05	613
2005–06	632
2006–07	672
2007–08	704
2011–12	1000
2016–17	1365
2021–22	1772
2031–32	2500

The per capita consumption (Table 1.4) has been calculated on the basis of UN method (Gross Electrical Energy Availability/Population).

The per capita consumption rate in India is very less when compared to the world average. The per capita consumption of developing country like China is also nearly two times the consumption in India. So methodology, policy and schemes in India should be developed in such a way that the energy consumption could be raised progressively. Because of these factors the power reforms in India were attempted.

POWER SECTOR REFORM

During the 90s the situation of power industry was very grim and tendering to the situation of heavy loss to the power sector. The prime causes were the tariff mechanism, which was not uniform and the SEBs adopting their own method of tariff rate, not considering the real-time problems. The losses to utilities were accumulating and the situation was becoming such that the compensation to the same was going beyond control. Moreover power in broad sense can be stored and hence generation and load consumption in real-time base should have to be balanced for the optimal use of the same. So reform was becoming the only solution to overcome the described situation.

Power sector reforms were basically understood with the concept of deregulation, decentralization and unbundling of electrical utilities from the vertical integrated monopolies into separate generation, transmission and distribution bodies. The main objectives were to initiate and encourage the private players to take part in the power generation function. So proactive efforts were initiated in the beginning of 90s and private utilities were invited to invest in the field of generation, transmission and distribution.

During 1991, GOI (Government of India) allowed some of the private bodies into the power sector, guaranteeing 16% profit, which in turn attracted the various national and international utilities to the power business. The payment to them was within the control of respective state government with counter-guarantee of the central government. During the initial days, the entry was very encouraging and many utilities entered the generation wing. But the biggest problem was that the private bodies had no role in collection of revenues, while the distribution sector actually collected the revenues from the consumers. So in due course, the private players started showing less interest on the investment and, planned to withdraw from the business slowly. Considering the problems and the situation, the programmes like APDRP (Accelerated Power Development and Reform Programme) and the Electricity Act 2003, were enacted. The other programmes like ABT (Availability-Based Tariff), OPEN ACCESS, UI (Unscheduled Interface) were planned to be implemented.

Difficulties During the Initial Stages of Reform Act

1. Coordination among the utilities became critical due to the development of many number of companies with different modalities.
2. During the initial days, the state utilities did not understand the concept of ABT and open access, although open access was implemented in the central transmission utilities (CTU).
3. With the increase in number of companies, the implementation of GRID CODE becomes difficult and in practice the coordination of the relays and other hardware also become difficult to attain.
4. Central Government formed MoU and monitored the CTU, for the system improvement. But the state government does not show interest for the positive side of the power sector reform.
5. Realization of tariff and wheeling charges by the DISCOMs become difficult to implement.

Power Sector Reforms in Different States

1. *Andhra Pradesh*

a) Andhra Pradesh State Electricity Board (APSEB) was unbundled with effect from 1st February 1999.
b) APGENCO and APTRANSCO originated from unbundling.

c) From 3rd April 1999, Andhra Pradesh Electricity Regulatory Commission became operational.

d) World Bank and DFID loans were sanctioned for the development of the restructured companies.

2. *Haryana*

a) SEB was unbundled into HVPNL (Haryana Vidyut Prasarana Nigam Ltd.) and HPCL (Haryana Power Corporation Ltd.) on 14th August 1998.

b) Distribution companies were also established as UHBVNC (Uttar Haryana Bijli Vitaran Nigam Ltd) and DHBVNL(Dakshin Haryana Bijli Vitaran Nigam Ltd.)

c) World Bank, adaptable programme loans and DFID loans were also sanctioned for the development of the restructured format.

d) SERC (State Electricity Regulatory Commission) came into force from 17th August 1998.

3. *Karnataka*

a) Karnataka Power Transmission Corporation Ltd. (KPTCL) and Visvesvaraya Vidyut Nigam Ltd. (VVNL) were the two companies that came into force from 1st August 1999.

b) Five different regional groups were established under KPTCL to look into the regional electrical business.

c) Regulatory commission came into effect from 15th November 1999.

d) Other restructured programmes like origin of distribution sectors, privatization issues and other related methodologies were taken with GOI (Govt. of India) for proper implementation.

4. *Madhya Pradesh*

a) During the initial days of the reform of State Electricity Board (SEB), it was proposed for the development of three power generation companies, one power trading company and nine power distribution companies.

b) The main aim of the reform was to reduce transmission and distribution losses, energy audit and improvisation of quality power.

c) For reform structure World Bank funds, help from central government was extended in phased manner.

5. *Orissa*

a) Orissa was the first state undertaken for reform programme.

b) In June 1995, the process of division into generation, transmission and distribution companies was studied by different consultants.

c) In January1996, OSEB(Orissa State Electricity Board) was split into three different entities like OPGC(Orissa Power Generation Companies), OHPC (Orissa Hydropower Corporation) and Gridco(Grid Corporation of Orissa).

d) The transmission and distribution sectors were with Gridco for few days.

e) Then during November 1997 Gridco got divided into two groups as Gridco and Distco. Four regional distribution companies were formed under Distco.

f) These companies are WESCO, NESCO, SOUTHCO, and CESCO.

g) Gridco, the parent which was looking into the transmission and electrical trading business, again divided into two groups, as Gridco and OPTCL (Orissa Power Transmission Corporation Ltd.) in the year 2004.

h) This time the trading part was taken care by Gridco and transmission part was taken care by OPTCL.

i) Orissa state Electricity Regulatory Commission (OERC) was constituted in the year July 1996.

6. *Rajasthan*

a) During January 2000, State Reforms Act was notified.

b) SERC was also planned to be constituted during January 2000.

c) Reform programmes were initiated by state government and State Electricity Board.

7. *Uttar Pradesh*

a) During January 2000, the SERC (State Electricity Regulatory Commission) became functional.

b) UPSEB (UP State Electricity Board) was divided for generation, transmission and distribution activities.

c) Generation part was handled by the entities like:
   - UPRVUNL (UttarPradesh Rajya Vidyut Utpadan Nigam Ltd.) and
   - UPJVNL (UttarPradesh Jal Vidyut Nigam Ltd.)

d) Transmission and distribution functions were controlled by the company called UPPCL (Uttar Pradesh Power Corporation Ltd.)

e) Different banking sectors and financial institutions came forward to help the reform programmes in the states. World Bank, DFID loans were also sanctioned for these programmes.

8. *West Bengal*

During March 1999, the SERC became functional and re-organized the WBSEB (West Bengal State Electricity Board) to the transmission to different companies.

9. *Other States*

Other states like Arunachal Pradesh, Assam, Bihar, Chandigarh, Delhi, Goa, Gujarat, Himachal Pradesh, Jammu and Kashmir, Kerala, Maharashtra, Manipur, Meghalaya, Mizoram, Punjab, Tamil Nadu, Tripura have also entered into the frame of reform structure. Some of the above states have already adopted the reform acts.

Impact of Reform

The major consequence of restructuring is the emergence of separate entities for generation, transmission and distribution. So, competition among the companies increases to provide customer a quality product at a competitive price. The awareness regarding the use of electricity among the consumers increased. People could know the tariff structure, rules, regulations and policy structures of the electricity reform. Overall the reform in the electricity brought the positive guidelines for the electrical industries.

Accelerated Power Development and Reform Programme (APDRP)

During 2000–01, the SEBs were in a great trouble. The financial position was a great concern and losses had reached to an alarming level of 26,500 crores, which was equivalent to about 1.5% of GDP. So the programme called APDP (Accelerated Power Development Programme) was introduced in the year 2000–01 with a view to restore the commercial viability of the distribution companies. In due course, this programme was converted to APDRP (Accelerated Power Development and Reform Programme) in the year 2002–03. The reasons for conversion are described in "Objectives of APDRP."

Objectives of APDRP

1. Improving financial viability of the power utilities
2. Reduction of AT and C losses
3. Developing reliable and quality power supply
4. Improving consumers' satisfaction by adopting new technology like computer application for energy audit and energy management issues.

Electricity Act, 2003

Before Electricity Act, 2003, came into effect, there were three major acts concerning the electricity sector. They were the Indian Electricity Act, 1910, the Electricity Supply Act 1948, and the Electricity Regulatory Commission Act, 1948. All the three acts described have their own limitations and on review it was decided to be replaced by the new act (Electricity Act, 2003). So this act was introduced on June 10, 2003 after sufficient technical and political discussions. The objective, as printed is as follows:

"An act to consolidate the laws relating to generation, transmission, distribution, trading and use of electricity and generally for taking measures conducive to the development of electricity industry, promoting competition therein, protecting interest of consumers and supply of electricity to all areas, rationalization of electricity tariff, ensuring transparent policies regarding subsidies, promotion of efficient and environmentally benign policies constitution of Central Electricity Authority, Regulatory Commissions and establishment of Appellate Tribunal and for matters connected therein or incidental thereto."

The practical reform objectives can also be described in a summarized manner as follows:

1. Provision of electricity at the best and affordable cost.
2. Cheapest power to the farmers to enhance the food security to the country.
3. Develop competitiveness to design the best in different sectors like generation, transmission and distribution.
4. Accessibity of electricity to backward classes and the concerned areas.
5. Power to all and all areas by the concept of open access.

Salient features of Electricity Act 2003

The central government prepared a National Electricity Policy in consultation with state sectors. Some of the salient features of the policy are as follows:

1. Completion of rural electrification by the formation of different sectors like Panchayat Samiti, Cooperative Societies and franchisees groups.
2. Except for hydro projects, all other generation projects have been cancelled of their license.
3. Open access to transmission and distribution system to be provided to all type of consumers, with some financial modalities.
4. Provision for private licensees in transmission and distribution system.
5. De-regulation of power trading was introduced and extended to all the transmission utilities with the concept of open access.
6. Appellate tribunals were formed to hear the appeals against the decisions of CERC and SERCs.
7. Provision for developing rules and regulations regarding the safeguard of public interest for the use of electrical power.

So, the Electricity Act, 2003 brings together all the regulations of generation, transmission, distribution, trading and use of electricity. Liberalization of generation, transmission and distribution and implication of penal action to the theft and defaulters for the power consumers were the main features of the act. Captive plants were developed and consumers enjoyed the concepts of open access and purchased power directly from the generation and transmission utilities. Non-government, local bodies and private concerns were able to enter the distribution sector for the power business and developed the competition in the sectors. So finally the consumers got the flexibility of using the electricity. The process of restructuring and reforming are still going on for obtaining better result for the consumers and different clients. The reform processes of different states are described here for ready reference of the data.

Availability-Based Tariff (ABT)

Methodology of energy accounting on the basis of available frequency of the system is defined as availability-based tariff (ABT).

Earlier tariff mechanism was not considered for the frequency of the system. The regional grids had been operating in a very undisciplined and haphazard manner. The utilities neither were getting encouraged nor penalized for the frequency variation in the system. It was profitable for the generators to generate even when the consumer demand had come down. So after long discussions, this concept of ABT originated. CERC notified ABT in the year 2000. But implementation to different regional level came into force in later stage (Table 1.5).

Table 1.5   ABT implementation

Sl. No.	Region	Date
1.	Western Region	01.07.2002
2.	Northern Region	01.12.2002
3.	Southern Region	01.01.2003
4.	Eastern Region	01.04.2003
5.	North-Eastern Region	01.11.2003

ABT has three components.

1. **Capacity charge (CC)**   It is the fixed charge considered on the basis of plant capacity in MW. The practical elements of the CC are as follows:

    i.   Interest on capital borrowed
    ii. Certain O and M charges
    iii. Salaries of higher officials
    iv. Depreciations of the equipments
    v.   Return on equity shares
    vi. Other fixed charges

But full fixed charges are recoverable for the cumulative target availability of 80% or more of the capacity. For availability of less than the target, the fixed charges would be reduced on pro rata.

2. **Energy charges (EC)**   It is the variable cost and is considered on the basis of operational charges like cost of fuels, lubricants, running accessories and consumable products.

The energy charges payable by the DISCOMs are variable charges as well as fuel price adjustment (FPA) charges. The energy charges are payable for the ex-power-plant drawal schedules issued by RLDC. The variable charges and the working FPA for state generators as notified by the CERC is the rate of energy charge × ex-bus scheduled generation.

3. Unscheduled interchange (UI) charges UI charges are considered on the basis of frequency and accounted for deviation in actual generation/drawal and scheduled generation/drawal. Frequency is the vital tool for the load-generation balance and indicates the index of UI rate. Low frequency indicates availability of less generation and high frequency for high generation to meet the demand on the system. So the UI rate is high when frequency is low and vice versa. The generators and beneficiaries, both become conscious of the generation and utilization of the electrical energy. Both the utilities adjust their operating practices to keep the allowed frequency within the band of 49.0 to 50.5 Hz.

UI is accounted for the 15 minute time block. The following table (Table 1.6) was effective from 01.04.2009.

Table 1.6 UI rates

Sl. No.	Average frequency	UI Rate
1.	50.3 Hz and above	0
2.	50.3–49.50 Hz	12.0 paise linear in 0.02 Hz
3.	49.5–49.2 Hz	17.0 paise linear in 0.02 Hz
4.	50.3–50.28 Hz	12.0 paise
5.	49.5–49.48 Hz	497 paise
6.	49.22–49.2 Hz	735 paise

Advantages of ABT

1. **Improvement in frequency profile** Due to awareness the frequency profile of the utilities increased and regional grids maintain the operating zone of 49 to 50.5 Hz.
2. **Improvement of the grid discipline** The grids associated in the network for transmission and distribution of energy, get accustomed to the discipline of the rules and regulation and maintain the proper method of the supply of electrical energy.
3. **Improvement in voltage profile** The reactive power pricing scheme along with ABT implementation has resulted into improvement in voltage profile in all regions.
4. **Trading** This mechanism provides a fair and transparent settlement system as well as real-time balancing market price (15 minutes time block basis) and healthy platform for trading bilaterally.
5. **Increase of inter-regional power links** The inter-regional power exchanges increased due to ABT mechanism. Power could be availed by deficit region from the power surplus region. The available surplus generation capacity could be transmitted by utilizing the margins in the existing transmission system. SEBs and DISCOMs put their seasonal demand in advance and balance their demand-supply position on day-to-day basis through day-ahead and same-day transactions subject to technical feasibility.

6. **The installation of generating plant** Flexibility in installation of generating units, the installation of power generating plants could be possible at the suitable position as per the availability of resources and accordingly the power could be transmitted from one region to the other. The cost of transmission is much less than the cost of transportation of the raw materials for generation.

7. **Suitable utilization of power plants** The generators in the plant could be operated as per the demand of the power. Based upon the fixed energy charges, the costlier generators were replaced by the cheaper ones as per the frequency-linked dispatch guidelines.

8. **Reliability** The reliability of the system increased due to proper discipline and regular transaction of the power among the region. Commissioning of SLDC (State Load Dispatch Centre), RLDC (Regional Load Dispatch Centre) also helped to control the power flow smoothly to different load centres.

Open Access System

Open access means the non-discriminatory provision for the use of transmission lines or distribution system or associated facilities with such lines or system by any licensee or consumer or a person engaged in generation, in accordance with the regulations specified by the appropriate commission. For all transactions of power either of long-term or short-term deal, the utilities have to take the help of open access through transmission or distribution network. Both CERC (Central Electricity Regulatory Commission) and SERC (State Electricity Regulatory Commission) together prescribed the Open Access Regulation for the inter-state and intra-state transaction of power. CERC had come out with the Inter-State Transmission Open Access Regulation in the year 2004 and successive revision in the year 2005. The process is still continuing for the evolvement of better solution for the open access.

The main problem in the de-regulated operation is the problem of transmission congestion. Sometimes the load and power generation does not match each other due to the sudden setting of the industries and delay in the transmission corridors for various reasons like forest clearance, funding to the projects, etc. The temporary transmission congestion may be caused due to many of the reasons like line outage due to system disturbance, generation problems due to abnormal rain or scarcity of fuels. The load pattern on the system also varies from day-to-day which results in the congestion of the distribution network. These problems disturb the concepts of open access. But certain provisions have been made in the open access to deal with these problems in the real time. But the actual objectives of the open access are as follows:

1. Promote competition in generation and supply
2. Facilitate consumer's choice to select supplier
3. Reduce costs (charges)
4. Protect distribution licensee's viability through surcharge.

CONCLUSION

Indian power sector had already been driven by different rules, acts and models, but the basic concept was to operate on the basis of subsidy regimes. So the burden and responsibility was inclined to the government and in turn the models were not so successful. Moreover these polices were framed during the time, when power sector was running with loss and the quality of power and availability was very poor. Introduction of Electricity Act, 2003 was the turning point in the history of power sector reform which revolutionized the methods of setting proper financial exchange for the actual and quality power. Frequency-linked availability-based tariff (ABT) also yielded good result for the improvement of grid discipline and better mechanism for the power exchange for intra-state and inter-state consumers. The framing of CERC and SERC also provides a platform for the control of power flow in different de-regulated regimes. They have taken various positive steps for the finalization of IEGC (Indian Electricity Grid Code), ABT, Open access mechanism, etc. In conclusion all the utilities, involved with power business should extend their positive thoughts for achieving the goal—"power for all by 2012".

2

STUDIES ON CURRENT TRANSFORMER

INTRODUCTION

Current Transformer is the important interface between the high level of power system and low level of protection, measurement and control circuit in terms of current. Wherever the values of current become too high, this instrument transformer (CT) is used in the system to produce a proportional value for a scaled-down replica to the secondary working circuit of a system.

Current transformer is used with its primary winding connected in series with the actual line current flow of the power system. The primary winding consists of a bar conductor or a conductor with a very few turns and causes no appreciable voltage drop across the winding. The secondary winding has large number of turns, the exact number being decided by the turns ratio of the CT. The instruments/equipments like indicating meters, relays, etc. are connected on the secondary windings, which have very low impedance circuit. So the secondary windings are regarded as a circuit that works nearly with short circuit condition.

The performance of the measuring transformers (CT and PT) during and following large instantaneous changes in the input quantity are to be considered seriously. The response of the electrical parameters of these transformers upon the secondary circuit should be well within the satisfactory limit for both under steady state and transient condition. For measuring and slow speed recording application, only the steady state accuracy is relevant, whereas for high speed protection and other application, accuracy under transient condition is also important.

The cores in the secondary circuit of the CT are designed according to the requirement of the secondary circuit. These are of three types

1. Measuring current transformers (Metering core)
2. Protective current transformers (Protection core)
3. Protective current transformers for special purpose of application (Protection core PS class)

Typical specification of a CT with these cores have been explained in Table 2.1.

Table 2.1 Typical specification of CT

Type of core	Metering	Protection special type	Protection type
Output (VA)	20 to 40	–	20 to 40
Accuracy class	0.5 Fs<5	PS	5P
$V_k (V)_{min.}$	–	600 to1200	–
$I_e @ V_k$ (mA) max.	–	10 to 25	–
R_{CT} at 75° C max*.	–	2.5 to 5	2.5 to 5

Fs—Safety factor-number of times normal current up to which accuracy is maintained.
max*—Resistance of CT differs as per the temperature on it.
The maximum temperature for reference is selected as 75°C and corresponding value of resistance is to be mentioned as the specification reference.

BASIC CONSTRUCTION

The construction of current transformer (CT) depends upon the arrangement of the primary winding, secondary winding and type of insulation used between them. Sometimes typical application of the CT also decides the change of the physical construction of the current transformer.

But from the basic construction of CT, it can be described here that this instrument has certain primary winding, which generally consists of a bar conductor or a conductor with a very few number of turns. The CT has also secondary winding wound over the primary winding and has large number of turns as compared to primary winding.

CLASSIFICATION OF CURRENT TRANSFORMERS

CLASSIFICATION BASED ON WINDING ARRANGEMENT

The use of windings and the arrangement of such windings decide the classification of current transformer. The basic arrangement of primary winding upon the core is decided as the main factor for the construction practice of the current transformer.

Wound Type Current Transformer

For this type of transformer (Figure 2.1), the primary winding of more than one turn are wound upon the core and provides the primary section of the transformer. Such primary winding is so designed that it could be able to carry the short-circuit current of the system. According to the design and turn-ratio requirement, the secondary winding is accordingly wound to provide the required current on the secondary circuit of the system.

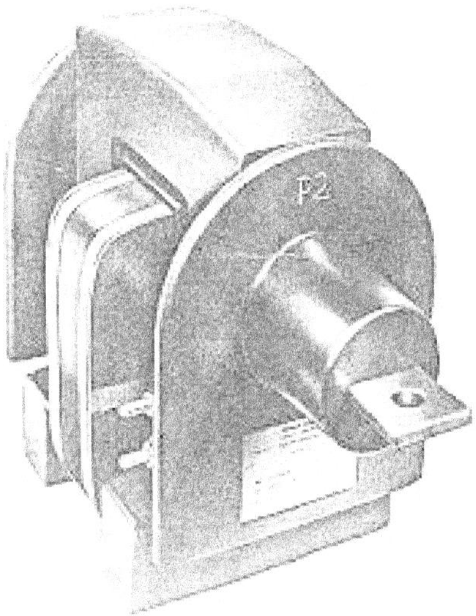

Figure 2.1   Wound type CT

Bar Type Current Transformer

For such current transformer (Figure 2.2), the primary winding consists of a bar conductor of suitable size and passes from one side of the CT to the other side. This conductor becomes an integral part of the current transformer. The primary winding is regarded as a conductor of a single primary turn and the secondary winding in the form of a toroid, occupies over the primary conductor, with proper insulation between primary and secondary winding.

Figure 2.2   Bar type CT

Bushing Type Current Transformer

This type of current transformer is similar to the bar type transformer. But in such type of CT neither any primary winding nor any insulation for such winding is provided.

Normally the bushing of the circuit breaker or the power transformer is used as a centrally placed primary winding for such CT. The necessary winding and the insulation provided over the winding, plays the role of primary part of the CT. Moreover the secondary winding being wound over the bushing primary provides necessary secondary current to the circuit.

Core-Balance Current Transformer

This category of CT is also similar to the bushing type and bar type CT. The core-balance CT is normally of the ring type, through the centre of which the three-core cable or three single core cables of three phases system is passed. The basic use of such CT is to monitor the residual current in the system. The sensitive earth fault relay is used on the secondary of this CT to operate for the unbalanced current in the system either due to the fault or unbalanced load in the system.

Core-balance transformers are normally mounted over a cable at a point near to the cable gland for the electrical equipments.

Ring Type Current Transformer

This type of transformer has no primary winding and looks like a toroidal ring that contains only secondary winding being wound over a core. The open space in the centre allows the active conductor to pass through the CT, i.e., this CT is taken through the primary conductor. This type of CT is also called **window type current transformer**. This CT has only one core and has only one CTR (current transformation ratio) and is generally used for small voltage rating system. The ring type CT is shown in Figure 2.3.

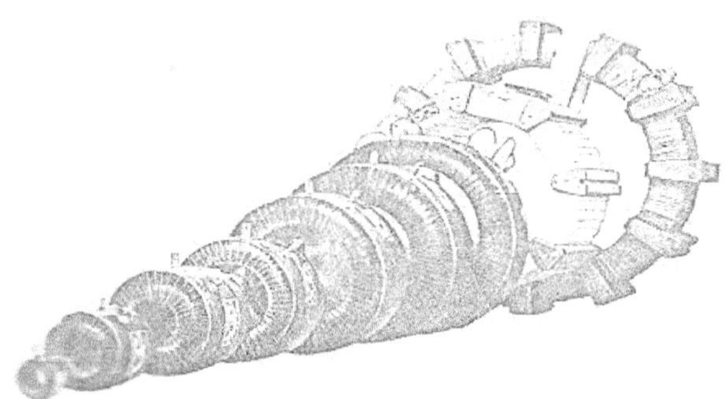

Figure 2.3 Ring type CT

CLASSIFICATION BASED ON APPLICATION

According to the use of secondary winding, the CT (Current transformer) is classified into different types as discussed below.

Measuring Current Transformer

This current transformer is basically applied to supply the current to the indicating instruments like ammeters, integrating meters like wattmeters, energy meters, etc.

The windings and design of the cores for this CT are considered on the basis of the requirement of the circuit. Metering circuit needs better accuracy performance, than the knee-point voltage margin. So the windings, that are used for this purpose should have better steady state accuracy factor, as compared to the transient accuracy factor.

The cost of the CT depends upon the metering core, accuracy limit and the cost becomes more for accuracy class nearer to zero. Nowadays precision CT of accuracy limit of 0.1 is also available for field use.

Protection Current Transformer

This current transformer is applied to supply the current to the protective relays in the circuit like current activated relays, power relays and impedance relays, etc.

For such transformer, the design of winding and core is considered on the basis of knee point voltage (k_n), excitation current at this knee point voltage. The accuracy limit factor of the CT is not that important as compassed to the metering class. But the action and response of this winding for both in transient and steady state condition should be active and ready to take action for operation of the relays in the circuit.

Protection Current Transformer (Special Purpose)

This type of current transformer is best suitable for the secondary circuit of relays, used for special purpose applications like balance protection scheme (differential circuit) and distance protection schemes. The knee point voltage, excitation current and secondary resistance of such transformer, plays the important role to decide the performance of the CT during the working operation on the circuit. These values should remain within the specification limit for both in transient and steady state condition up to the maximum through-fault current. The impedance of the CT is low as compared to other type of CTs.

Dual Purpose Current Transformer

This type of current transformer is used for connection to the secondary having both metering circuit and protection scheme in the system.

The design of the winding and core is decided in such a way, so that the accuracy factor remains within the specification limit for all the stages of current loading on the CT.

To maintain such accuracy limit, the relevant factors like design, cost, space and withstanding capability of the instrument (CT) are accordingly considered and becomes more in comparison. But nowadays such type of CTs are rarely in use.

Single Ratio-Multiple Core Transformer

For economical practice of the use of different number of secondary circuits from a common primary winding of the current transformer, a unit comprising more than one core can be designed. Such type of CT is designated as single ratio-multiple core CT. According to the characteristics of the circuit requirement, the core of the CT is designed separately. So from a single CT, different cores for different purpose like metering core, protection core and protection core for special purpose, etc. can be designed.

Multiple Ratio, Multiple Current Transformer

By the use of a single CT and with the control of CT circuiting on primary and secondary side, more than one current transformation ratio (CTR) can be obtained. These ratios are obtained by the use of taps on the primary or secondary windings or both, or by series/parallel connection of separate primary or secondary windings on a common core.

CLASSIFICATION BASED ON INSULATION

The insulation material that is used in the current transformer plays a vital role in the working of CT. According to the type and class of insulation used in the current transformer, the voltage rating of the CT is decided and the classification is done accordingly. Following are a few modes of CT classification on the basis of insulation used.

Air Support Current Transformer

These are small current transformers in which a small air gap is included in the core to produce a secondary voltage output proportional in magnitude to current on the primary winding. Sometimes they are termed as "transactors" or "quadrature CT". Such type of CT is rarely used in main power system circuit, because of its limitations.

Dry Type Current Transformer

The insulation used for this type of transformer is of solid and dry nature (Figure 2.4(a)). It does not require the use of any liquid or semi-liquid material as the insulation of the CT. The solid insulation may be of epoxy resin bound material or polycrete type material. But polycrete instruments have better strength in comparison to other solid insulations.

For the case of few small voltage rating transformers, paper insulation is also provided as the solid insulation and the range of such class is restricted to 13 kV range. The epoxy resin-filled bound and polycrete instruments (Figure 2.4(b) and 2.4(c)) have ranges even up to 72 kV class and the performance is well within the satisfactory limit.

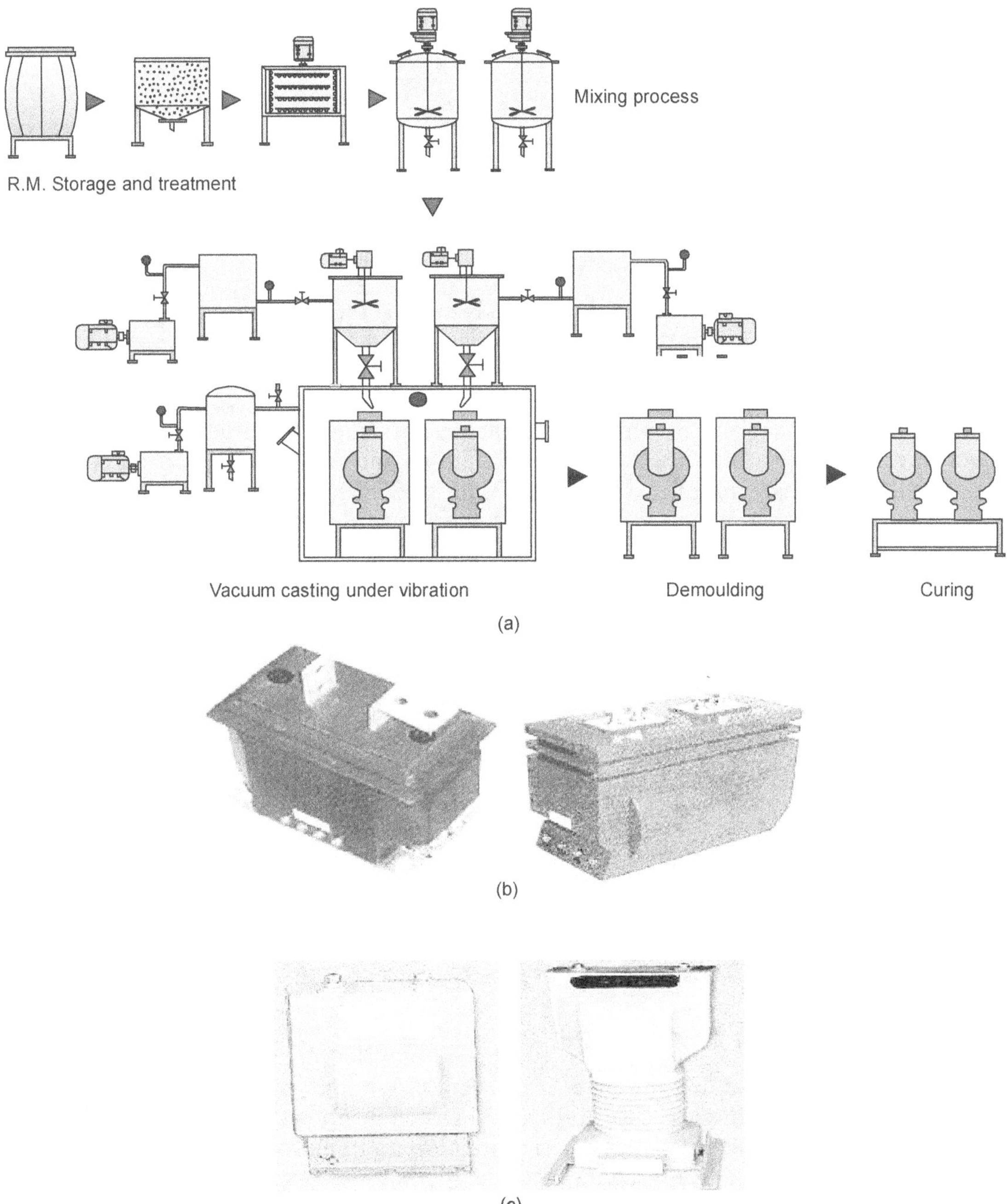

Figure 2.4   Different dry type CTs (a) Manufacturing method of dry type CT (b) Epoxy resin filled CT (c) Polycrete (dry type) CT

Such equipment provide the following features.
1. Compact and solid insulation design.
2. High mechanical strength.
3. Excellent outdoor performance under extreme ambient conditions.
4. Maintenance-free and longer life.
5. Simplified transport and erection.
6. Environment-friendly.

Oil Insulated Current Transformer

For this type of CT shown in Figure 2.5, oil is used as the insulation between primary and secondary winding. In practice high dielectric strength oil-impregnated paper being dried and degassed is used as the insulation material. The oil is filled over to this paper insulation. Because of this design insulation, it is called paper-oil insulated CT. To protect the paper-oil insulation from ambient temperature, maximum CTs are hermitically sealed by bellows arrangement. Porcelain insulator stack of suitable and required voltage range is used to provide paper insulation for the working limit of CT. This insulator stack should also be mechanically strong enough to withstand the physical weight of the top tank and the tensions of the conductor to the CT connection.

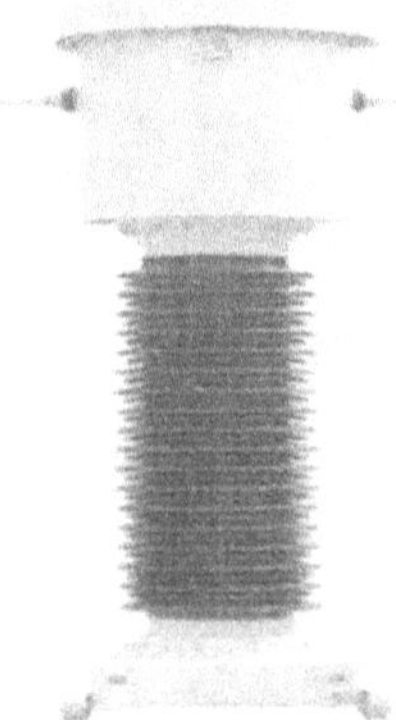

Figure 2.5 Oil insulated current transformer

SF$_6$-filled Current Transformer

Nowadays some Indian manufacturers have started manufacturing current transformers with SF$_6$ gas as the insulating medium instead of conventional oil insulation. This type of instrument transformer provides better insulation and due to use of SF$_6$ gas, the type of paper insulation that is required for the construction of CT can be avoided in design. The rating of the insulation voltage level could also be increased. But the performance of this type of CT has not yet been popularly experienced by Indian utilities, for which the same has not yet been accepted openly. CGL in India has successfully manufactured and tested the CT up to 245 kV range.

CLASSIFICATION BASED ON TANK DESIGN

The design factors of HV and EHV CTs depend upon the following points also.

1. Use of HV insulation in CT.
2. Positioning of primary and secondary winding in the system.

Generally post-type current transformer is used in HV/EHV system. The basic construction of this type of CT consists of a common primary conductor and an insulator (the post). One or more cores are arranged either at the bottom of the insulator around the primary conductor or at the top of the insulator housing with primary conductor passing straight through the CT. So according to the use of insulation and positioning of primary and secondary winding, this post-type CT is classified into two types.

1. Hairpin or pendulum design CT (Dead tank type)
2. Top tank or inverted design CT (Live tank type)

Hairpin or Pendulum Design (Dead Tank CT)

In hairpin type design, the primary winding is taken up to the bottom box, where the secondary winding is wound over the primary winding to obtain required ratio of the CT.

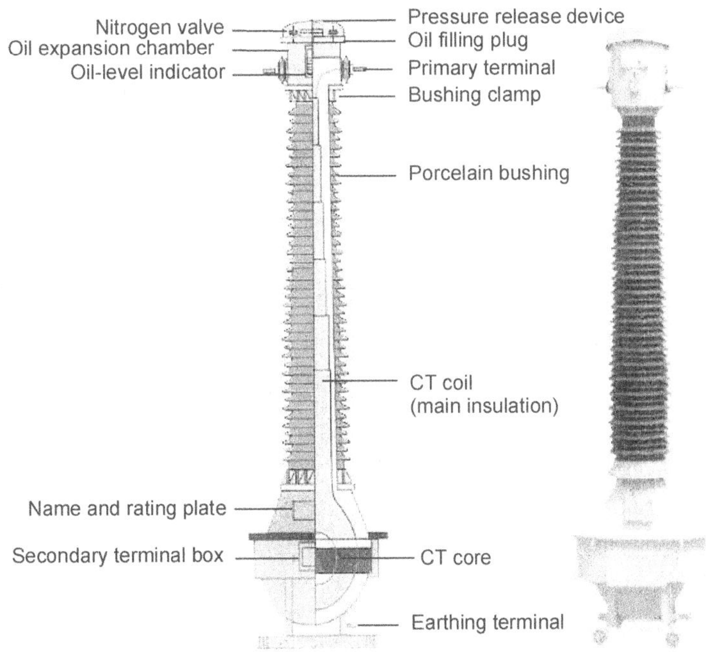

Figure 2.6   Dead tank CT with cut-section view

The major insulation is provided on the primary winding and housed properly in the bottom tank, which is earthed or dead. Because the housing of the conductor is connected to the earth, such type of CT is called "dead tank CT". This CT also looks like a pendulum and hence called "pendulum design CT". The dead tank CT along with its cut-section view is shown in Figure 2.6.

Top Tank or Inverted Design CT (Live Tank CT)

In top tank type design, the primary winding is spread in a uniform and symmetrical way around the core and available in the top tank of the CT. The secondary windings are wound over the primary winding inside this box. In live tank CT, the core and secondary winding is insulated against the high voltage and major part of the insulation is kept on the core and secondary winding. This transformer looks like an inverted CT, and is so called "inverted CT" and as it is housed in the tank, which is the live part of the system, it is called "Live Tank CT". The live tank CT with cut-section view in shown in Figure 2.7.

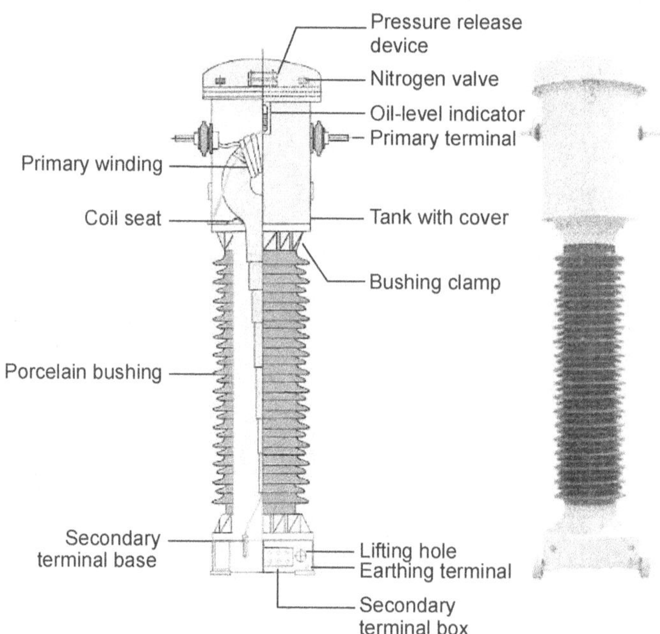

Figure 2.7   **Live tank CT with cut-section view**

Comparison of Live Tank and Dead Tank CT

Considering the practical design pattern and use of insulation in the system, some comparisons can be discussed as described in Table 2.2.

Table 2.2   Comparison of live and dead tank CT

Live tank	Dead tank
The core and secondary windings are kept in the top tank, which is live (charged with system voltage).	The core and secondary windings are kept in the bottom tank, which is earthed or dead.
The core and secondary assembly is insulated against the high voltage in this live tank.	In dead tank, the primary winding is brought down to the bottom tank and is insulated from earthed tank.
Insulation is on the core and secondary winding.	Insulation is on the primary winding.
This transformer is compact and economical.	This transformer is bulky and costly.
Because the primary winding is shortest and directly passes through the live tank, it offers high strength against the short-time current dynamic force.	Primary winding passes through the porcelain insulator stack and larger length of primary conductor produces maximum mechanical force during short-time dynamic current, which might damage the insulation.
Due to minimum length of primary winding, the I^2R loss is minimum and thermal stability would be better.	Since the length of primary winding is more and its insulation design is complex, the heat dissipation is critical. Heat generated is dissipated into two stages, i.e., winding to insulating material and then to the oil.
Because the single HV conductor passes directly, the volume of insulation and oil is comparatively less.	Because of two primary conductors the volume of the insulator and oil is comparatively more.
Due to top tank design, the mechanical strength in normal condition is also not stable.	Due to dead tank being preponderant at the base, its mechanical stability is comparatively more.
Heat generated between terminal connectors and primary terminal joints, affects the primary winding, as it is mounted very near to the terminals (High voltage part).	Primary winding is lengthy in comparison and takes sufficient path for heat dissipation. So, heat generation in jointing terminals is comparatively less.
In the case of insulation failure, top chamber and insulator are scattered and the surrounding equipments are damaged.	The main insulation is at the bottom part, so the failure causes less damage to contents in the bottom tank.

OTHER SPECIAL TYPE CT

Sometimes, according to the requirement of the secondary current in the electrical circuit, the conventional CT designs are modified. Basically for some of the protection and metering scheme, the current availability from the existing secondary CT does not match to the current requirement of the available scheme. For such practices, intermediate CT is provided in the scheme to match the current for the circuit. Considering the above factors in system, two different types of CTs are realized.

1. Summation current transformer
2. Interposing current transformer

Summation Current Transformer

Sometimes, it becomes necessary to calculate the total current drawn through a set of parallel feeders emanating from a common bus system. To achieve this, summation CTs (Figure 2.8) are used. From the associated line feeders, the secondary windings of line CT are wound on the primary side of summation CT, forming the primary winding of the CT. This CT has a single secondary winding, which is connected to the circuit (burden). The ratios of the feeder CT may or may not have same ratio. If individual feeder CT ratios are the same, number of turns of each primary winding shall be identical. If however, the feeders CTR are different, it shall be necessary for the manufacturer to alter the number of turns of each primary with an ultimate objective of obtaining current on the secondary side which is proportional to the summated load current. In such cases the individual feeder CTR must be specified.

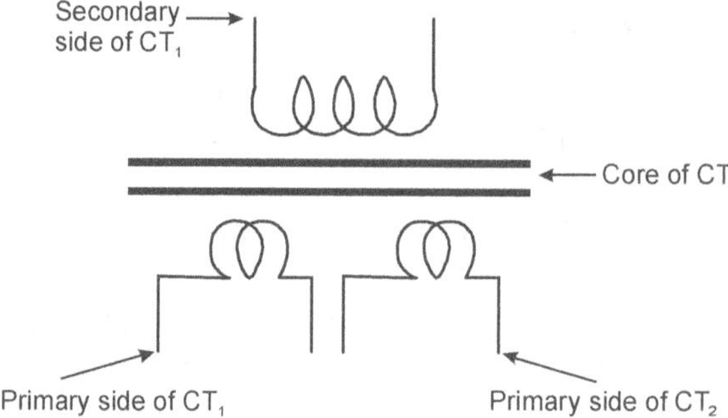

Figure 2.8   Circuit diagram of summation CT

Interposing Current Transformer (ICT)

This current transformer is interposed between the main CT and the secondary electrical circuit (burden) of the system (Figure 2.9). The secondary circuit (winding) of the main CT is connected as the primary winding of the ICT. The CTR and number of turns on the winding to feed the correct current magnitude to the burdens depends upon the correction factor. These transformers are of small rating transformers and become the burden for the main CT in the system.

Some typical applications of ICT

1. For differential circuit connection in power transformer protection, the current magnitude required for either side of the transformer should be same. But sometimes,

the magnitudes do not match due to use of different CTR and phase shifting of winding connection. So for such practice ICT is used.

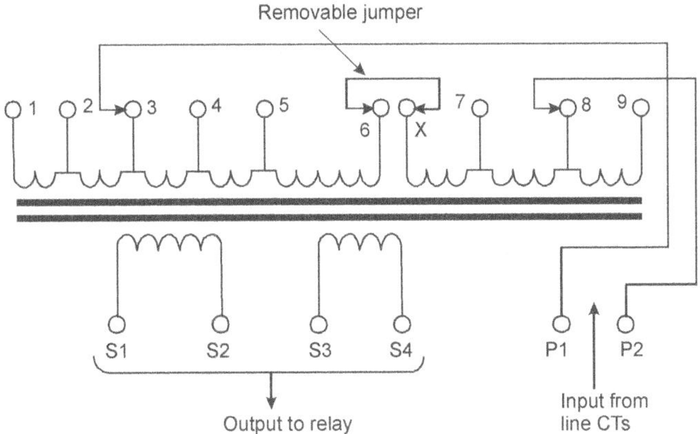

Figure 2.9   Circuit diagram of interposing CT

2. When main CT is having secondary current of 1 A with requirement of burden of 5 A, this ICT of ratio 1 : 5 is selected in the scheme. Similarly with main CT having secondary current 5 A and requirement of burden 1 A, demand ICT is of 5 : 1 ratio.

3. For the high ratio main CT, with low-rated load current, ICT has to be used to reduce the ratio effectively.

   Example   Suppose main CTR is of 2000/5 or 2000/10, then suitable ICT of 5/1 or 10/1 can be used to get suitable overall ratio.

4. When it is difficult to design a main current transformer with low ISF, because of various reasons, designed ISF can be achieved by using a suitably designed ICT.

5. For controlling the saturation factor of the main CT and to restrict the maximum current flowing through the burden, this ICT is also used in the system.

Advantages

1. For correct operation of electrical circuitry and to match the main CTR with burden load current, the use of ICT is the only choice to be provided in the scheme.

2. To achieve the desired ISF for the main current transformer, ICT becomes suitable for use in the system.

3. Even to limit the very high primary current for suitable range, ICT is interposed in the system.

Disadvantages

1. The accuracy class of the circuit, particularly for metering scheme is raised. Suppose, an ICT is used with a main CT of having accuracy class 0.2 and 0.5 respectively, then net accuracy would be 0.7 for metering circuit. So the error in measurement would be raised, for the use of ICT in the circuit.

2. As ICT is connected on the secondary circuit of main CT, so this CT becomes burden for the main CT.

3. As ICT is connected in between the main CT and burden of the circuit, so the possibility and chance of opening of the circuit becomes more.

4. It increases or reduces the size of the connected burden by the square of the ratio used in the system.

In all of the above conditions, it can be concluded that the use of ICT in the scheme causes error with extra burden to the system. So, unless otherwise desired by the system or circuit, the use of ICT should be avoided.

CONSTRUCTION PRACTICE OF 245 kV LIVE TANK CT

For a detailed idea upon the construction practice of the current transformer, a typical description has been explained for a live tank CT of rating 245 kV class. The cut-section view of 245 kV live tank CT is shown in Figure 2.7.

Components of a 245 kV Live Tank CT

For such type of CT, following are regarded as the main components for the construction of the transformer.

1. *Primary winding*   The primary winding is of either rigid aluminum or copper sections (rod, tube, flats, etc.) present on the top tank and passes directly with the externally accessible primary connecting links. A metal bellow made of stainless steel as the expansion element for the temperature-dependent oil volume changes its level according to the volume of oil. The oil level and operating mode of bellow can be monitored by means of a level marker for maximum and minimum level.

2. *Secondary winding*   The secondary windings are of toroidal type winding, insulated copper wire uniformly distributed over the circumference of the core. The leads of the secondary windings are taken through a condenser bushing and porcelain to the bottom housing for external connection. Now the entire cavity is filled with impregnated oil.

3. *Insulation*   Major part of the insulation is provided between primary and secondary winding and between winding to earth also. This insulation is of impregnated kraft paper. In general practice, the complete assembled CT is dried and filled with specific quantity of oil under vacuum. Hollow porcelain insulator provides the space for secondary leads to pass through the aluminum tube. This insulator also helps to provide proper insulation between HV terminal and earth base.

4. *Bottom housing box*   This box is used to take out the secondary leads for external connections. This box also contains grounding terminal pad with external oil-drain screw, lifting lugs/holes and mounting holes being the other accessories of the CT.

Working Principle

Primary winding of the current transformer is connected in series with the line conductor whose current is required to be monitored. Secondary winding is connected to the burden of measurement and protection circuit of the system. Primary current (I_1) results in an alternating flux in the core and this flux causes induced emf across the secondary winding. According to the use of primary and secondary number of turns in the CT, the secondary current (I_2) passes through the circuit. For theoretical practice due to quality of primary and secondary ampere-turns, the following equation holds good for CT.

i.e., $$N_1 I_1 = N_2 I_2 \text{ (ideal condition)}$$

But in practical current transformer due to the effect of magnetic circuit certain excitation current (also called no-load current) flows in the primary winding of the circuit.

Such excitation current is the resultant of I_m (magnetizing component) and I_w (iron loss for working component). So for actual practice, the primary current responsible for the flow of secondary current I_2 is considered as I_1', but not I_1 as explained in ideal condition. To explain the above theory, the following vector diagram as shown in Figure 2.10 can be drawn.

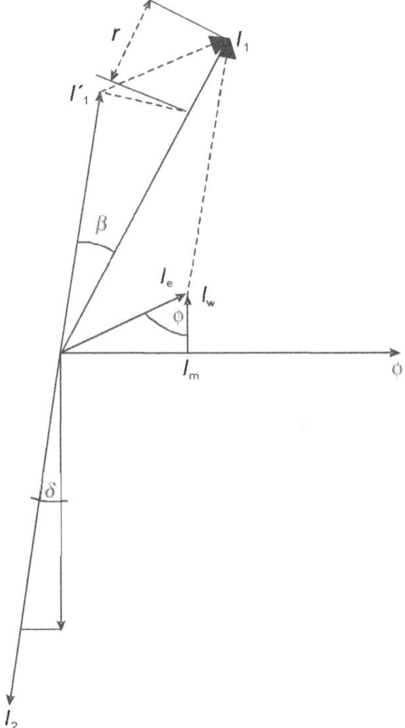

Figure 2.10   Vector diagram of CT

Derivation from vector diagram

$$\% \text{ Ratio error} = r = -100[(I_m \sin \delta + I_w \cos \delta)\,/\,I_1]$$

$$\text{Phase error} = \beta = \frac{180}{\pi}[(I_m \cos \delta - I_w \sin \delta)\,/\,I_1]$$

In the vector diagram,

I_1 = Actual primary current
$I_1{}'$ = Reflected primary current
I_e = Excitation current
I_m = Magnetizing component of exciting current
I_w = Working component of exciting current
δ = Overall secondary power factor angle
β = Phase angle error
γ = Ratio error
ϕ = Excitation angle

NOTES ON ERRORS IN CT

In actual practice under working condition of CT, the flow of current on primary and secondary side do not respond to the transformation ratio fully. Certain difference results between actual transformation ratio and rated transformation ratio. Similarly the phase angles of both the current do not match to the actual phase angle. Certain errors result in both current ratio and phase angle.

1. Current error (ratio error)
2. Phase angle error
3. Composite error

Current Error (Ratio Error)

During actual monitoring or measuring of current flow in the CT, the deviation observed in comparison to the rated transformation ratio, is expressed in the form of ratio error.

$$r = \% \text{ Current error}$$

$$= (K_n I_2 - I_1)100\,/\,I_1$$

$$= \left(\frac{K_n}{K} - 1\right) \times 100$$

where,

r = % Ratio error

K_n = Nominal or rated transformation ratio (rated CTR)

K = Actual transformation ratio (obtained CTR during measurement) = I_1/I_2

I_1 = Actual primary current

I_2 = Actual secondary current during measurement

Phase Angle Error

For a perfect CT, the phase angle difference between primary and secondary current is considered zero. But during actual working operation, this phase angle differs as compared to the actual and contributes certain error, called phase angle error.

This error is said to be positive when secondary current vector leads the primary current vector and negative when secondary current vector lags behind the primary current vector.

Composite Error

During actual operation of CT, the current wave form does not become purely sinusoidal, due to non-linearity nature of exciting impedance and loading impedance.

Harmonics in the system also play the role to distort the original wave form. So, the current error and phase angle error contribute an error in composite form, called composite error.

The composite error is generally expressed as a percentage of RMS values of primary current.

$$E_c = \frac{100}{I_1} \times \sqrt{\frac{1}{T}\int_0^T (K_n\, i_1 - i_2)^2\, dt}$$

where,

E_c = Composite error

K_n = Nominal or rated transformation ratio

I_1 = RMS value of primary current

i_1 = Instantaneous value of primary current

i_2 = Instantaneous value of secondary current

T = Duration of one cycle

dt = Time derivative

NOTES ON CONNECTION PRINCIPLE

Polarity and Connection

The primary and secondary terminals of the CT are identified with polarity markings by the symbols like (P_1 and P_2) for primary and (S_1 and S_2) for secondary. It is marked with a common convention that when primary current enters the 'P_1' terminal, secondary current leaves the 'S_1' terminal to the load circuit. So primary 'P_1' terminal corresponds to the secondary of 'S_1' terminal. It is regarded as "DOT" convention. Its significance is in showing the direction of current flow relative to another current or to a voltage as well as to aid in making the proper connection. The same is explained in the Figure 2.11.

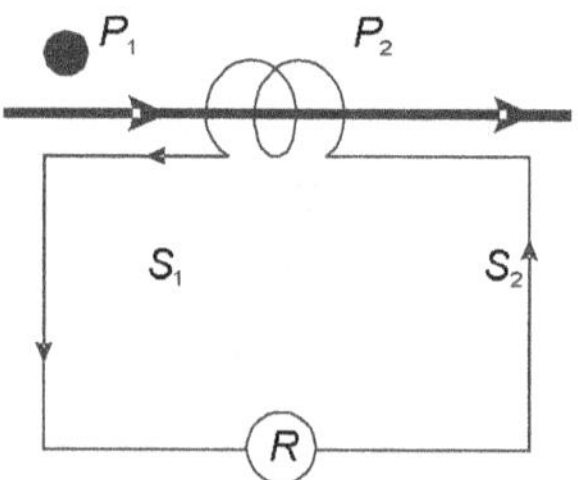

P_1— primary marking at one end S_1—secondary marking at one end

P_2— primary marking at the other end S_2—secondary marking at the other end

R—relay

Figure 2.11   Polarity marking of CT

Concept of CT Circuit Connection

Star connection circuit (Y-connection)   CT secondary windings are connected either in star (Y) or in delta (Δ) connection as per the requirement of the circuit.

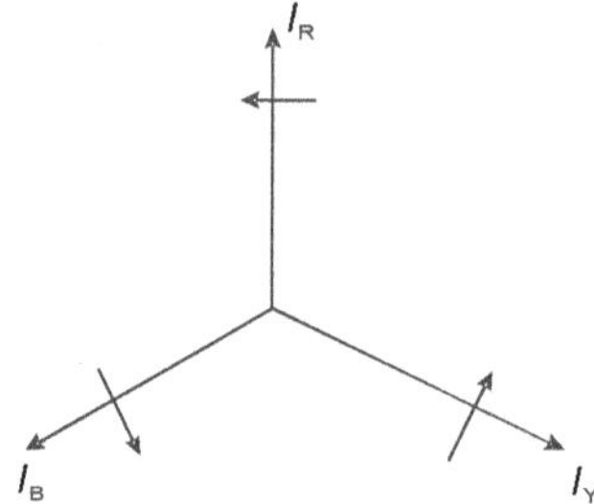

Figure 2.12   Vector diagram of star connection

For star connection (Figure 2.12), the currents on each phase (I_R, I_Y and I_B) are related vectorially and expressed in sequence components as follows.

$$I_R = I_{R1} + I_{R2} + I_{R0}$$
$$I_Y = I_{Y1} + I_{Y2} + I_{Y0} = a^2 I_{R1} + a I_{R2} + I_{R0}$$
$$I_B = I_{B1} + I_{B2} + I_{B0} = a I_{R1} + a^2 I_{R2} + I_{R0}$$

So, $I_R + I_Y + I_B = 3 I_{R0} = 3 I_{Y0} = 3 I_{B0}$

where, 1, 2, 0 designate +ve, −ve and zero sequence components and a, a^2 are the operators.

Note Current vectors are in +ve sequence only with current on R phase taken as reference.

Vectorial expression

$$I_R = I\,\underline{(0°)}$$
$$I_Y = I\,\underline{(-120°)}$$
$$I_B = I\,\underline{(120°)}$$
$$I_R + I_Y + I_B = I_N$$

Delta connection circuit (Δ-connection) For delta connection of the CT circuit, the pattern of connection can be made by two possible ways. In one type of connection secondary S_1 of one phase is connected to the secondary S_2 of next phase in regular sequence (R, Y, and B). In other type of connection secondary S_1 of one phase is connected to the secondary S_2 of next phase in opposite sequence (R, B, and Y). The connections of the windings are shown in Figures 2.13a and b and 2.14a and b.

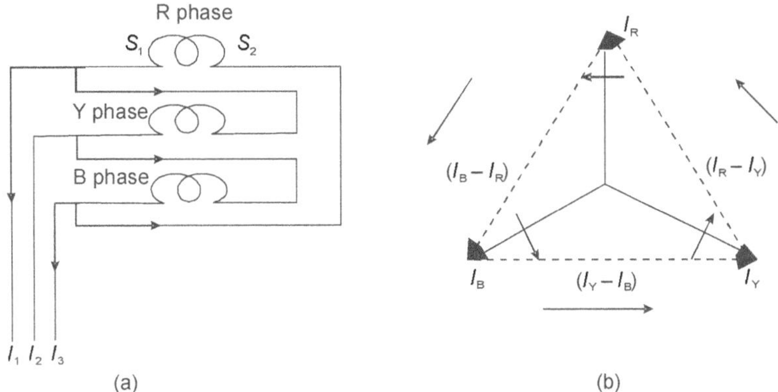

Figure 2.13 Positive sequence delta connection

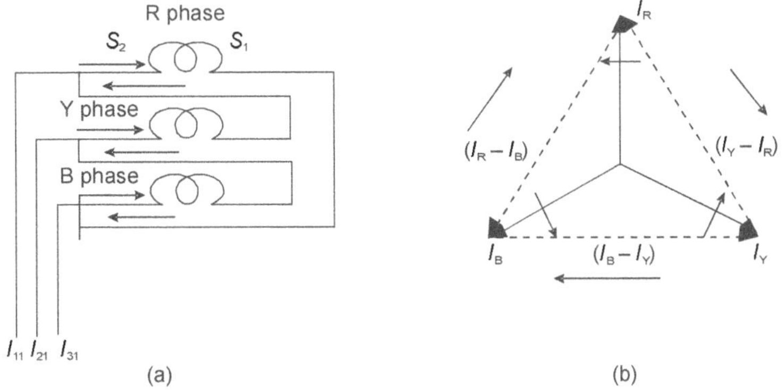

Figure 2.14 Negative sequence delta connection

Considering vector analysis principle with "R" phase as reference and balance load connection, the following results are expressed.

For Figure 2.13(b) the vectorial expression is

$$I_R = I\angle 0°, I_y = I\angle -120°$$

$$I_B = I\angle 120°$$

So, $I_R + I_Y + I_B = I_N = 0$

$$I_1 = (I_R - I_Y) = \sqrt{3}\, I \angle 30°$$

$$I_2 = (I_Y - I_B) = \sqrt{3}\, I \angle -90°$$

$$I_3 = (I_B - I_R) = \sqrt{3}\, I \angle 150°$$

For Figure 2.14(b) the vectorial expression is

$$I_R = I\angle 0°, I_y = I\angle -120°$$

$$I_B = I\angle 120°$$

So, $I_R + I_Y + I_B = I_N = 0$

$$I_{11} = (I_Y - I_R) = \sqrt{3}\, I \angle -150°$$

$$I_{21} = (I_B - I_Y) = \sqrt{3}\, I \angle 90°$$

$$I_{31} = (I_R - I_B) = \sqrt{3}\, I \angle -30°$$

CT Connection Related to the Ratio

To obtain multi-CT ratio in a common CT, the windings of the primary and secondary side are controlled by different connections. These connections are of three different types.

1. Ratio by primary control
2. Ratio by secondary control
3. Ratio by both side control

Ratio by primary control   By the connection of available primary windings in different ways like all in series or all in parallel or combination of series and parallel, the ratios of the CT are changed. The detailed connections are shown in Figure 2.15. The connection sequences are described in Table 2.3.

Ratio by secondary control   By the use of tapping terminals on the secondary side of the windings, different CT ratios can be obtained. By this principle primary connection remains fixed and ratios are controlled on secondary terminals. The connection diagram is shown in the Figure 2.16. The connection sequences are described in Table 2.4.

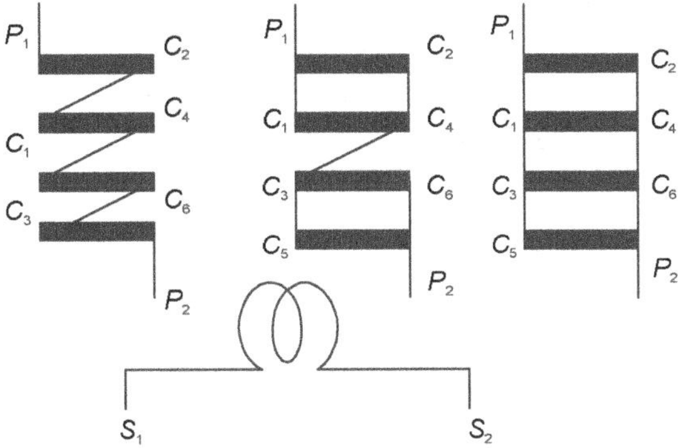

Figure 2.15 Primary control ratio CT

Table 2.3 Primary control CTR

Connection sequence		Current ratio
Primary	**Secondary**	
$(C_1 + C_2)$	$S_1 - S_2$	Lowest ratio (CTR$_1$)
$(C_3 + C_4)$		
$(C_5 + C_6)$		
$(P_1 + C_1)$	$S_1 - S_2$	Middle ratio (2CTR$_1$)
$(C_2 + C_3 + C_4 + C_5)$		
$(P_2 + C_6)$		
$(P_1 + C_1 + C_3 + C_5)$	$S_1 - S_2$	Highest ratio (4CTR$_1$)
$(P_2 + C_2 + C_4 + C_6)$		

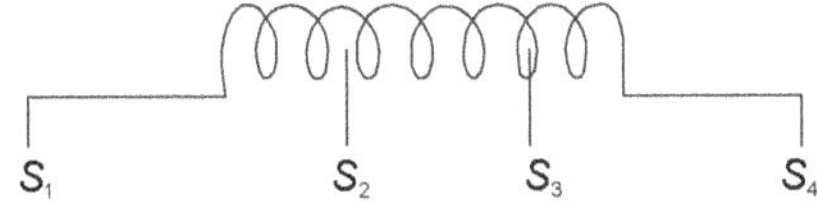

Figure 2.16 Secondary control ratio CT

Table 2.4 Secondary control CTR

Connection sequence		Current ratio
Primary	**Secondary**	
$P_1 - P_2$	$S_1 - S_2$	Lowest ratio (CTR$_1$)
$P_1 - P_2$	$S_1 - S_3$	Middle ratio (2CTR$_1$)
$P_1 - P_2$	$S_1 - S_4$	Highest ratio (4CTR$_1$)

Ratio by both side control   Both primary and secondary winding can be connected with the tapping terminals to obtain different CT ratio from a common CT. By this type of control, the connection stability and flexibility increases and ratios are obtained as per the suitability. Nowadays maximum CTs are designed with CTR control from both primary and secondary control. The detailed connections are shown in Figure 2.17. The connection sequences are described in Table 2.5.

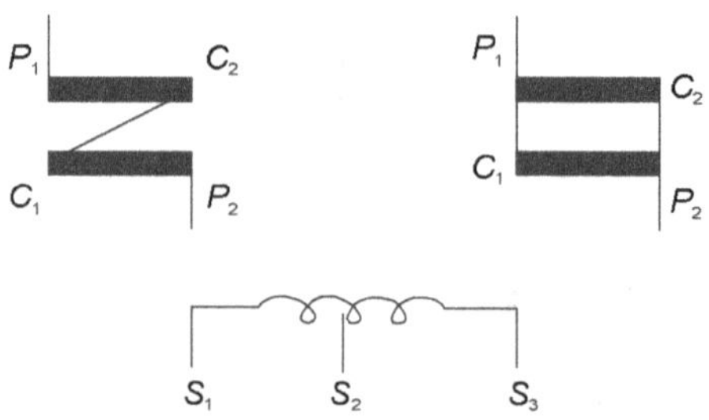

Figure 2.17   Both side control CT

Table 2.5   Both side control CTR

Connection sequence		Current ratio
Primary	**Secondary**	
$(C_1 + C_2)$	$S_1 - S_2$	Lowest ratio (CTR_1)
$(P_1 + C_1)$, $(P_2 + C_2)$	$S_1 - S_2$	Middle ratio $(2CTR_1)$
$(C_1 + C_2)$	$S_1 - S_3$	Middle ratio $(2CTR_1)$
$(P_1 + C_1)$, $(P_2 + C_2)$	$S_1 - S_3$	Highest ratio $(4CTR_1)$

Mathematical Analysis for Availability of Different CT Ratio

CTR availability due to primary link control only   Consider the resistance of each portion between two consecutive terminals being R ohm (Figure 2.15). So, value of resistance for different patterns of connection will be as follows. It can be noted here that the VA value becomes same for all patterns because of same CT. Now value of current in each pattern can be calculated as follows (Table 2.6) for availability of CT ratio.

Table 2.6   CTR calculation table for primary control CT

Pattern of connection	Value of resistance	VA value
All in series	$4R = R_1$	$V_1 \times I_1 = I_1 R_1 \times I_1 = I_1^2 R_1$
Series and parallel	$R = R_2$	$V_2 \times I_2 = I_2 R_2 \times I_2 = I_2^2 R_2$
All in parallel	$R/4 = R_3$	$V_3 \times I_3 = I_3 R_3 \times I_3 = I_3^2 R_3$

VA value being same

$$V_1 \times I_1 = V_2 \times I_2 = V_3 \times I_3$$

$$4I_1^2 R = I_2^2 R = I_3^2 R/4$$

$$I_1/I_2 = 1/2 \text{ and } I_2/I_3 = 1/2 \quad \text{(As shown in Figure 2.15)}$$

So, $CTR_2 = 2CTR_1$ and $CTR_3 = 4CTR_1$

CTR availability due to secondary tap control only　For such condition primary side has a particular number of turns (say N_1) but the secondary turns are tapped at different points to provide different ratios according to involvement of number of turns in the winding (Table 2.7).

Total number of secondary turns for tapping between S_1 and $S_2 = N$

Number of turns for tapping between S_1 and $S_2 = 2N$

Number of turns for tapping between S_1 and $S_4 = 4N$

Now $T_1/T_2 = 2$ and $T_1/T_3 = 4$

In a transformer, current ratio is inversely proportional to the turns ratio

So, $CTR_2 = 2CTR_1$ and $CTR_3 = 4CTR_1$

Table 2.7　**CTR calculation for secondary control CT**

Connection	No. of turns	Turns ratio $= \dfrac{\text{Primary}}{\text{Secondary}}$
S_1 and S_2	N	$T_1 = N_1/N$
S_1 and S_3	$2N$	$T_2 = N_1/2N$
S_1 and S_4	$4N$	$T_3 = N_1/4N$

CTR availability due to primary and secondary control　Mathematical calculation for CT ratio is the combination of the cases as described for CTR availability due to primary and CTR availability due to secondary (self-explanatory).

BASIC INSTALLATION PRACTICES

Transportation and Shipping

The current transformers are generally packed in wooden crates and securely packed with supports and reinforcement are provided in packing to protect the primary terminals and porcelains. Transportation is generally done in horizontal position, with necessary shipping marks on the packing, as shown in Figure 2.18.

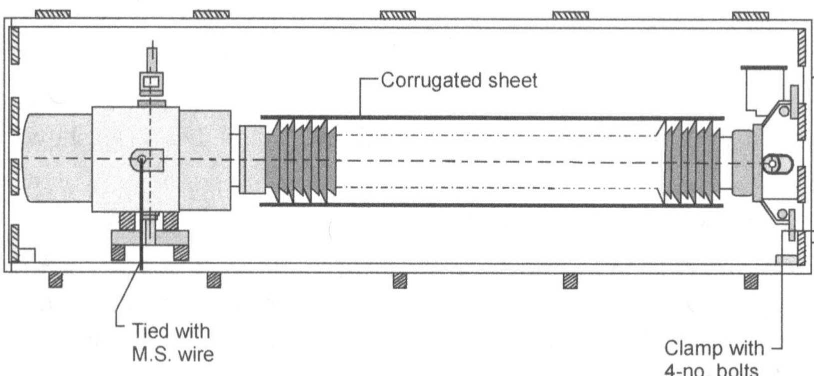

Figure 2.18   Horizontal position of CT

Unloading and Handling Practice

Every manufacturer provides proper procedures for unloading and handling of the current transformer. These instructions are strictly to be followed during the time of unloading and handling procedure. But in general, unloading and handling of CT is done in two different methods (horizontal and vertical). For horizontal handling the lifting point should be chosen such that the tension on the string should be uniform. The details are shown in Figure 2.19. This method is generally adopted during erection of the CT.

For vertical handling, the string is tied and fastened through the bottom holes, encircling the two ropes on each side as shown in Figure 2.20(c). This method of CT handling is done for installation of the instrument on the proposed platform.

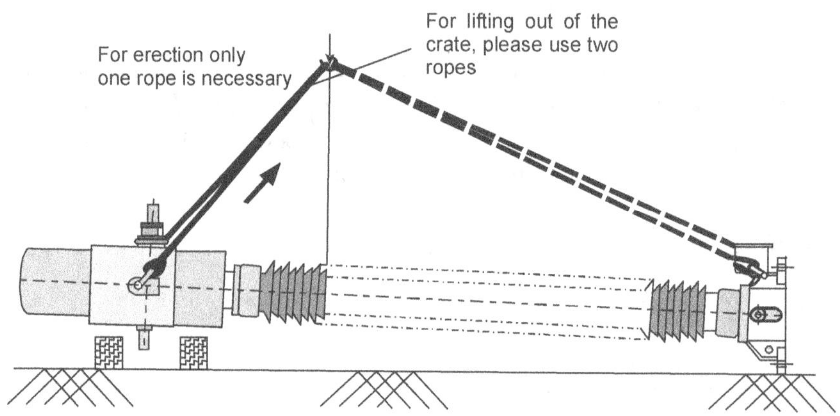

Figure 2.19   Horizontal handling of CT

Transportation and Erection

Horizontal handling   On removing from packing case CT should be kept on a levelled ground as shown in Figure 2.20a. Ensure wooden supports are kept in position as shown, to prevent damage to CT primary terminal.

Erection   During the whole erection phase, the rope must be at 90° angle to earth as shown in Figure 2.20b. Ensure the wooden support is firm on the ground to avoid heavy swinging of the CT while making it vertical.

Note   The top of the CT is heavy and so should be handled with care.

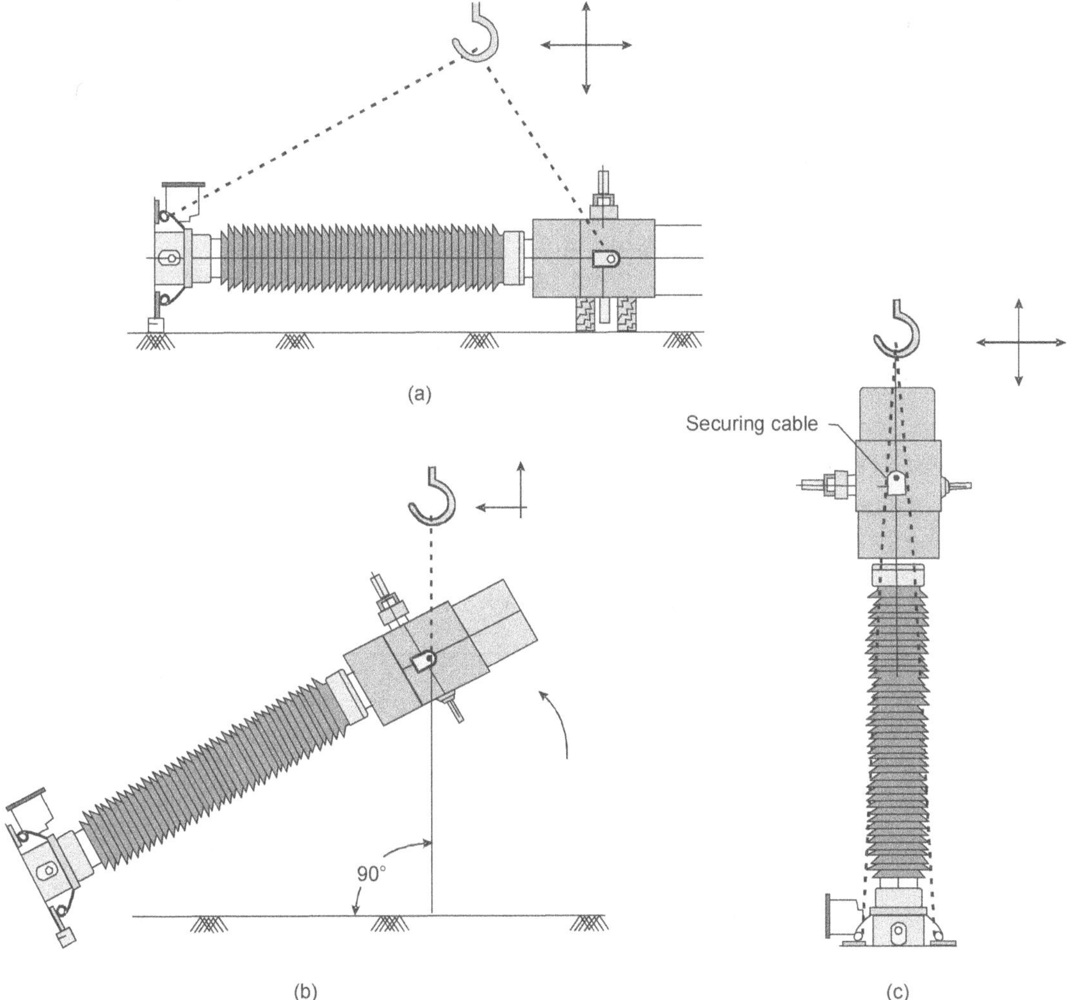

Figure 2.20 Transportion and erection of CT (a) Placing CT before erection (b) Erection method of CT (c) Vertical hadling of CT

Vertical handling   Ensure securing cable is tied through the lifting hole, encircling the two ropes on each side (Figure 2.20c). Ensure that the lifting ropes do not damage the hood.

Installation Practice

1. By the use of suitable hoist or crane, the CT is handled vertically and lifted to a suitable height for the installation on the platform. Mounting holes are matched according to the foundation drawing and fastened to the support properly.
2. Proper earthing of the transformer is important and necessary earthing bolts are provided for this purpose.

Pre-commissioning tests   The following preliminary tests are to be done before installation of the CT on the platform to confirm the CTR and the insulation condition.

1. The selected CTR has to be confirmed by primary injection testing kit.
2. The insulation resistance value of primary to secondary, primary to earth and secondary to earth for all the available cores have to be checked, by suitable range of megger.
3. Ensure the polarity of each core of the CT.
4. Measure the resistance value of each secondary winding.
5. If possible and available, then tan delta measurement has to be conducted for reference.
6. Maintain the record of the above test values in the prescribed format for future reference.

Caution   For measurement of insulation resistance (IR) of secondary windings to earth, low-voltage rating megger preferably 500 V/1000 V range instrument has to be used.

Pre-commissioning check   Following points are to be checked before charging of the CT into the system (Table 2.8).

Table 2.8   Pre-commissioning check of CT

Attributes	Check
Tightness of bolts/screws upon structure	Done/not done
Tightness* of primary connections	Done/not done
(i) Stud/pad to line conductor	
(ii) Primary links	
Tightness of secondary connections	Done/not done
Shorting of spare (unused) secondary core of the CT	Done/not done
No secondary winding is open**	OK/not OK
Correct and suitable secondary tappings in use	Yes/no
Firmly earthing of base structure	Done/not done

Table 2.8   (Continued)

Attributes	Check
Measurement lead (C_x) if available, is connected to earth inside the secondary box	Done/not done
Correct and suitable primary links in use	OK/not OK
Connected burden value of each core is within the rated value	OK /not OK
Polarities of primary*** and secondary terminals are as per the schematic diagram	OK /not OK
Desired ratio is correctly selected as per the scheme	OK /not OK

For CT with bellow system
  (i)   Bellow position normal
  (ii) Protective cover around the bellow is removed and bellow is visible through windrow
  (iii) All the CTs for a 3 phase system have approximately same bellow level
	OK /not OK

For CT with oil level windrow
  (i) Oil level normal
  (ii) Windrow is clean and clear
  (iii) All the CTs have approximately same level
	OK /not OK

Secondary box cover is tightly closed	Yes/no
Noting down of Sl.No, P.O. No, and Ref. No	Yes/no

*   to avoid corrosion, do not connect aluminum terminals with copper cables/tubes directly. Use weather-resistant protective coating and suitable washers and wire ribbon, etc.

**   for the case of tapped secondary winding [IS_1, IS_2, IS_3, IS_4] if secondary circuit is connected to any section say (IS_1, IS_2), then the other sections (IS_3, IS_4) of the same winding must not be shorted.

*** in general convention "P_1" terminal of CT is installed towards the substation bus conductor.

STANDARDS AND PRACTICES

Standards and practices are the limitation terms, to which manufacturers, purchasers, customers and other utility members, refer to maintain the required performance of the CT in India. For the formulation of this standard, assistance has been obtained from different institutions like CPRI, ERDA, NPL, ERTL, IDIMI, CBIP, IEEMA, SEBs, PSU, etc. Our national standard is IS 2750 (Part I-to IV-1992) for CT, which is at par with international standards.

But it has been observed over the years that the purchasers, customers while placing their order with the manufacturers, often over-specify the specification than is actually required, keeping the margin of safety more. This practice not only increases the size and dimension but also increases the avoidable cost of the instrument. The purchasers should refer the present standards and recommendations, instead of copying the old specification and standards. They should specify the information as per the requirement and suitability of the situation. It is known that the CT technology has experienced a considerable development in various applications and accordingly the standards have been modified to meet the requirements.

The standard of the CT basically covers the information on the following factors.

1. Ratio
2. Accuracy class
3. Rated burden
4. Basic insulation level and voltage class
5. Service conditions
6. Testing of CT
7. Marking of CT

RATIO OF CT

Ratio of CT is usually mentioned as the ratio between the rated primary current to the rated secondary current. The selection of ratio of current transformer depends upon the selection of primary and secondary current.

Rated Primary Current

As per IS 2705 (Part-I/1992), the standard value in amperes of rated primary currents are 10, 12.5, 15, 20, 25, 30, 40, 50, 60, 75 and their decimal multipliers or fractions. The underlined values are generally preferred for application in practice.

Points for consideration

1. For the indoor application, single-ratio CT is preferable for voltage rating up to 33 kV class. (It is due to the size and cost factor of the CT)
2. For outdoor application, multi-ratio CT can be preferred for voltage rating above 33 kV level.
3. Selection of rated primary current becomes preferable if it is decided according to the rated short-time current for 1 sec, and also according to the availability of minimum load current in the system. The design of the relay becomes economical if both factors are considered properly.
   i. Rated primary current = (Rated short-time current for 1 sec)/150.
   ii. Value obtained from (i) should be more than the load current and a multiple of the underlined values as mentioned in the standard.

Notes on rated short-time current Every CT has its own limitation of highest current flow in the primary winding with the requirement of thermal rating and dynamic rating.

i. Rated short-time thermal current   The maximum value of primary current up to which the CT withstands for a rated time of one second with secondary circuit being short-circuited is called **rated short-time current.**

Note 1   Rated short-time current (I_{ST}) is taken as $150I_p$ for 1 sec, depending upon the value of fault current I_F. If I_F is less than $150I_p$ then the short time current may be I_F with duration t such that $\sqrt{t} = 150I_p / I_F$.

Note 2   In general, duration time is selected "1 sec" for this rated short-time current.

ii. Rated dynamic current   The peak value of primary current (normally of 2.55 times the rated short-time current) is considered as the **rated dynamic current**.

Note   For testing of CT with rated dynamic current, the CT secondary circuit is to be short-circuited for 1 sec.

Rated Secondary Current

Standard rated secondary current as per IS 2705 (P-1/1992) are of 1 A or 5 A. For delta connected group the standard values are $(1/\sqrt{3} = 0.578 \text{ A})$ or $(5/\sqrt{3} = 2.89 \text{ A})$.

Points for consideration

1.   The preference of secondary current depends upon the burden on the circuit.
2.   It also depends upon the type and ratio of relays and instruments to be connected in the circuit.

CLASS OF ACCURACY

The accuracy class is decided according to the application of the secondary winding and the type of core used in practice. Basically CTs are used for metering and protection purpose. For metering purpose, the accuracy class is required to be considered seriously as compared to protection class. Protection class needs robust performance even during saturation region.

Class of Accuracy for Metering Core

Standard accuracy classes for measuring current transformers shall be (0.1, 0.2, 0.5, 1, 3, and 5) and (0.2s, 0.5s). In practice, all the mentioned accuracy class of CT has certain limits of errors. These limits are considered as the standard of the accuracy class. IS 2705 (Part-I/1992) is the Indian standard for this metering core CT. The importance and significance of the core is judged on the basis of accuracy class. Lower the limit of error, more the accuracy and vice-versa.

Class of Accuracy for Protection Core

For protection purpose, limiting value of error is not so important as compared to the metering class. The nature of magnetization and demagnetization characteristics is required to be designed correctly to meet the purpose of protection. Standard accuracy classes for protection in current transformers shall be (5P, 10P, 15P, etc.) and for special purpose it is given as "PS". This expression of percentage error like 5P, 10P etc. are restricted and intended to maintain so within the declared limits of error up to certain primary current (multiple of rated current).

ALF (Accuracy limit factor)   This is a certain standard multiplying factor of rated primary current up to which the declared limits of composite error in CT is maintained. The factors are generally of 5, 10, 15, 20 and 30 and expressed with accuracy class of CTs like the term 5P10, 10P20, etc.

Meaning of 5P10   This term 5P10 corresponds to the accuracy class of protection core and indicates that 5% of composite error in CT is limited up to 10 times the primary current flow in the CT. Similarly, the expression 10P20 indicates 10% of composite error in CT up to 20 times flow of primary current.

ISF (Instrument safety factor)   Every CT has certain limit of allowable primary current to flow in it up to which the instrument can work safely. This factor is the ratio of instrument limit primary current to the rated primary current. General standard is taken as the value less than or equal to 5. But choice and decision depends upon the purchaser and manufacturer. But general recommendation should be as low as practicable.

"PS" Class protection core   For this type of protection core, the error due to ratio and phase angle is not that important. But the other factors like knee-point voltage (V_K), permissible magnetizing current and allowable secondary burden, etc. are required to be mentioned in the circuit.

Minimum knee-point voltage (V_k)   It is the point over and above which, increase of 10% in excitation voltage causes increment of exciting current of 50%. The corresponding voltage at that point is called knee-point voltage (V_k). Refer Figure 2.26.

Minimum knee-point voltage is specified by a formula $V_k = KI_s (R_{CT} + R_L)$

where,

K = A parameter that depends upon the system fault level and characteristics of the relay

I_s = Secondary reflected current

R_{CT} = CT secondary resistance at 75°C

R_L = Resistance of secondary circuit with lead and

V_k = Minimum knee-point voltage.

Note 1   The knee-point voltage should not be less than the minimum knee-point voltage.

Note 2   The value of V_k should be left to the manufacturer with assumption of R_{CT} also.

Notes on "k" for knee-point voltage   In usual practice the value of "k" is selected on the basis of fault current "I_F" and considered as 2 I_F.

Example 1

Consider the CT core for differential relay with following data.

20 MVA, 132/33 kV Transformer, % Imp = 9%, $CTR_1 = 100/1$ A, $CTR_2 = 400/1$ A, Fault MVA = 2000 MVA

Calculation

H.T full load current = 87.5 A

L.T full load current = 350 A

Corresponding fault current on the basis of fault MVA (2000 MVA) for L.T side = 35 kA

Now,

I_F = Reflected fault current to relay = $35 \times 10^3/400 = 87.5$

So

$$K = 2\,I_F = 2 \times 87.5 = 175$$

On the basis of % impedance and rating of transformer

Fault MVA　=　Transformer rating/per unit impedance (Considering zero source impedance)

$$= 20 \times 100/9 = 222.22 \text{ MVA}$$

Corresponding fault current = 3.888 kA

I_F = Reflected fault current = 9.72

So,

$$K = 2 \times 9.72 = 19.44 = 20 \text{ (say)}$$

So,

K can be taken as "20" instead of "175".

Note　Refer appendix Z for selection of CT as per different manufacturers.

Maximum excitation current (I_{em})　CT core needs to be identified by certain parameters for "PS" class core, magnetization nature of this core is referred with the current called maximum magnetization current at minimum level of knee-point voltage (V_k). The value in practice should be less than the maximum magnetization current.

$$I_{em} = p_{mA} \text{ at } V_k/F$$

where,

p_{mA} = Permissible magnetization current in mA and

F　= Factor usually specified by the relay manufacturer (2 or 4 depending upon application).

RATED BURDEN

The loads connected on the CT secondary circuit are called as burden. The loads include relays, instruments and other devices that are connected in the secondary circuit. Burdens are expressed in VA (volt ampere) at the rated secondary current with some particular power factor up to the specified accuracy limit.

So the rated output (VA) is the apparent power, which the CT can deliver to the secondary circuit at the rated current and rated burden while maintaining its accuracy within the specified class.

Standards on Burden

Some standard instruments have certain burdens that are connected to the metering core, the burden of these instruments have been expressed in the Table 2.9.

Table 2.9 Burden of instruments

Loads	Burdens (VA)
Ammeter	1.0
Current coil of Watt/Var meter	1.5
Current coil of energy meter	2.0
Current coil of P.F meter	2.5
Current coil of TV meter	3.0
Leads between CT and instruments	2.0

Note 1   The value of burdens expressed in the above table is considered on the basis of maximum limit. But nowadays the numerical instruments, that are used have reduced burden as compared to the old equipments.

Note 2   The standard values of rated output are 2.5, 5, 7.5, 10, 15 and 30 VA.

Relationship (Burden and ALF)

The satisfactory operation of the scheme for protection core, depends upon the development of voltage across the winding to drive the current through the relays or instruments during the time of fault.

So the selection of burden and ALF of the core is considered on the basis of voltage developed during the fault condition. It is expressed as follows.

$$V = (\text{Burden} \times \text{ALF})/\text{Rated secondary current}$$

Example 2

Consider a CT of following parameters:

I_F = Fault current = 10 kA

CTR = 400/1

Accuracy class = 5P20

Burden = 15 VA

R_{CT} = 3Ω, R_L = (2 × 1)Ω

R_R = 0.2 Ω (Use of numerical relay)

1. Actual voltage during the time of fault

    $$V_A = I_F' \times (R_{CT} + R_L + R_R)$$

    where, I_F' = Reflected current on secondary side = $10 \times 10^3/400$ = 25

    So, V_A = 25 (3 + 2 + 0.2) = 137.5 volt = Actual voltage

2. Calculation of voltage on the basis of burden and ALF

    V_D = (15 × 20)/1 = 300 volt = Driving voltage

    Since $V_D > V_A$ this CT is practically accepted for design.

Note   According to the burden of the core the cost of CT varies. So during selection of CT, according to the use of CT, secondary current, equipments and burden should be selected.

Selection of VA Output of Bushing Current Transformer

Considering the assumption that diameter of the bushing insulator cannot exceed 90 mm for 33 kV classes CT, the relationship between rated primary current and burden is expressed in the following Table 2.10.

Table 2.10   Relationship between primary current and burden

% Rated primary current	Measurement core						Protection core	
	0.1	0.2	0.5	1	3	5	5P10	10P10
60			2.5	5	5		–	7.5
100			5	10	15	15	10	10
200	2.5	5	15	30	30	30	15	15
300	5	15	30	30	30	30	30	30
400	10	30	30	30	30	30	30	30
500	15	30	30	30	30	30	30	30

BASIC INSULATION LEVEL (BIL)

To identify CT for its withstanding voltage level, different voltage ranges are generally mentioned in name plate details. Some of the voltages like NSV (nominal system voltage), HSV (highest system voltage), power frequency withstanding voltage and lightning impulse withstanding voltage are provided to study the insulation level of the equipment. IS 2705/1992 (Part 1 1992) provides Tables 2.11 & 2.12 for various voltage ranges from 0.66 kV to 765 kV.

Table 2.11   Rate of insulation level for HSV from 0.66 kV to 145 kV

Nominal system voltage kV (RMS)	Highest system voltage kV (RMS)	Power frequency withstand voltage kV (RMS)	Lightning impulse withstand voltage kV (Peak)	
			List 1	List 2
Up to 0.6	0.66	3	–	–
3.3	3.6	10	20	40
6.6	7.2	20	40	60
11	12	28	60	75
33	36	70	145	170
66	72.5	140	325	325
110	123	185	450	
		230	550	
132	145	230	550	
		275	650	

Table 2.12   Rate of insulation level for HSV from 245 kV to 765 kV

Nominal system voltage kV (RMS)	Highest system voltage kV (RMS)	Power frequency withstand voltage kV (RMS)	Lightning impulse withstand voltage kV (Peak)
220	245	360	850
		395	950
		460	1050
400	420	950*	1175
		1050*	1300
		1050*	1425
525	524	1050*	1425
		1175*	1550

*   Switching impulse withstand voltage in kV (Peak)

SERVICE CONDITION

Some standards are followed to put the CTs in service condition. These standards are generally of ambient temperature condition, altitude, earthing to the system, etc.

Ambient Temperature Condition

1. Maximum ambient temperature $\leq$ 45°C
2. Maximum daily average ambient temperature $\leq$ 35°C
3. Minimum ambient temperature $\geq$ 5°C

Note   The values mentioned are as per the Indian standard conditions.

Limits of temperature rise   Temperature is one of the important factors for the service condition of CT. Due to climatic conditions, temperature rises on CT. The limiting values of temperature rise have been mentioned in Table 2.13. Such values have been mentioned on the basis of rated thermal primary current with rated burden and unity power factor on the system.

Table 2.13   Limits of temperature rise of windings

Class of insulation	Maximum temperature rise
All classes immersed in oil	55°C
All classes immersed in oil and hermetically sealed	60°C
All classes immersed in bituminous compound	45°C
Classes not immersed in oil of bituminous compound	
Y Class	40°C
A Class	55°C
E Class	70°C
B Class	80°C
F Class	105°C
H Class	130°C

Note 1   The values mentioned in Table 2.13 correspond to ambient temperature mentioned.

Note 2   Reference ambient temperature values higher than those given in Table 2.13 are to be reduced by 40°C.

Note 3   For positioning of transformer above 1000 m sea level the values are to be reduced by a % as mentioned below for each 100 m of exceed.

| Oil immersed | : | 0.4% |
| Dry-type | : | 0.5% |

Altitude

Sometimes few users insist upon the manufacturers to declare altitude factor for the installation of CVT. But this standard is not so important for the installation of CT in the system. However for standard practice, this factor can be chosen up to 1000 m above mean sea level.

System Earthing

For both protection and safety of the system, earthing to the equipment is considered as one of the most important factors. The following points are to be followed as the normal standard and practice regarding the earthing of the equipments.

1. The structure, framework upon which equipment is installed should be earthed with two different terminals. The equipment base, marked with earth point has to be connected to solid earth point.
2. The earthing terminals should be of required size and protected against corrosion.
3. The earthing of secondary star terminal should be done at one point only and preferably it is to be done at switchyard instead of at control or relay panel.
4. The star terminal of different core available in CT should be separately earthed.

TESTING OF CT

It is discussed in detail under testing procedures.

MARKING OF CT

Marking of CT includes the details about CT regarding the ratings, burden ratio, etc.

1. Identification number like serial number, designation, type, etc.
2. Ratio of primary and secondary current with number of cores,
   e.g., [400-200-100/1-1-1-1]: This ratio indicates that it is multi-core CT having three possible ratios and of four number of secondary cores.
3. Rated working frequency.
4. Nominal system voltage and highest system voltage.
5. Rated insulation level (mentioned with power frequency withstand voltage and lightning impulse withstand voltage).
6. Rated short-time current with time duration.
7. Reference to the standard.
8. Name of manufacturer and their details.

9. Other details
    i. Weight of oil
    ii. Weight of core
    iii. Total CT weight
    iv. Reference drawing, Purchase order, etc.
    v. Other caution remarks like secondary terminals must be shorted of spare cores, power factor terminals must be earthed during operation, etc.
10. Detailed core identification with its ratings.
    i. *Metering core* Rated burden, accuracy class and instrument safety factor has to be mentioned.
    ii. *Protection core* Rated accuracy class with ALF has to be mentioned. The rated burden in "VA" is to be declared also.
    iii. *Protection core* (special purpose)

Following values has to be mentioned.

   o Rated minimum knee-point voltage
   o Maximum exciting current at rated knee-point voltage or a fraction of it.
   o Secondary winding resistance at 75°C.

11. Ratio connection diagram or detail.

The typical name plate details and circuit diagram of a 33 kV CT are shown in Table 2.14, 2.15, 2.16 and Figure 2.21 respectively.

Typical Name Plate Details of 33 kV CT

Table 2.14 33 kV CT detail

Particulars	Rating/value
Make	–
Reference standard	IS 2705-1992
Ratio	400-200-100/1-1-1A
Insulation level	70/170 kV
Serial number	1019/1981
Type	OCT/33
Frequency	50 Hz
Highest system voltage	36 kV
Short-time current kA/sec	25 kA/1

Table 2.15　33 kV core detail

Core	Rated VA	Accuracy class	SF	V_k (V) (min.)	$I_e @ V_k$ (mA) (max.)	ISF	R_{CT} at 75°C (max.)
I	30	5P	10	–	–	–	–
II	30	0.5	–	–	–	< 5	–
III	–	PS		1000–1200	25/15		<3 /<6

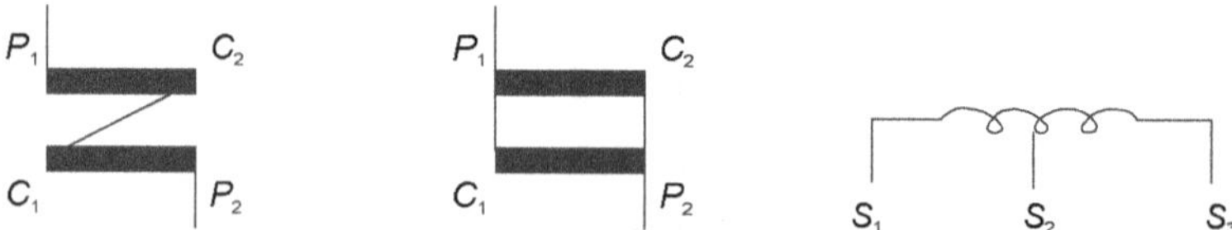

Figure 2.21　Circuit diagram

Table 2.16　Connection diagram of 33 kV CT

Ratio	Primary connection	Secondary connection
100/1	$C_1 + C_2$	$S_1 - S_2$
200/1	$P_1 + C_1$ and $P_2 + C_2$	$S_1 - S_2$
400/1	$P_1 + C_1$ and $P_2 + C_2$	$S_1 - S_3$

The typical name plate details and circuit diagram of a 132 kV are represented in Table 2.17, 2.18, 2.19 and Figure 2.22.

Typical Name Plate Details of 132 kV CT

Table 2.17　133 kV CT detail

Particulars	Rating/value
Make	–
Reference standard	IS 2705-1992
Rated primary current	600-300-150 A
Insulation level (kV)	275 RMS/650 Peak
Frequency	50 Hz
Minimum creepage	3625 mm
Serial number	0-325/B/1
Type	CTO/145

Table 2.17 (Continued)

Particulars	Rating/value
Nominal system voltage	132 kV
Highest system voltage	145 kV
Short-time current kA/sec	18.2/3
Weight of oil/CT kg	120/550
Drawing no. suitable for hotline washing	100325/145/CTO

Caution

1. Secondary terminals must be shorted before burden is disconnected.
2. Power factor testing terminal to be earthed during operation.

Table 2.18 Core details of 132 kV CT

Core	Ratio/ ampere	Secondary connection	Rated VA	Accuracy class	V_k (v) (min.)	$I_e@V_k$ (mA) (max.)	R_{CT} @ 75°C (max.)
1	600/1	1s1-1s3	–	PS	1200	10	5.0
	300/1	1s1-1s2	–	PS	600	20	2.5
	150/1	1s1-1s2	–	PS	600	20	2.5
2	600/1	2s1-2s3 ($S_1 - S_3$)	15	0.5 Fs<5	–	–	–
	300/1	2s1-2s2 ($S_1 - S_2$)	15	0.5 Fs<5	–	–	–
	150/1	2s1-2s2 ($S_1 - S_2$)	15	0.5 Fs<5	–	–	–
3	600/1	3s1-3s3	–	PS	1200	10	5.0
	300/1	3s1-3s2	–	PS	600	20	2.5
	150/1	3s1-3s2	–	PS	600	20	2.5

Connection Diagram of 132 kV CT

Table 2.19 Ratio connection table

Ratio	Primary connection	Secondary connection
150/1	$C_1 + C_2$	$S_1 - S_2$
300/1	$P_1 + C_1$ and $P_2 + C_2$	$S_1 - S_2$
600/1	$P_1 + C_1$ and $P_2 + C_2$	$S_1 - S_3$

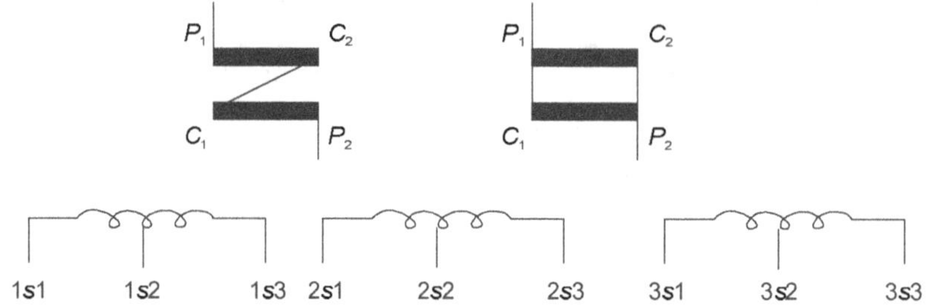

Figure 2.22   CTR control of CT

The typical name plate details and circuit diagram of a 220 kV CT are shown in Tables 2.20, 2.21 and Figure 2.23 respectively.

Typical Name Plate Details of 220 kV CT

Table 2.20   220 kV CT detail

Particulars	Rating/value
Make	–
Reference standard	IS 2705-1992
BIL	1050/460 kV
Frequency	50 Hz
Making capacity	100 KAP
Serial number/year	5643/1997
Type	IT-245
HSV/NSV	245/220 kV
Short-time current	40/1 kA/sec
Weight of oil	350 kg
Total weight	1200 kg

Table 2.21   Core details of 220 kV CT

Ratio	400-200-100/1-1-1-1-1		
Primary/ secondary current (A)	400/1	200/1	100/1
Primary connection	$P_1C_1 - P_2C_2$	$C_1 - C_2$	$C_1 - C_2$

Table 2.21 (Continued)

Ratio	400-200-100/1-1-1-1-1				
Secondary connection					
Core 1	$1S_1-1S_3$	$1S_1-1S_3$	$1S_1-1S_2$		
Core 2	$2S_1-2S_3$	$2S_1-2S_3$	$2S_1-2S_2$		
Core 3	$3S_1-3S_3,S'$	$3S_1-3S_3,S'$	$3S_1-3S_2,S'$		
Core 4	$4S_1-4S_3$	$4S_1-4S_3$	$4S_1-4S_2$		
Core 5	$5S_1-5S_3$	$5S_1-5S_3$	$5S_1-5S_2$		
Core	1	2	3	4	5
Output*	–	–	40	–	–
Accuracy class*	PS	PS	0.5	PS	PS
ISF/ALF*	–	–	≤ 5	–	–
V_k (V) min*	1200	1200	–	1200	1200
$I_e@V_k$ *(mA) max *	25	25	–	25	25
R_{CT} at 75°C max*	5	5	–	5	5

* At 400/1 and 200/1 ratio only

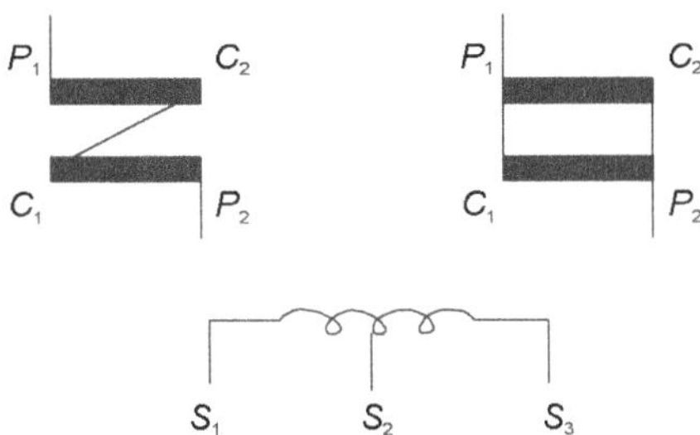

Figure 2.23 CTR control diagram of 220 kV CT

GTP AND SPECIFICATION

GTP stands for "guaranteed technical particulars". Sometimes it is also called as general technical particulars. The manufacturer of the equipment declares the values of different technical parameters and satisfies the customer by testing the equipment in their presence, during the time of inspection. The sheet thus prepared with these particulars is called GTP. It is similar to the rating plate, but values are mentioned in detail.

Specification is an agreement term basically designed by the customers. Purchaser specifies the values of the technical parameters as per the requirement and suitability condition to the reference standard that is used in practice. Explanation regarding the terms used in specification have already been described under the title standards and practices.

Table 2.22   Specification detail for 220 kV CT

Particulars	Technical values
1. Manufacturer's name and address	
2. Type	Oil-filled, live tank
3. Rated voltage (kV) or normal voltage (kV)	220
4. Highest system voltage (kV)	245
5. Rated primary current (A)	1200-600-300
6. Rated secondary current (A)	1-1-1-1-1
7. Number of cores	5(Five)
8. Details of core	

Core	Ratio	Burden (VA)	Accuracy class	ALF/ISF	k_n (min.) (volt)	I_e (max.) at $V_k/2$ (mA)	R_{CT} at 75°C (Ω)
1		–	PS	–	600 for 300/1	25	2.5 for 300/1
2	1200–600 –300/1	–	PS	–	1200 for 600/1 and 1200/1		5.0 for 600/1 and 1200/1
3		30	0.2	<5	–	–	–
4		–	PS	–	600 for 300/1	25	2.5 for 300/1
5		–	PS	–	1200 for 600/1 and 1200/1	–	5.0 for 600/1 and 1200/1

9.	Rated short-time current (kA/sec)	40 kA/sec
10.	Rated dynamic current (kA_p)	100 kA
11.	Power frequency voltage (1minute) kV(RMS)	
	(i) Primary	460
	(ii) Secondary	3
12.	1.2/50 μsec impulse voltage(kV_p)	1050
13.	Temperature rise	55°C
14.	Reference standard for oil	IS 335
15.	Reference standard for CT	IS 2705
16.	Quantity of oil	350 kg
17.	Total weight	1050 kg
18.	Maximum creep age of porcelain insulator	6125 mm
19.	Frequency	50 Hz
20.	Outdoor/indoor	Outdoor
21.	Treatment of external ferrous surface	Epoxy painted
22.	Facility of tan δ point measurement (C_x)	Yes/external
23.	Facility of oil sample collection value	Yes
24.	Monitoring status	Bellow indicative
25.	CT ratio control	Both primary and secondary

GTP of a Typical 220 kV CT—Specification Inquiry of CT

While placing the order of the CT, the following technical informations have to be provided (Table 2.22).

1. Type of CT (Outdoor/indoor, ring type/cast resin/oil immersed)
2. Rated system voltage and highest system voltage
3. Rated insulation level
4. Rated short-time current
5. Rated CTR
6. Rated core details
   i.   Number of cores
   ii. Type of cores (Metering, protection, etc.)
   iii. Burden, accuracy class, k_n (min.), I_e (max.), secondary resistance
7. Reference standard
8. Facilities to be provided like monitoring status, collection of oil sample, measurement of tan δ, etc.
9. Frequency other than 50 Hz
10. Other particulars as per suitability and requirement.

MAINTENANCE PRACTICE

The current transformer does not need any special maintenance. Periodically it requires some scheduled checking and some testing practice as per demand. Following maintenance schedule may be followed for current transformer (Table 2.23).

Note 1   For the periodicity as mentioned (daily, weekly and monthly), the maintenance schedules as described are of visual checking of the CT in service. If any abnormality is observed, then planned shutdown of the equipment can be availed to do the needful.

Note 2   The maintenance practices/checkings/testings as mentioned under the yearly schedule are done during the time of annual shutdown of the system. So, detailed checking and corrective actions can be done under this schedule.

Note 3   Sometimes, due to the fault on the system, forced shutdown is availed for rectification. During this period and according to the availability of shutdown duration, some of the checking as mentioned under the yearly schedule can also be attended.

Table 2.23 Maintenance Schedule of CT

Periodicity	Checking/testing	Actions to be taken
Daily	1. Oil leakage. 2. Abnormal noise. 3. Other visual checking. 4. Level of metal bellow [@]	Follow-up action.
Weekly	1. Oil level. 2. Current reading of secondary circuit. 3. Earthing of secondary terminals at CT console box.	1. Fill up oil. 2. Record for corrective action. 3. Follow-up action.
Monthly	1. Analysis of CTR from working range of currents. 2. Secondary circuit terminal checking at terminal box. 3. Analysis of level of metal bellow.	1. Follow-up action. 2. Tighten the terminal. 3. Follow-up action.
Yearly	1. Earthing of base plate. 2. Connection checking of secondary circuit at junction box and panel end. 3. Checking of CTR. 4. Cleaning of porcelain insulators. 5. Checking of corrosion of metal part. 6. Measurement of IR value. 7. Measurement of earth resistance of the earth pit used in the system. 8. Checking of primary terminals and primary links. 9. Dielectric strength of oil used.	Follow-up action.
Five-yearly	1. DGA of oil used. 2. tan δ measurement of the CT. 3. Checking of typical characteristics of oil sample.	Compare the result with commissioning value and accordingly take follow-up action.

[@] Status of insulation level as per the bellow position. Following Table 2.24 can be referred to obtain the status of insulation of CT.

Table 2.24   Diagnosis of CT through bellow position

Bellow position	Results and measures to be taken
All the CTs in the same station should have approximately same bellow position.	CT insulation OK.
At about 10°C, the position is approximately at half between maximum and minimum.	CT insulation OK.
Different bellow position for the same identical CTs in the station.	Observe the CT and check the oil leakage on minimum bellow position CT.
Minimum bellow position.	Check oil leakage and take corrective action.
Bellow position is independent of temperature.	Bellow is not sealed properly and bellow might be jammed.
Bellow exceeds the maximum position.	Problem in the CT due to internal gas formation, immediately disconnect the CT.

TESTING PROCEDURES

Different tests are carried out to check or compare the quality and design of the current transformer with the reference standards. These tests are classified into three types. They are type test, routine test and optional test.

TYPE TESTS

To compare and confirm the major parameters of the electrical equipment, some tests, are required to be done by the suitable methods, available in different standards. These tests are called "type tests". Following are few tests, categorized in type test.

1. Short-time current tests
2. Temperature-rise test
3. Lightning impulse test
4. Switching impulse test
5. High voltage power frequency wet withstand test
6. Determination of errors and other characteristics

Short-Time Current Test

During this test, the secondary windings are short-circuited and current is applied at a value of 'I' ampere for time 't' second, with a condition that

$$I^2 t \geq I_{st}^2 t_r$$

where,

I_{st} = Rated short-time current

t_r = Rated duration = 1 sec (standard)

I = Current to be applied for testing and

t = Duration of testing time

= (0.5 to 5) seconds.

Dynamic current test This test is also done with short-circuited secondary winding and application of primary current as per the condition

$$I^2 t \geq I_{dyn}^2 \times t_r$$

where,

I_{dyn} = Dynamic value of current $\cong$ 2.25 I_{st}

Note 1 The applied primary current should have the peak value, not less than the rated dynamic current for at least one peak.

Note 2 Sometimes both manufacturer and purchaser may mutually agree to conduct this test with a specified connected burden also.

Pre-set and post-set conditions

1. Before conducting the tests (short-time and dynamic current tests), the routine tests have to be carried out for reference.
2. After conducting the tests, the same routine tests have to be checked for comparison and conclusion.
3. After testing, the core has to be demagnetized and the CT to be cooled to ambient temperature for post-test checking. The following points are to be checked.
   i. Physical condition of CT.
   ii. The magnetic characteristics do not differ from those recorded before and after the tests conducted. If so varying, it should not be more than half of the permissible limit.
   iii. Particularly the ratio error, exciting current value, knee-point voltage, etc. should be checked.
   iv. It withstands the dielectric test, with the test voltages or current reduced to 90% of those specified.
   v. The insulation next to the surface of the conductor does not show significant deterioration. This examination can be avoided if the current density in winding corresponding to the rated short-time thermal current of one second, does not exceed 160 A/mm^2.
   vi. The nature of winding conductivity should not be less than 97% of the value given in IS 613 :1984.

Temperature-Rise Test

For testing of temperature rise in CT, the following conditions are to be satisfied.

1. Rated continuous thermal current to be allowed to flow in primary winding.
2. Rated burden on secondary winding with unit power factor.
3. Mounting of transformer as per the mounting in service.
4. Ambient temperature shall not exceed 40°C.

By satisfying these conditions, the temperature rise is measured and compared with the standard for conclusion.

Various methods are adopted to measure the temperature rise in the winding.

Some methods among them are:

1. Variation of resistance method
2. Thermometer or thermocouple method

Note 1   Temperature rise of winding is measured by the variation of resistance method, or thermocouple method as per suitability.

Note 2   The temperature rise of parts other than winding shall be measured by thermometer or thermocouples.

Lightning Impulse Test

Every current transformer is rated with certain impulse voltage on its rating plate. During the lightning impulse test of the CT, this rated impulse voltage is referred for the testing. Application of this voltage under test is considered for withstanding capacity of the CT.

According to the standard IS 2071 (part-I and II); 1974, the test voltage shall be applied between the primary winding and earth (framework, secondary winding terminals are to be earthed together). The peak value and wave shape of impulse voltages shall be recorded for comparison of the results with the testing standards for finalization of the result.

Application of voltage   Depending on the type and rating of voltage class, the number of full wave impulses of each polarity either with or without correction for atmospheric condition are applied during the time of testing. The Table 2.25 can be referred for impulses for application.

Confirmation of the test   CT shall have passed the test if the following conditions are satisfied:

1. No disruptive discharges occur in the non-self-restoring insulation.
2. No flashovers occur along the non-self-restoring with external insulation.
3. No more than two flashovers occur across the self-restoring external insulation.
4. No other evidence of failure is detected.

Table 2.25   Voltage impulse application for impulse test

CT	Number of impulse	Remark
Indoor ≤ 36 kV	5 consecutive full wave impulses of each polarity with correction for atmospheric condition.	
Outdoor ≤ 245 kV	15 consecutive full wave impulses of each polarity without correction for atmospheric condition.	For external insulation it is 3 in number and without correction.
420 kV	3 consecutive full wave impulses of each polarity without correction for atmospheric condition.	

Switching Impulse Test

This testing procedure is similar to the impulse test method as described in lightning impulse test. The CT of voltage class 420 kV and above is generally considered for testing with this switching impulse test. The test voltage application is similar to the voltage mentioned before in BIL table.

High-Voltage Power Frequency Wet Withstand Test

For such test, the application of voltage is as per the values mentioned in BIL table. The external insulation shall be tested for one minute to the power frequency voltage having peak value equal to $\sqrt{2}$ times the value specified in table and test is applicable for outdoor type CT up to 245 kV class.

Determination of Errors and Other Characteristics

Every CT is used mainly for two different secondary purposes like metering scheme and protection scheme.

For the measurement of errors and characteristics of these cores, some of the following points are to be considered during the type test of CT.

1. Test shall be made at the rated frequency.
2. Temperature for testing to be at room temperature, also for both extreme temperatures as decided by the manufacturer and user.
3. Rated burden to be connected as per the rating plate.
4. The accuracy result should confirm the limits of errors.

Determination of errors for measurement core   According to the accuracy class of CT, the burden on secondary side is connected and CT is tested with the percentage of current as per the standard (IS-2705) (Table 2.26 and 2.27).

CT of class (0.1, 0.2, 0.5, 1.0, 0.2s *and* 0.5s)

Table 2.26   Errors in CT (Metering core)

Accuracy class	± % Current ratio error at % of rated current					± Phase angle displacement error in minutes at % of rated current				
	1	5	20	100	120	1	5	20	100	120
0.1	–	0.4	0.2	0.1	0.1	–	15	8	5	5
0.2	–	0.75	0.35	0.2	0.2	–	30	15	10	10
0.5	–	1.5	0.75	0.5	0.5	–	90	45	30	30
1.0	–	3.0	–	1.0	1.0	–	–	–	–	–
0.2s	0.75	0.35	0.2	0.2	0.2	30	15	10	10	10
0.5s	1.5	0.75	0.5	0.5	0.5	90	45	30	30	30

Note 1   Test to be done at 25% to 100% of rated burden.

Note 2   Power factor of 0.8 lag for all cores above 5 VA and unit power factor for 1 VA to 5 VA to be used for test.

Table 2.27   Error in metering CT of class 3 and 5

Accuracy class	± % Current ratio error at % of rated current	
	50	120
3	3	3
5	5	5

Note 1   Test to be done at 50% to 100% of rated burden.

Note 2   Power factor same as above class, as mentioned in Table 2.26.

Instrument security current test

For this test, primary winding is kept open and secondary winding is energized at rated frequency of RMS voltage equal to the secondary limiting emf. By application of this voltage the excitation current is measured. It should be of at least 10% when compared with respect to the rated secondary current being multiplied by ISF.

$$\frac{I_e}{I_{sec} \times FS} \times 100 \geq 10\%$$

where,

I_e = Excitation current obtained
I_{sec} = Rated secondary current and
FS = Instrument security factor.

Determination of errors for protection core   For protection core, three different errors current error, phase angle error and composite error are considered for accuracy class measurement.

Current error and phase angle error   The testing is done according to the method described in the previous section and the values obtained are compared with the values in the following Table 2.28.

Table 2.28   Errors in protection core

Accuracy class	Current error at rated primary current (%)	Phase displacement at rated primary current (minutes)	Composite error at rated accuracy primary current (%)
5P	± 1	± 60	5
10P	± 3	–	10
15P	± 5	–	15

Note   Burden to have power factor of 0.8 lag above 5 VA and unit power factor for 1 to 5 VA

Composite error   Similarly for test comparison, the values in the above table are referred but here tests are done by both direct and indirect methods. For detailed testing under directed and indirect method, different standards can be referred [IS 2705 Part-3, 1992].

Determination of errors for PS class protection core   This test is generally grouped under routine test.

ROUTINE TESTS

The normal tests generally conducted with the CT to know the condition and performance of CT are called "routine tests". Following are few tests that have been grouped under routine test of CT. They are:

1. Terminal and polarity marking
2. Power frequency dry withstand test on primary winding
3. Power frequency dry withstand test on secondary winding
4. Over-voltage inter-turn test
5. Partial discharge test
6. Determination of errors and of other characteristics

Terminal and Polarity Marking

The primary terminals and secondary terminals are generally marked in CT. For confirmation of these markings, the polarity test is done as per the method described below.

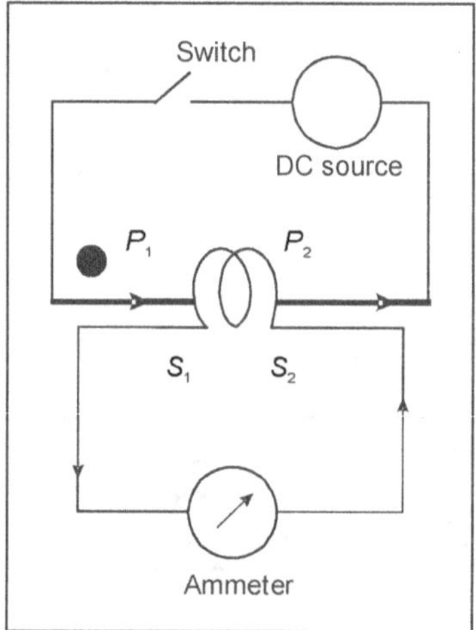

Figure 2.24   Polarity test of CT

A small DC voltage source of magnitude 6 V to 9 V is applied to the primary side of the CT with a switch arrangement as shown in Figure 2.24. On the secondary circuit an ammeter is connected to read the deflection of the current. According to the deflection of current on the ammeter, the confirmation of the polarity is determined. For correct polarity by the momentary switching application, the current on the ammeter should deflect towards right, indicating the flow of current from left to right on the secondary circuit.

Power Frequency Dry Withstand Test on Primary Winding

The test method is similar to the power frequency test as described in type test. But for CT having HSV 420 kV and above, the following points are to be considered during the test.

CT with HSV ≥ 420 kV

Type (1)   For this type of CT, the value of appropriate voltage should be decided according to the rated lightning impulse withstand voltage and power frequency short duration withstand voltage as mentioned in the Table 2.29.

Type (2)   In another method, the appropriate power frequency pre-stress voltage can be injected for 10 secs. The reduced voltage corresponding to partial discharge test voltage without interruption can also be considered for injection for 5 minutes. The maximum partial discharge shall be 10 pC during last minute at the specified test voltage (Table 2.30).

Table 2.29 High voltage test values of CT

Rated lightning impulse withstand voltage kV (peak)	Power frequency short duration withstand voltage kV (RMS)
1175	510
1300	570
1425	630
1550	680
1800	790
2100	880
2400	975

Table 2.30 Standard for withstand test of CT

HSV	Power frequency pre-stress voltage	Partial discharge test voltage
420	510	315
525	630	395
765	790/880[$]	575

[$] To be determined by the rated lightning impulse withstand voltage

Note 1. For CT with sectionalized winding each section shall withstand for one minute a test voltage of 3 KV (RMS) between section and earth.

2. If for any reason, it is necessary to repeat the test, then the test voltage shall be reduced to 80% of the original test voltage.

Power Frequency Dry Withstand Test on Secondary Winding

For this test, the test voltage is of 3 kV (RMS) for 1 minute between secondary windings (connected together) and earth. During this test the frame, body and primary windings are to be shorted and earthed.

Similarly for multiwinding or sectionalized winding, the same 3 kV (RMS) can be applied for 1 minute withstand, other windings and frameworks to be connected together, and earthed.

Over-voltage Inter-turn Test

Two different methods are adopted for testing of over-voltage inter-turn test.

***Method* 1** By this method secondary winding is made open-circuited and voltage at power frequency is applied to the primary winding. But this method of testing is generally not adopted due to the possibility of permanent damage to the CT.

***Method* 2** By this method, primary winding is kept open and voltage at a frequency limited to 5 times the rated frequency is applied to the secondary winding to produce the rated secondary current or the current corresponding to a value of 4.5 kV peak, whichever is lower. The applied voltage should be withstood for one minute without resulting in any considerable disruptive discharge.

Partial Discharge Test

Partial discharge is always associated with the degradation of insulation systems in high voltage equipment. To know the status of insulation, this test is conducted. Two different methods are adopted for this test.

Electrical detection method By the application of power frequency voltage, called PDIV (partial discharge inception voltage), certain charge is allowed to develop across the insulation. Then the developed charge is discharged by decreasing the voltage called PDEV (partial discharge extinction voltage). According to the pattern of PDIV and PDEV, the status of insulation is determined.

Acoustic method Because of the fault in the current transformer, the solid insulation part being heated, produces acoustic waves. By this method the sensors that are fixed on the inner wall of the transformer, receive the signals for analysis of the fault by the computing machines outside.

***Note* 1** Test is applicable for CT with solid insulation for HSV 7.2 kV and above.

***Note* 2** For liquid insulation from HSV 72.5 kV and above.

***Note* 3** IS 11322 : 1985 standard to be followed for PD Test.

Determination of Errors and Other Characteristics

The methods of testing for determination of errors and other characteristics have been discussed under type test. But for routine test, some conditions are only changed with similar test principle.

Determination of errors for measurement core Test should be carried out only at % of rated current as mentioned in Table 2.31.

Determination of errors for protection core

Current error and phase angle error

***Note* It is similar to type test as mentioned before.

Table 2.31 Percentage application of current for routine test

Class	Percentage of rated current			
0.1 and 0.2	120	100	20	5
0.2s	120	20	5	1
0.5, 1.0	120	20	–	–
0.5s	120	20	–	–
3, 5	120	50	–	–

Composite error It is also same to that of type test. But for low reactance current transformer, the correction factor should be applied to the results to compare the values.

Determination of errors for PS class protection core For PS class protection core, the determination and testing of following points/factors are important.

   i. Knee-point voltage (V_k)
  ii. Maximum excitation current at rated knee-point voltage or fraction of the same.
 iii. Secondary winding resistance at 75°C.

i. Knee-point voltage By keeping primary winding open, sinusoidal voltage at rated frequency is applied to the secondary terminals.

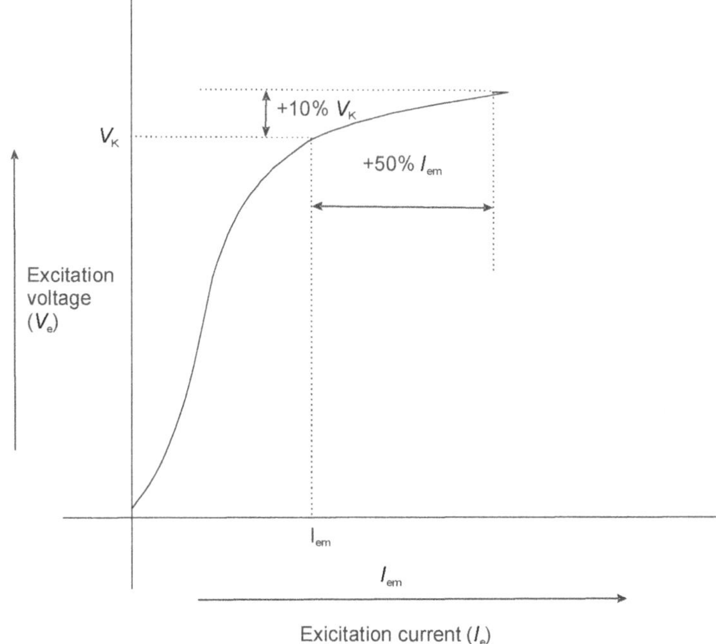

Figure 2.25 Excitation curve for knee-point voltage

The supply voltage and corresponding exciting currents are noted down to obtain the curve for excitation voltage and excitation current. During this voltage application, the point just above which 10% rise in voltage causes 50% rise of excitation current, is called **knee-point voltage** (Refer Figure 2.25).

ii. Maximum excitation current   Similarly for obtaining the maximum excitation current the same procedure and same curve can be referred.

iii. Secondary winding resistance   By the prescribed and suitable method/technique the resistance value of CT secondary can be measured. The value of this resistance should be corrected to the temperature of 75°C to compare with the specified value.

SPECIAL TEST/OPTIONAL TEST

These tests are generally conducted based on a mutual understanding between manufacturer and purchaser. The following tests come under the special/optional test category.

1. Chopped lightning impulse test.
2. Dielectric dissipation factor of oil immersed CT.
3. Commissioning test of new CTs selected from lot.

Chopped Lightning Impulse Test

This test is carried out with lightning impulse full-wave voltage with negative polarity. Following points are also considered for different conditions.

   i. For HSV $\leq$ 245 kV with testing requirement of external insulation
- One 100% full impulse
- Two 100% chopped impulses
- Fourteen 100% full impulses

   ii. For HSV up to 245 kV where external insulation need not be checked by 15 full-wave impulses and for HSV 420 KV and above.
- One 100% full impulse
- Two 100% chopped impulses
- Two 100% full impulses

Note   Standard lightning impulse shall be chopped after 2 to 5 μ second and overswing of positive polarity to be limited to 30% of chopped impulse.

Dielectric Dissipation Factor of Oil-immersed CT

The measurement of dielectric dissipation factor (tan δ) is conducted by the use of Schering Bridge in the system. The ambient temperature plays the important role for the variation of tan δ value. During this measurement, corresponding compensation factor has to be incorporated to compare the value at 20°C. However, the voltage application for tan δ measurement ranges from 2 kV to 12 kV and standard value is taken as 10 kV.

Commissioning Tests on New Transformers

Sometimes purchasers decide to test the CT from the selected lot of CT at the manufacturer's premises. The type of tests, standards to be followed for those tests, etc. are decided in pre-hand with mutual agreement between purchaser and manufacturer. Then the selected CTs from the ordered lot are chosen randomly and taken for the commissioning tests as per the standards decided.

CASE STUDIES

CASE STUDIES ON CT CIRCUITRY

INTRODUCTION

Current transformers (CT) and its circuitry play the most vital role for the protection, control and metering of HT and EHT lines. During the time of commissioning or modification of the CT circuitries, mistakes are even overlooked by the protection engineers during the time of stability test (primary injection test, sensitive test, etc.). Sometimes the testing procedures are approximated to reduce the testing duration, which develops problems in real practice. After commissioning and energization of the said circuit, these mistakes result problems in protection, control and metering circuit. The problems are listed as follows.

1. False tripping of charged feeder with certain rise of load current, before the settable limit of overload relays.
2. False tripping of transformers even for external feeder fault.
3. Wrong metering, may cause discrepancies in operation control and commercial billing among utilities.
4. Difficulties in relay coordination for interconnected lines.
5. Problems in load assessment in any loading feeders.
6. Wrong assessment of tripping analysis.

So some practical case studies with real-time occurrences have been described with supportive analysis in circuit.

Case Study 1

Situation/problem   Tripping of one 132 kV feeder on E/F relay was observed at a 220/132 kV grid substation, during peak load conditions for the rise of load current above a particular load.

Steps attempted   During off-peak load condition, the currents on the secondary circuits, used for back-up relays were measured by means of clamp-on ammeter. The results are tabulated in Table 2.32.

Table 2.32   Results of case study 1

Phase/ wire no.	Current (in mA)	Reference from vector diagram	Remarks
R phase C_{11}	61	OA	OK
Y phase C_{31}	62	OB	OK
B phase C_{51}	60	OC	OK
Neutral C_{71}	120	BD = 2OB	Doubt?
$C_{11} + C_{31}$	105	AB = $\sqrt{3}$OA	Doubt?
$C_{31} + C_{51}$	104	BC = $\sqrt{3}$OC	Doubt?
$C_{11} + C_{51}$	62	CD = OA	OK

Conclusion of the readings   From the readings of back-up core as described in the Table 2.32, it was concluded that "Y" phase secondary CT terminals have been altered and connected in the circuit. The same can be confirmed from the vector analysis (Figure 2.26a and b).

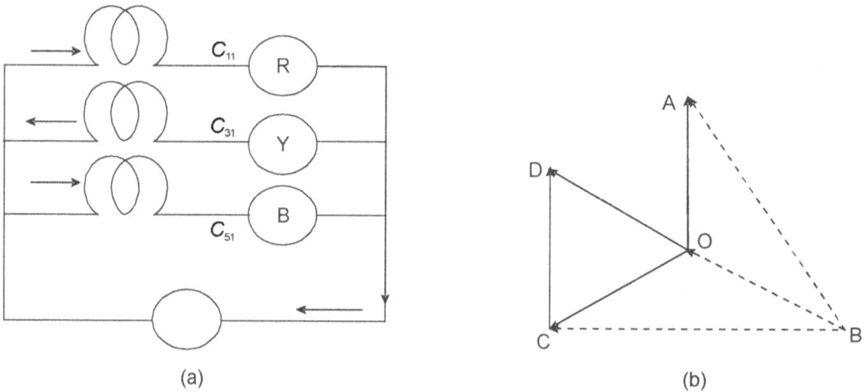

Figure 2.26   a) Relay circuit connection b) Vector diagram of case study 1

Case Study 2

Situation/problem   In one 132/33 kV S/S, the indicating instruments (wattmeter, ammeter, etc.) on 132 kV incomer feeder were recording erroneous readings.

Steps attempted　During loading condition of the said feeder, the currents were measured in the metering core circuit by means of clamp-on ammeter. The results were obtained as follows in the Table 2.33.

Table 2.33　Results of case study 2

Phase/wire no.	Current (mA)	Reference from vector diagram	Remarks
R phase D_{11}	0	OA	Doubt?
Y phase D_{31}	80	OC	OK
B phase D_{51}	82	OE	OK
Neutral D_{71}	81	OD = OC	Doubt?
$D_{11} + D_{31}$	83	OC	OK
$D_{31} + D_{51}$	81	OD	OK
$D_{11} + D_{51}$	82	OE	OK
$D_{71} + D_{11}$	81	OD	Doubt?
$D_{71} + D_{31}$	140	$OG = \sqrt{3}OC$	Doubt?
$D_{71} + D_{51}$	142	$OH = \sqrt{3}OE$	Doubt?

Conclusion of the readings　From the readings of metering core as described in the Table 2.33, it was concluded that "R" phase secondary CT terminals might have been shorted. So the detailed physical connections of the R phase were checked, but no such short circuiting of the R phase was found. Instead of short-circuiting, mixing of R phase winding was observed with other core of same phase as shown in Figure 2.27a.

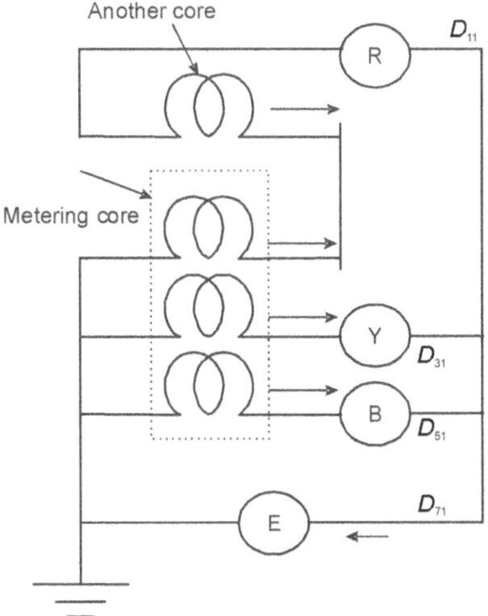

Figure 2.27a　Secondary connection for case study 2

The analysis was confirmed by drawing the vector diagram in Figure 2.27b.

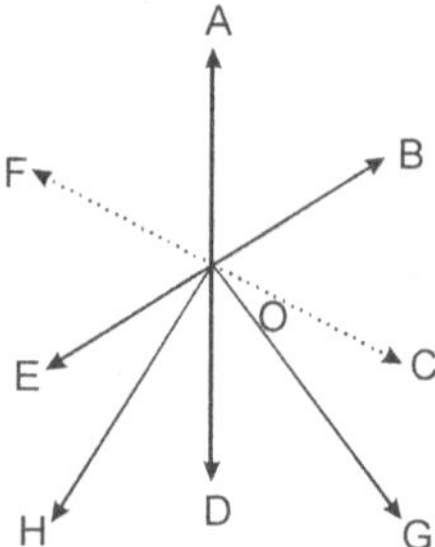

Figure 2.27b   Vector diagram for case study 2

Case Study 3

Situation/problem   One 33 kV feeder was tripping on E/F relay frequently for the load current more than approximately 45 A. Line CTR = 200/1, setting of E/F PSM= 0.1

Steps attempted   Load current was restricted to 30 A for measurement of secondary current in the back-up core. The currents are measured and tabulated as in Table 2.34.

Table 2.34   Results for case study 3

Phase/wire no.	Current in mA	Remarks
R phase C_{11}	184	Doubt?
Y phase C_{31}	151	OK
B phase C_{51}	153	OK
E/F C_{71}	38	Doubt?
$C_{11}+C_{31}$	172	Doubt?
$C_{31}+C_{51}$	152	OK
$C_{11}+C_{51}$	176	Doubt?

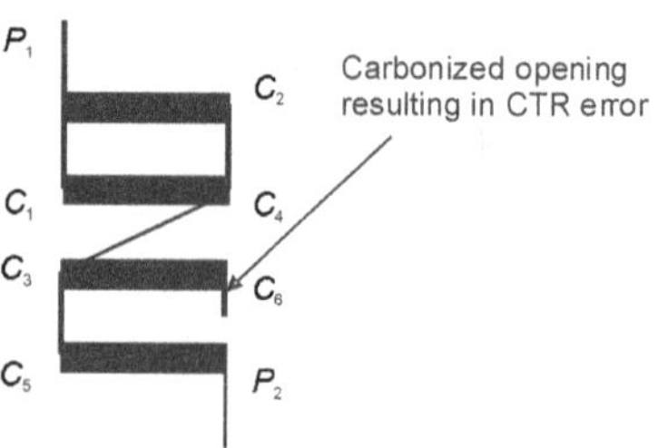

Figure 2.28   Primary side CT problem

Conclusion of the readings   From the readings of back-up core as described in the Table 2.34, it was concluded that "R" phase CT is associated with wrong CTR, which may be due to saturation of core or wrong primary connection.

The detailed secondary circuit of R phase was checked and found OK. But on physical verification, the error was found with the carbonized opening of one link on primary side of the CT as shown in Figure 2.28.

Detailed analysis   The carbonized opening between the link C_6 and P_2, due to loose contact and sparking, has developed an erroneous CTR.

For correct connection of links, the CTR is 200/1 with equivalent resistance (R Ω). But opening of one link as shown in Figure 2.28, has resulted in the rise of equivalent resistance (1.5 R Ω).

$$\text{Since } I_1^2 R_1 = I_2^2 R_2$$
$$200^2 \times R = I_2^2\, 1.5\, R \text{ and } I_2 = 163.3$$

So, new ratio becomes 163.3/1 instead of 200/1.

The CT link was replaced by a new one.

Case Study 4

Situation/problem   One newly commissioned 220/132 kV auto transformer was tripping in REF relay for the external fault on any 132 outgoing feeder. HT CTR = 300/1, LT CTR = 600/1.

Steps attempted   During load condition of the transformer, the currents on various windings were measured. The values were obtained as mentioned in Table 2.35.

Table 2.35   Results of case study 4

Phase/wire no.	Current in mA	Remarks
HT R phase C_{11}	450	OK
HT Y phase C_{31}	454	OK
HT B phase C_{51}	455	OK
HT E/F C_{71}	18	OK
REF (restricted earth fault) circuit	21	Doubt?
Secondary Neutral CT	12	OK

Detailed analysis   Net REF (Restricted Earth Fault)

$$\text{Current} = \text{Residual current} \pm \text{secondary NCT (neutral CT) current}$$

$$\text{Residual current} = (\text{Secondary 220 kV current} \sim \text{secondary 132 kV current})$$

$$\text{Residual current} = (18 \sim 18 \times 220/132) = 12 \text{ mA}$$

So, Net REF (Restricted Earth Fault) current = (12 ± 12) mA

For additive REF (Restricted Earth Fault) current = $(12 + 12) = 24$ mA

For subtractive REF (Restricted Earth Fault) current = $(12 - 12) = 0$ mA

When the measurement value of 21 mA is encountered, so suspecting the reverse connection of NCT secondary terminals, the polarity was changed. But the problem was not solved, so the detailed circuit was checked and found with unequal CTR for HT and LT. The residual path of the restricted earth fault was connected from the auxiliary CT connection of differential core.

Rectification and modification

1. The differential core was separated from the REF circuit.
2. Separate core was used for REF circuit and same CTR was used for all the windings (HT CTR = 300/1, LT CTR = 300/1, Neutral CTR = 300/1). For connection refer Figure 2.29.

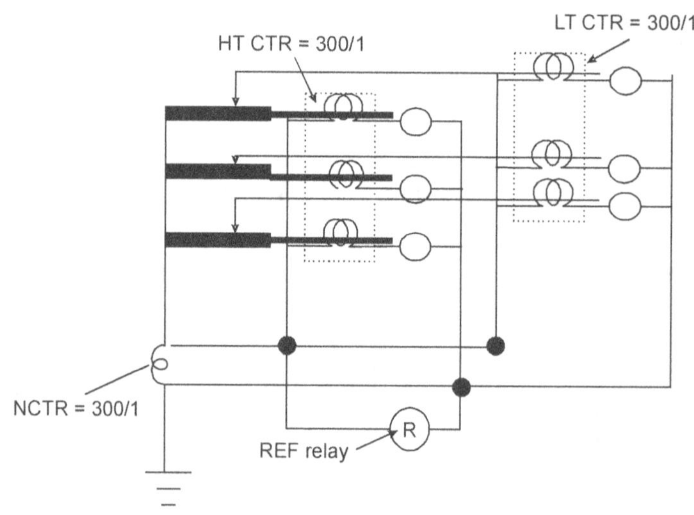

Figure 2.29   REF relay connection

Case Study 5

Situation/problem   During stability test (load balancing test) of a 132/11 kV transformer, the following currents were obtained in differential circuit. Auxiliary CTs are used on both side of the CT secondary winding.

Steps attempted   From the readings as obtained in Table 2.36 it was concluded that,

1. HT side CT connections are correct in nature.
2. LT side CT connections are expected with faulty connection. Y phase terminals might have been shorted with other connection.

Table 2.36   Results of case study 5

Phase	HT side	LT side	Operating coil
		Currents in mA	
R phase	27	23	12
Y phase	27	0	27
B phase	27	23	12

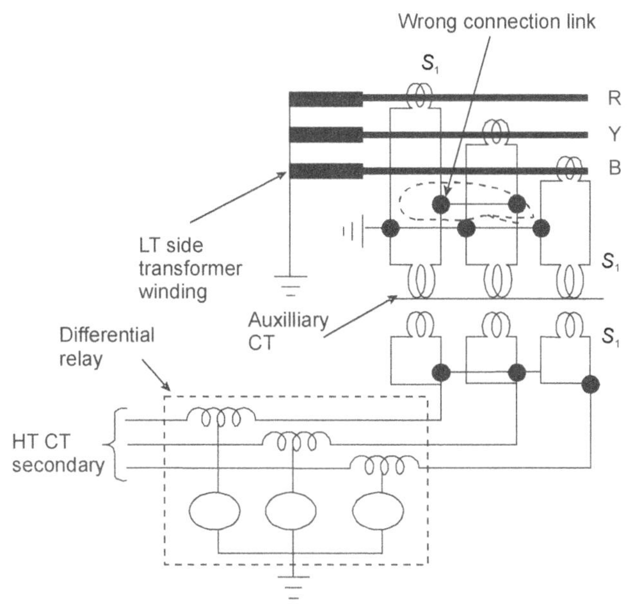

Figure 2.30   Circuit analysis of case study 5

Physically the LT CT secondary side was checked and found with a wrong connection link of S_1 terminal of both Y phase and B phase on the primary side of auxiliary CT as like shown in Figure 2.30.

Analysis of current flow   For the above wrong connection, the current flow in different circuits are explained in Figure 2.31.

Vector analysis of current flow   The current on operating coils of the differential relay are obtained due to the combination of currents from both HT and LT side secondary values. The values are analysed in the Table 2.37 and vector diagrams is shown in Figure 2.32. The results of case study 5 are analysed in Table 2.36.

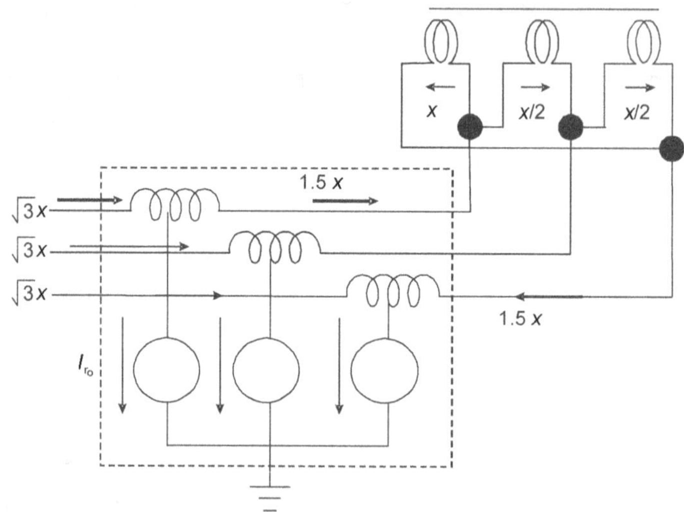

Figure 2.31   Analysis of current flow case study 5

Table 2.37   Current calculation-case study 5

Phase	HT side	LT side	Operating coil
R phase	$BA = \sqrt{3}x(30°)$	$AC = 1.5x\ (180°)$	$BC = DB/2(-90°)$
Y phase	$DB = \sqrt{3}x(-90°)$	0	$DB = \sqrt{3}x(-90°)$
B phase	$AD = \sqrt{3}x(150°)$	$DE = 1.5x\ (180°)$	$EA = DB/2(-90°)$

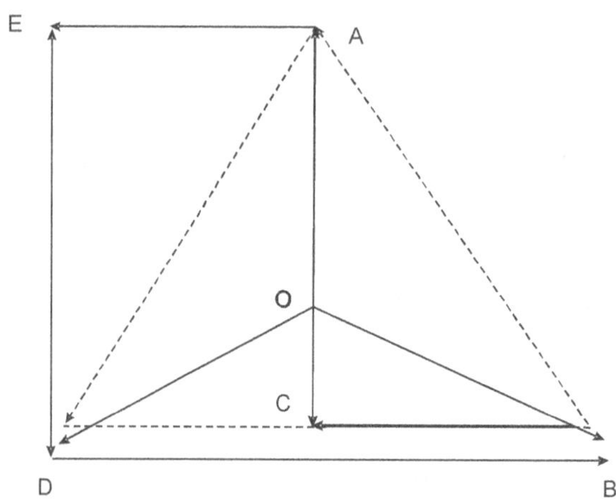

Figure 2.32   Vector diagram for case study 5

Remarks on stability test for transformer The transformer secondary circuits contain both star and delta connections. So behaviour of current flow and analysis of the same becomes difficult for delta connection. So, for easy study of the current behaviour, single phase supply connection should be used instead of three phase connection.

CASE STUDIES ON FAILURE OF CT

Case Study 1

Situation/problem One of the 33 kV CT failed with violent explosion. The flying debris and boiling oil was scattered and it damaged the adjacent equipments, resulting in power interruption in the system.

Post-incident report The damaged CT was investigated and the following report was submitted.

1. Sparking at the joint between stud with the primary winding.
2. Rupture of primary insulation.
3. Enbulging of primary box in scattered form.
4. Splitting of bushing porcelain.

Analysis

1. The joint between the stud with the primary winding was found improper due to bimetallic material.
2. It is apprehended that the insulation material might have suffered with degradation and deterioration trends, which ruptured when the condition become severe.
3. Due to post-incident effect and separation of the insulation, both the bushing porcelain and primary box were scattered.

Case Study 2

Situation/problem In one of the 132/33 kV sub-station, 33 kV CT failed and burst during the time of initial charging before allowing current flow to the CT.

Post-incident report The damaged CT was inspected and following were observed.

1. Post-insulator porcelain was found splitted into pieces.
2. Spilling of insulting oil.
3. Complete damage of secondary terminal box.

Analysis The secondary circuit, connections, etc. were checked and found in order. The primary links, connections were ascertained also. No specific reason was concluded upon the analysis.

But from experiences and observations it was concluded that the insulations including oil had been in deteriorating condition. The testing and maintenance personnel were enquired about the pre-commissioning test results of IR (insulation resistance) and PI (polarisation index) values. As reported IR value was around 70 MW at 2500 V range of megger. But PI value was neither taken nor recorded.

It is apprehended that PI value might have been in declination trend. It is seen from experience that HV equipment like CT may fail during initial energization with voltage only, if PI value becomes less than 1 (one).

Case Study 3

Situation/problem   In one of the 220/33 kV substation, a series of 33 kV CTs (8 Nos.) failed in regular interval during working condition.

Observation   The following observation were found with the CTs .

1. Maximum number of failed CTs (6 nos.) were observed with spilling of oil and enbulging of primary box.
2. Rest of the CTs (2 nos.) were found burst with explosion, flying of debris.

Investigation and analysis   All the necessary checking of the circuits were done and found in order. It was suspected that the problem was regarding earthing of CT secondary circuit.

Some of the earth terminals were found rusted and corroded. The earth pit resistance was measured and found with certain higher value of approximately $\cong 10\Omega$. So, during faulty condition, due to delayed clearance of the fault from the system, the insulation such provided might have been deteriorated, leading to final rupture and failure of the CT.

Remedy   When the earth terminals and earth pits condition were improvized, then failure of CT was completely reduced.

Case Study 4

Situation/problem   One of the 33 kV CT failed with burning of CT secondary circuit on the relay panel at control room.

Observation

1. Enbuldging of primary box of the CT.
2. Spilling of oil insulation from the CT.

Analysis   From the detailed inspection of the available healthy circuit on adjacent panel it was found that:

1. Definite O/C relays are in use with minimum time set of 0.3 sec.
2. The CT secondary wires have been terminated with 'I' shape TBs (terminal blocks) at relay panels.
3. The CT circuits have been earthed at both switchyard console box and relay panel end.

Explanation   Use of definite O/C relay results in the tripping of the circuit according to the definite time set in the relay, irrespective of magnitude of current. Sometimes due to close fault occurrence, the tripping time gets delayed and waits for the definite time set of the relay. So, persistence of the fault current affects the secondary circuit, in turn the insulation of the CT also.

The joints, connecting points of the CT circuit also play the role for increase of the secondary burden in due course of time. For every fault occurrence on the system, certain mechanical forces develop on both primary and secondary circuits. Due to use of 'I' pin TBs, the terminal point on the terminal block is loosened, which may be of very slightest in nature, but succeeding fault occurrence in the system causes prominent loosening of terminals and finally leads to the opening of the CT secondary circuit. This nature of opening of CT secondary circuit causes deterioration of insulation of the CT and final failure of the CT. Moreover earthing at both switchyard console box and relay panel causes sharing of fault current in parallel paths to earth. So, delayed de-energization might also be considered as the cause of CT failure.

Remedy

1. The definite O/C relays were replaced by IDMT O/C relays.
2. The "I" pin TBs were replaced by ring and nut bolt TBs to avoid the loosening due to fault.
3. The earthing at relay panel was removed and the earth connection was strengthened at CT console only.

Case Study 5

Situation/problem   One of the 220 kV CT failed, with fire and spilling of oil damaging the porcelain of adjacent CT due to flying debris on a drizzling day.

Observation

1. The porcelain insulation was broken and scattered into pieces, causing damage to the adjacent CT.
2. The twin primary conductor with paper insulation caught fire.

Analysis

1. This CT during summer season was drawing overload current of 25% extra continuously for around 6 to 8 hours daily. So, temperature rise is one of the reasons of CT failure.
2. The CT was causing minor oil leakage also at the CT clamp point, as reported by the maintenance engineer, but due to non-availability of shut down, the same was left unattended.
3. A defect was noticed in the connection of line conductor with CT clamp at the CT stud. The conductor was bent downward and connected to the clamp with tension

(Figure 2.33). This tension might have caused the crack of the insulator porcelain ring, causing minor leakage of oil.

During summer season the insulating oil might have deteriorated due to contraction and expansion. For the first few showers, the ingress of rain water caused further damage of the oil. But the connection of line conductor in downward slope and the ingress of rain water becoming more due to surface tension, have resulted in the final failure of CT. Burnt CT is shown in Figure 2.34.

Figure 2.33 Bent line conductor

Figure 2.34 Burnt CT

Remedy The height of CT support was raised for connection of the line conductor in upward slope and less tension was applied on the clamp.

COMPARISON STUDY WITH THRESHOLD VALUES

Calculation of Primary Current

The detailed study for the calculation of the rated primary current has already been covered earlier. Again the same is repeated by comparison with limit value.

Factors to be considered

  i. Rated continuous thermal current.
 ii. Maximum load current on the system with a provision of 25% extra flow.
iii. Based on the short-time current for allowable time period.
 iv. Suitable multiple of factors 10,12.5,15, 20,25,30,40,50,60,70.

Points for calculation

1. Comparison is done between rated continuous thermal current and (Maximum load current + 25% extra) current in the system. The load current should be within the limit of rated continuous thermal current.
2. Calculation on the basis of short-time current is also taken into account for selection of primary current.

   Rated primary current = (Rated short-time current for 1 sec)/150

The lower value as such obtained by comparison of primary current calculation on the basis of short-time current and maximum load current, is considered. Now the data (value) thus obtained is taken to the nearest multiple of underlined factors for calculation of final primary value.

Calculation of CT Secondary Drive Voltage

Sometimes due to incorrect choice of CTR, burden and ALF for protection class, the secondary terminal voltage becomes insufficient to operate for secondary circuit. Now comparison can be done between the design value and actual value by stating the following example.

Example Consider CTR 800/1, 5P10, 20 VA, R_{CT} =5 Ω R_L= (2 × 1) Ω R_{relay} = 0.2 Ω (numerical), Fault current = 20 kA.

Calculation During fault occurrence, the reflected fault current on CT secondary passes through the relay for its actuation.

Actual voltage = I_{SF} × $(R_{CT} + R_L + R_r)$ = [20 × 10³/800] (5 + 2 + 0.2) = 180 volt

Design voltage = Burden ALF/I_s = 20 × 10/1 = 200 volt

Now design voltage is more than actual voltage. So the parameters can be selected for CT.

Calculation of Knee Point Voltage (V_k) and Excitation Current

The calculation of knee-point voltage (V_k) and excitation current is important for the protection class core for special purpose, used for balanced current schemes.

Value of K_n The knee-point voltage on CT plays the role of threshold value of CT saturation. It is the point at which CT starts saturating. So the significance of this voltage becomes prominent for fault condition. During fault condition, the voltage value should be such that CT can drive current for operation of the relay. Particularly for external fault condition and with balance current scheme protection (differential protection), the requirement of voltage becomes as follows:

$$V_0 = 2I_F (R_{CT} + R_L)$$

From comparison of values it can be concluded that min. $V_k > V_0$

Note 1 Sometimes for designing CT of required and economical size, the standard values chosen for calculation are 20 times the rated primary current or system fault current, whichever is low.

Excitation current (I_e) Excitation current also plays a role for incorrect CTR and mal-operation of relay in current balance scheme. So it is relevant only during normal operating condition to ensure stability and prevent false tripping and is not relevant for fault condition.

$$V_k \text{ (min.)} = 2I_F (R_{CT} + R_L)$$

$$V_0 = I_R (R_{CT} + R_L)$$

$$\text{So } V_0 = V_K \text{ (min.)} \times (I_R/2I_F)$$

Example Say $I_R = 800$ A, $I_F = 10$ kA

$$V_0 = V_k \text{ (min.)} \times (I_R/2I_F) = V_k \text{ (min.)} \times (800/10 \text{ kA}) = 8\% \text{ of } V_k \text{ (min)}$$

So, value of excitation current expression at 8% is even suitable and for easy limit calculation, it can be chosen at 25% or 50% of V_k ($V_k/4$ or $V_k/2$)

Note 1 Limiting value of $I_e < 30$ mA at $V_k/4$ for 1 A CT

$$I_e < 150 \text{ mA at } V_k/4 \text{ for 5 A CT}$$

Note 2 Lesser the value of excitation current, indicates the CT of better quality.

CONCLUSION

This topic related to the study on current transformer provides information regarding the basic construction, working principle, maintenance practice, etc. of the instrument transformer. Some views, comments, etc. are restricted to Indian culture. Particularly case studies on CT circuitry cover the fundamental approach of vector analysis of the currents from three phase supply source. Combination of the vector analysis and study of the current value measurement in different circuits provides the conclusive idea regarding the faults in the system. The various case studies as described above are the physical examples of the practical occurrences. Moreover the table described with the fault finding study for star-connected circuitry becomes quite helpful to trace the possible faults directly, without the study of vector analysis and fundamental approaches.

APPENDIX

APPENDIX 1

Fault Finding Study for Y (star) connected CT Protection Circuitry

Sl. no.	Current in the CT secondary	Expected faults
1.	$R = Y = B = x$ A, $N = 0$ A	No fault in the circuits
2.	$R = Y = B = x$ A, and $N = 2x$ A	Any one of the phase 'CT' polarity reversed
	i. If $(R + Y) = (Y + B)$ $= \sqrt{3}x$, and $(B + R) = x$	Y phase reversed
	ii. If $(R + Y) = (B + R)$ $= \sqrt{3}x$, and $(Y + B) = x$	R phase reversed
	iii. If $(Y + B) = (B + R)$ $= \sqrt{3}x$, and $(R + Y) = x$	B phase reversed
3.	i. If $R = 0$ A, and $Y = B = N = x$ A then check for all other R phase CT secondary cores, if values obtained are in same pattern, then	R phase primary side open
	ii. Similarly for Y phase and B phase also.	Corresponding phase primary side open
	iii. If $R = 0$ A, and $Y = B = N = x$ A for only in one core, then	R phase secondary is shorted or R phase is mixed with other cores or with use of auxillary CT any one of the side might be shorted.
	iv. Similarly for Y phase and B phase also.	Corresponding phase
4.	$R = Y = B = x$ A, $N = 3x$ A	All phases have been connected to one CT only instead of different cores, as 1st, 2nd, 3rd cores, etc. as R phase cores and Y phase cores and B phase cores or primary side has been connected from a single source.

Sl. No.	Current in the CT secondary	Expected faults
5.	i. R = Y = $x/2$ A, B = x A, N = 0 A	R and Y phases of CT secondary similar polarities have been shorted.
	ii. Y= B = $x/2$ A, R = x A, N = 0 A	Y and B phases of CT secondary similar polarities have been shorted.
	iii. B = R = $x/2$ A, Y = x A, N = 0 A	B and R phases of CT secondary similar polarities have been shorted.
6.	i. R= x A, Y = B = 0 A, N = x A	Y and B phases of CT secondary have been shorted.
	ii. Y= x A, R = B = 0 A, N = x A	B and R phases of CT secondary have been shorted.
	iii. B = x A, R = Y = 0 A, N = x A	R and Y phases of CT secondary have been shorted.
7.	R = Y = B = N = 0 A	All the 3 CTs are shorted.
8.	If the values are resulted other than the above readings as described.	1. CTR may be different. 2. Wrong primary link connection. 3. Phase angle problem. 4. CT saturation problems

APPENDIX 2

Relay Detail for Selection of Instrument Transformers

Transformers Differentials

(a) Alstom make

1. Relay type: DTH 31/32

    $V_k > 40 \times I (R_{CT} + 2R_L)$; Example: $V_k > 40(1)(3 + 4) > 280$ V

2. Relay type: MBCH 12/13

$$V_k > 24\, I_n\, (R_{CT} + 2R_L)$$

where,

V_k = Knee-point voltage

I_n = Relay rated current

R_L = Total lead resistance

I_e = <3% I_n at $V_k/4$ for both above types of relays, i.e., 0.03 I

i.e., 30 mA at $V_k/4$

3. Relay type: KBCH, MiCOM P630 (Numerical)

Application	Knee-point voltage (V_k)	Through-fault stability X/R	I_f
Transformers, generators, generator transformers, motors, shunt reactors, series reactors also.	$24\,I_n\,(R_{CT} + 2R_l)$	40	$15\,I_n$
Overall generator-transformer units.	$48\,I_n\,(R_{CT} + 2R_L)$	120	$15\,I_n$
Transformers connected to a mesh corner, having two sets of CTs each supplying separate relay inputs.	$48\,I_n\,(R_{CT} + 2R_l)$	40 120	$40\,I_n$ $15\,I_n$

(b) *ABB make*

1. Relay type : RADSB (Static)(Medium impedance)

$V_k > 30(R_{CT} + 2R_L + R_{re})\,I_n$, $>30(4 + 4 + 3)1$, >330 V

Note   Over-current factor of 30 recommended

Excitation current: Not applicable*

2. Relay type: SPAD346C (Stabilized differential relay)

$V_k > 4 \times I_{max}\, x\, (R_{in} + R_L)/n$,

where,

n = Transformation ratio of CT $> (R_{CT} + R_L + 0.5/\text{sq.of } I_{sn})$

R_{in} = Secondary resistance of CT

$2R_L$ = Control cable ("to and fro") resistance

I_{max} = $I_d/I_n >>$set on relay (range available 5 to 30, default set is 10)

3. Relay type: RET316 (Stabilized differential relay)

$n' = n\,(Pr + Pe)/(Pb + Pe)$,

where,

n = ALF,

n' = Effective over-current factor, is a function of fault current (I_k) frequency and time constant of network, and read from graph in RET manual,

Pb = Connected burden at rated current,

Pe = CT losses of secondary windings and

Pr = Rated CT burden, DC time constant assumed is 300 ms.

* Relay provided with "magnetizing inrush restraint" based on second harmonic content of the inrush current and hence "I_{mag}" calculation is not applicable.

(c) *Easun rerolle*

1. Relay type: 4C21 (Static) (Low impedance)
    CT Class: PS, $V_k > 2 I_f (R_{CT} + R_L + R_{ICT}$ (P)) + ($I_{CT} V_k \times I_{CT}$ ratio)
    Example $V_k > 2 \times 10.9375 (2 + 3 + 1) + (14.43/0.875) \times 0.577 > 140.75$ volts
    R_{CT} = Main CT resistance,
    R_{ICT} (P) = ICT primary winding resistance and
    R_L = Lead resistance and I_f = Maximum through-fault current.
2. Relay type: Duo bias M(Numeric), (Differential and restricted earth fault)

$$V_k > 4 \times I(A + C),$$

where,

I = Either maximum 3-phase through-fault current referred to secondary (as
    limited by transformer impedance) or high-set setting, whichever is greater,

A = Secondary winding resistance of each star connected CT,

C = CT secondary loop resistance for internal faults and

CT Class recommended-PS, X to BS 3938, TPS to IEC-44.

Generator Differential Protection

(a) *Alstom*

1. Relay type: CAG34 (High impedance scheme)
    $V_k > 2I_f (R_{CT} + 2R_L)$
    Example $V_k > 2 \times 10(3 + 4) > 140$ V
    where,
    I_f = Secondary equivalent of fault current
    $I_e = I_s - I_r = (0.15 - 0.10)/2 = 25$ mA at $V_k/2$
2. Relay type: LGPG, MiCOM 340 (Numerical)
    i.   For voltage-dependent, over-current, field failure and negative phase, sequence
         protection.

$$V_k > 20 I_n (R_{CT} + 2R_L)$$

    ii. For stator earth fault protection:

$$V_k > I_s (R_{CT} + 2R_L + R_R)$$

    iii. For generator differential protection:
         Low impedance differential $V_k > 50 I_n (R_{CT} + 2R_L)$
         High impedance differential $V_k > 2 V_s$
    where,

$$V_s = 1.5 I_f (R_{CT} + 2R_L) \text{ and } R_s = V_s/I_s$$

3. Relay type: YTGM15, YCG15AA, ZTO11(Generator Backup)

$$V_k > 2I_f\,(R_{CT} + 2R_L + M + CM),$$

where, CM = Connected burden.

(b) ABB

1. Relay type: RADHA/RADHD (high impedance)

$$V_k > 2I_k\,(R_{CT} + R_L),\ >2 \times 25(4 + 3),\ >350\ V,$$

R_L in case of generator is longer, i.e., $2R_L = 6\ \Omega$

I_k will be higher considering Xd (0.2 pu) and CT secondary of 5 A.

Excitation current: Not applicable*

Excitation current is kept low for increasing the primary sensitivity

  * Relay provided with "magnetizing inrush restraint" based on second harmonic content of the inrush current and hence I_{mag} calculation is not applicable.

(c) Easun reyrolle

1. Relay type: 4B3 (EM)/DAD 3 (Static)/Argus-1 (Numeric)(High impedance scheme)

$V_k > 2I_f\,(R_{CT} + 2R_L),$

Example: $V_k > 2 \times 10(3 + 4) > 140$ volts

CT Class: PS,

I_f = Maximum through-fault current,

R_{CT} = Main CT resistance and

R_L = Lead resistance between CT and relay.

2. Relay type: GAMMA (Numeric) (High impedance)

For two of 3 phase inputs (line end and neutral end) and for neutral earthed CTs.

In case of low impedance bias differential functions

  i. $V_k > 50 \times I_n(R_{CT} + 2R_L + R_R)$

    where,

    Maximum through-fault current = $10 \times I_n$ with maximum $X/R = 120$.

  ii. $V_k > 30 \times I_n(R_{CT} + 2R_L + R_R)$

    where,

    maximum through-fault current = $10 \times I_n$ with maximum $X/R = 60$

      I_n = Rated current sec. $X/R = X/R$ ratio for maximum through-fault condition

    R_{CT} = Secondary resistance of CT,

      R_L = Lead resistance between CT and relay and

    R_R = Resistance of any other protection functions sharing the CT.

Bus Differential Protection

(a) ABB

1. Relay type: RADHA/RADHD (High impedance scheme)
   $V_k > 2I_k (R_{CT} + R_L)$, $>2 \times 40 (4 + 4)$, >640 V
2. Relay type: RADSS (Medium impedance scheme)
   Depending on differential ratios, for 1 A CT, V_k shall be 500 V.
   Excitation current: Not applicable*

*   Relay provided with 'magnetizing inrush restraint' based on second harmonic content of the inrush current
    and hence I_{mag} calculation is not applicable.

(b) ALSTOM

1. Relay type: CAG34 (High impedance scheme)
   $V_k > 2I_f (R_{CT} + 2R_L)$, Example: $V_k > 2 \times 10(3 + 4)$, >140 V
2. Relay type: DIFB –DIFBCL
   $V_k > K \times I_n (RTCP + RF + R_d/n_2)$,
   where,

   K $= (1.2/40) \times (I_{CC}/I_N)$

   I_n $=$ Main CT primary rated current,

   I_{CC} $=$ Maximum short-circuit current delivered to bus bar via the input where MCT is installed,

   $RTCP =$ Resistance of secondary of MCT,

   RF $=$ Resistance of link loop between MCT and auxiliary CT,

   n $=$ Ratio of auxiliary CT and

   $R_d/n_2 =$ Value of differential resistance transposed to ACT primary.

3. Relay type: MCTI 34 (Numerical)
   $V_k > 1.6V_s$, $V_s = 1.25 \times I_f (R_{CT} + 2R_L)$
   where,

   $R_{CT} =$ CT resistance,

   R_L $=$ Maximum lead resistance from CT to common point and

   I_f $=$ Maximum internal secondary fault current.

(c) Easun rerolle

1. Relay type: B3 (EM)/DAD3 (Static)
   CT Class: PS, $V_k > 2I_f (R_{CT} + R_L)$
   Example $V_k > 2 \times 10(3 + 4)$ >140 V
   I_f $=$ Maximum through-fault current,

R_{CT} = Main CT resistance and

R_L = Lead resistance between CT and relay.

Distance Protection

(a) *Easun reyrolle*

1. Relay type: THR (Static)

$$\text{CT Class: PS, } V_k > I\left[R_L + R_2 + \frac{X}{R}(R_3 + R_2)\right]$$

Example $V_k > 10(3.8 + 7 + 4(1.2 + 7)) > 436\text{ V}$

where,

R_L = Burden of relay (3.8 W max.)
R_2 = Resistance of leads plus resistance of CT secondary
X/R = Ratio of reactance to resistance of the system for fault at the end of zone 1
R_3 = Constant depending on impedance setting of zone 1. (1.2 W maximum)
I = Secondary fault current for fault at end of zone 1

Note $X/R = 4$ for 132 kV system

$X/R = 7$ for 220 kV
$X/R = 11$ for 400 kV

2. Relay type: Omega (Numeric)
    V_k should be equal to or greater than the higher of the following two expressions.
    i. $V_k > K \times (I_P/N(1 + X_P/R_P)) \times (0.03 + R_{CT} + R_L)$ for phase-phase faults
    ii. $V_k > K \times (I_e/N(1 + X_e/R_e)) \times (0.06 + R_{CT} + R_L)$ for phase-earth faults

I_P = Phase fault current calculated for X_P/R_P ratio at the end of zone 1
I_e = Earth fault current calculated for X_e/R_e ratio at the end of zone 1 N-CT ratio
X_P/R_P = Power system reactance to resistance ratio for the total plant including the feeder line parameters calculated for phase fault at the end of zone 1
X_e/R_e = Similar to above ratio but calculated for an earth fault at the end of zone 1,
R_{CT} = CT resistance,
R_L = Lead burden CT to relay and
K = Factor chosen to ensure adequate operating speed which is >1.0

(b) *Alstom*

1. Relay type: Micromho, Quadramho
    $V_k > I_f (X/R)(M + R_{CT} + nR_L)$
Example $V_k > 10(4)(10.2 + 3 + 4) > 40(17.2) > 688\text{ V}$
    $I_e < 3\% I_n$ at $V_k/2 < 30$ mA at $V_k/2$

where, M = Relay resistance (phase fault).

2. Relay type: MiCOM 430/441/442, EPAC, LFZR, LFZP, PD521, PD932 (Numerical)
$V_k > I_f \times (1 + X/R) \times (R_{CT} + R_L + R_b)$
where,

X/R = Primary system ratio

R_b = Relay burden

R_L = Rest of cable connecting CT to relay (lead and return for ground faults, lead only for phase faults)

(c) ABB

1. Relay type: RAZAO/REL511 (Static)(Numerical)
Secondary limiting voltage $> (I_k \times I_{sn}/I_{pn}) \times a \times (R_{CT} + R_L + 0.5/(I_{sn}/I_{pn})2)$
where,

a = Factor for the DC time constant (approximately 10 for about 100 msec) and excitation current $<0.2\ I_{sn} <0.2$ A <200 mA.

Feeder Differential Protection

(a) Easun reyrolle

1. Relay type: Solkor-M and Microphase-FM (Numerical) (Current differential)

$$V_k > K \times X/R \times I_f /N \times (R_{CT} + 2R_L + R_b)$$

where,

K = Stability factor = 0.8 for Microphase-FM and

X/R = Ratio for the maximum through-fault conditions.

(The value of this transient factor depends upon the sum of the source and transmission circuits impedances.)
R_b = Burden of relay,
The AC burden of the relay per phase is
0.05 VA at 1 A, tap = 0.05 Ω and 0.30 VA at 5 A, tap = 0.012 Ω.
The values of magnetizing currents of CTs at two ends should not differ by more than $I_n/20$ for output voltages up to $50/I_n$ volts.

(b) Alstom

1. Relay type: MBCI
Translay 'S' differential (for feeder and transformer)
(A) For plain feeders:
i. $V_k > 0.5 \times N \times K_1 \times I_n (R_{CT} + XR_L)$
where,

V_k = KPV of CTs for through-fault stability,

R_L = Resistance of CT secondary circuit,

X = 1 for core wire connections between main CT and the relay,

= 2 for six wires connection

N = Relative neutral turns on summation transformer winding and

K_1 = The selected time-dependent constant.

ii. For all application at or above 220 kV where X/R ratio are large:

$V_k > N \times K_1 \times I_n (R_{CT} + X_{RL})$

Magnetizing current $< 0.05 \times I_n$ at $10/I_n$ V

(B) For transformer feeder differential:

a. $V_k > 50 \times I_n (2.2/I_n2 + R_{CT} + R_L)$ for star-connected CTs.

b. $V_k > 50 \times I_n/\sqrt{3} \, (9.7/I_n2 + R_{CT} + R_L)$ for delta-connected CTs.

2. Relay type: MiCOM P540 (Numerical)

$V_k > K \times I_n \, (R_{CT} + 2R_L)$

where,

K is a constant depending on I_f = The maximum value of through-fault current for stability and is determined as follows:

For relays set at Is_1 = 20%, Is_2 = 2 I_n, K_1 = 30%, K_2 = 150%:

$K = 40 + (0.07 \times (I_f \times X/R))$and

$K = 65$. This is valid for $(I_f \times X/R) < 1000 \, I_n$

For higher $(I_f \times X/R)$up to 1600 I_n:

$K = 107$

For relays set at Is_1 = 20%, Is_2 = 2 I_n, K_1 =30%, K_2 =100%:

$K = 40 + (0.35 \times (I_f \times X/R))$and

$K = 65$. This is valid for $(I_f \times X/R) < 600 \, I_n$

For higher $(I_f \times X/R)$up to 1600 I_n:

$K = 256$

Over-current and Earth Fault Relay

(a) *Alstom*

1. Relay type: CDG11 (IDMT)

   This relay has 3.5 VA burden. So total VA burden requirement is 10 or 15 VA.

   Note   ALF factor of 10 is sufficient. If backup protection scheme is envisaged, ALF of 15 is required.

   The time current setting characteristic of IDMT relay becomes a straight line after 15 times setting; therefore, time discrimination is ineffective after 15 times current setting.

2. Relay type: MiCOM P120,P140 (Numerical)
   Class: 5P10, Burden: 5 VA

(b) *ABB*

1. Relay type: SPAJ 140 (Numerical)
   This relay generally requires CT with 5P10/5P20 CT with very low burden, e.g., 0.1 VA

(c) *Easun reyrolle*

1. Relay type: ARGUS/MIT (Numerical)
   Class: 5P10, Burden: 5 VA

2. Relay type: Solkor-R/RF (Pilot wire differential protection)
   CT Class: PS $V_k = 50/I_n + (I_f/N)(R_{CT} + R_L)$
   I_n = Rated current
   I_f = Maximum primary steady state through-fault current
   N = CT ratio, R_{CT} = CT resistance and R_L = Lead resistance.

Accuracy Class as per Canadian Standard CAN3-C13/1983(Fr. 60 Hz)

Class designation	Burden PF	Error limits		Composite error	Corresponding I_{ec} accuracy class
		Current error	Phase angle error		
0.3 0.6 1.2	0.6...1.0	According to the metering core as per CAN3-C13/1983 (Fr. 60 Hz)		– – –	0.2 0.5 1.0
2.5L 10 2.5L 20 2.5L 30	0.9	–	–	2.5	5P20
2.5L 100 2.5L 200 2.5L 400 2.5L 800	0.5	–	–	2.5	5P20
10L 10 10L 20 10L 50	0.9	–	–	10	10P20
10L 100 10L 200 10L 400 10L 800	0.5	–	–	10	10P20

Limits of Accuracy Classes of CT for Metering Core as per U.S.A, ANSI C53.13-1993 (fr 60 Hz)

		Current error	Phase angle error lagging and leading in minutes
Accuracy class	**0.3**	0.991	−30
		0.994	−20
		0.997	−10
		1.000	0
		1.003	+10
		1.006	+20
		1.009	+30
	0.6	0.982	−60
		0.986	−40
		0.994	−20
		1.000	0
		1.006	+20
		1.012	+40
		1.018	+60
	1.2	0.964	−120
		0.976	−80
		0.988	−40
		1.000	0
		1.012	+40
		1.024	+80
		1.036	+120

Limits of Errors According to Australia Standard–AS 1675–1986

(Class-MCTs) Metering Core

Class	Current error (per cent)					Maximum variation in error with change in primary current	
	Primary current (per cent)					Between 5% and 10%	Between all combination of 10, 20, 100 and 120%
	5	10	20	100	125		
0.1M	±0.15	±0.1	±0.1	±0.1	±0.1	0.05	0.05
0.2M	±0.30	±0.20	±0.20	±0.20	±0.20	0.2	0.1
0.5M	±0.75	±0.5	±0.5	±0.5	±0.5	0.4	0.25
1M	±1.5	±1.0	±1.0	±1.0	±1.0	–	–
2M	–	–	±2.0	±2.0	±2.0	–	–
5M	–	–	±5.0	±5.0	±5.0	–	–

(Class-MCTs) Protection Core

Class	Current error (per cent)					Maximum variation in error with change in primary current	
	Primary current (per cent)					Between 5% and 10%	Between all combination of 10, 20, 100 and 120%
	5	10	20	100	125		
0.1M	±0.30	±0.2	±0.15	±0.15	±0.15	0.09	0.12
0.2M	±0.60	±0.20	±0.30	±0.30	±0.30	0.15	0.23
0.5M	±1.2	±0.9	±0.75	±0.60	±0.60	0.3	0.45
1M	±2.0	±1.5	±1.0	±0.9	±0.9	–	–
2M	–	–	±5.0	±3.5	±3.5	–	–
5M	–	–	–	–	–	–	–

APPENDIX 3

STANDARDS USED IN OTHER COUNTRIES

Accuracy Class According to [U.S.A ANSI C57.13-1993 (Fr. 60 Hz)]

Class designation	Burden PF	Error limits		Composite error	Corresponding I_{ec} accuracy class	Notes
		Current error	Phase angle error			
0.3		**According to the metering core as per USA, ANSI C57.13-1993 (Fr. 60 Hz)**		–	0.2	
0.6	0.6…1.0			–	0.5	
1.2				–	1.0	
T10						
T20	0.9	–	–	10	10P20	Note 1
T30						Class T must
T100						be verified by
T200	0.5	–	–	10	10P20	testing
T400						
T800						
C10						Note 2
C20	0.9	–	–	10	10P20	Class C must
C50						be verified by calculation
C100						
C200	0.5	–	–	10	10P20	
C400						
C800						

APPENDIX 4

Standard Burdens According to CAN3-C13/1983(Fr. 60Hz)

Designation	Burden impedance	Output with 5 A (sec 9 VA)	Power factor
B-0.1	0.1	2.5	0.9
B-0.2	0.2	5.0	0.9
B-0.5	0.5	12.5	0.9
B-0.9	0.9	22.5	0.9
B-1.0	1.0	25	0.5
B-1.8	1.8	45	0.9
B-2.0	2.0	50	0.5
B-4.0	4.0	100	0.5
B-8.0	8.0	200	0.5

APPENDIX 5

LIST OF INDIAN STANDARDS

No.	Title
IS 335 : 1983	New insulating oil
IS 1885 : 1993 (pt. 28)	Electro-techno vocabulary of instrument transformer.
IS 2099 : 1986	Bushing for alternating voltages above 100 volts.
IS 2165 (pt.1) : 1977	Insulation coordination (phase to earth insulation coordination).
IS 3716 : 1978	Application guide for insulation coordination.
IS 4201 : 1983	Application guide for CT.
IS 11322 : 1985	Method for partial discharge (PD) measurement in instrument transformer.

QUESTIONS AND ANSWERS

1. Why should CT secondary star connection be earthed?

   Answer

   In an electrical system, every winding should be connected in such a way that in case of fault in the system, the current could be circulated in the winding to get easy access to earth. CT is used for two different purposes (measurement and protection) in the system. For these two purposes CT is connected with different burdens. For any abnormality in the primary side of the CT like unbalanced currents, fault currents, etc. the reflected current, on the secondary circuit becomes unbalanced/abnormal. Now this current needs certain path to flow to earth.

   Moreover every winding (R phase, Y phase, B phase) requires certain return path for circulation of current. This path is provided by connecting either side (supply side and load side) of star connections by a common wire, called neutral wire. The earth connection of star point plays the following roles:

   1. It becomes the reference point for circulation current in the circuit.
   2. For the case of fault/abnormality, etc. earth provides the path for flow of this current.

   Note

   1. CT secondary star point should be solidly earthed.
   2. Every individual core in CT secondary should be separately star-connected and earthed.
   3. The spare cores of CT secondary should be shorted and earthed.
   4. Earthing used in CT secondary must be checked intermittently. Failure of earth connection causes the problem in CT, regarding the burden to the insulation due to delayed subsidisation fault current.

2. Why should the star point in CT circuits be earthed at one point only?

   Answer

   The star point earthing is used for reference in the circuit and quick flow of reflected fault current to the earth terminal. So during fault condition though available protection clears the fault within the minimized setting time, still the de-energization of the circuit needs certain extra time to settle down. So for quick de-energization of the circuit, earthing plays the important role. Now the discussion is to be concentrated whether it is to be earthed at one point or at multi points. CT circuits must be earthed at one point only, because multipoint earthing provides fault current to share through the parallel paths.

3. Why is it preferred to earth the star point near CT console at switchyard instead of at control panel?

   Answer

   CT circuit starts from the CT secondary terminals and the driving voltage for the current to flow is regarded as reference that starts from the primary side of the CTs. The following points can be summarized regarding the earthing of state point.

i.   Generally fault in the circuit results on primary side of the CT. So for quick reference of the current flow the nearer end of CT secondary should be earthed.

ii. As fault current is associated with the earth-terminal of the circuit, during fault condition if protection used in the system gets delayed or does not actuate, then earthing plays the role for flow of this current. So for safety of the working personnel, the earthing is to be done at the switchyard instead of control/relay panel.

4. What happens if metering core of the CT is connected to the protection circuit and vice-versa?

Answers

It is essential to study the characteristics of different class of the CT cores. The following table can be considered as the reference data for studying the characteristics.

Type of core	Metering	Protection special type	Protection type
Output	20 to 40	–	20 to 40
Accuracy class	0.2 Fs<5	PS	5P
V_k (V) min.	–	600 to1200	–
$I_{exc}@V_k$ (mA) maximum	–	10 to 25	–
R_{CT} at 75°C max	–	2.5 to 5	2.5 to 5

From the characteristics of the table, it can be concluded that the metering core needs high accuracy class core with certain burden across it. The role of the knee-point voltage and the value of excitation current do not become important for such core.

But for the protection core, the fault current on the secondary circuit plays the role to clear the fault by the actuation of protective device (relay). Here knee-point voltage is to be considered for the working operation of the protection device instead of accuracy class.

Now as per the discussion, if the cores are interchanged, then metering scheme will be affected due to less accuracy class core and will result in errors while reading. Similarly the protection scheme will be affected due to quick approach of knee-point voltage during fault condition. In such a possibility, the cores may be saturated in due course of time.

5. What happens to the CT ratio for the case of saturation of CT core?

Answer

For the saturation of CT core, the development of flux resulting in induced emf across the winding becomes saturated quickly. So, with reduced flow of primary current the rated secondary is obtained. This behaviour results with reduced CTR.

6. Why does CT bursts for opening of secondary circuit?

Answer

Instrument transformers (CT, PT, etc.) work on the principle of mutual induction. For CT, when primary current is injected certain flux is developed in the core, which in turn results with induced emf across the winding of the primary. The same flux being linked to the secondary winding also induces certain emf across it. Now if secondary side is kept open then due to absence of flow of secondary current, no counter flux for the primary flux would be available in the core. So the forward flux due to large current on primary being linked with more number of turns of secondary winding will cause large emf even more than the allowable value. This effect develops both electrical and mechanical stress on the winding and insulation. So after a certain condition the CT starts bursting severely.

7. Why is it preferred to have the CT ratio availability by the connection control on both primary and secondary?

Answer

CTR availability can be obtained by different control methods. However the CTR availability by both primary and secondary control method becomes suitable due to its stability and flexibility of CT connection. By this method the required CTR can be obtained, with the CTs being on charged condition. The secondary terminals are available in tappings and to obtain the required ratio, only the tappings are changed.

8. Why does it become preferable nowadays to have twin core secondary cable connection to the terminals instead of single core cable above 500 m of secondary connection?

Answer

The performance of CT depends upon the availability of secondary burden on it. The calculation of knee-point voltage for different relaying device depends upon the value of CT resistance and lead resistance.

Example   For differential connection

$$V_k \text{ (knee-point voltage)} > 40I \ (R_{CT} + R_l)$$

If lead resistance would be more, then the type of CT required should have higher knee-point voltage, which in turn will not be economical and sometimes will not be feasible in practice. So the cable to be used for the connection between the secondary terminals and controlling devices should be such that the lead resistance is to be limited to some value. On basic standard of lead resistance (copper cable from CT to control room), the value of resistance is taken as (0.6 ohm to 0.8 ohm for 100 m of cable). So for 500 m it is to be taken as $(0.8 \times 5) = 4$ ohm. But for above 500 m the value of lead resistance would be more than the allowable standard of 4 ohm. So above 500 m the secondary cable should be taken in double (twin core) to limit the resistance. Moreover for the use of twin core, the chance of opening of CT secondary due to delink or damage of lead can be reduced. So twin core is preferable above 500 m run of secondary CT cable.

9. Why is top tank primary design CT preferred to old type hairpin CT?

Answer

In hairpin type design, the primary windings are taken up to the bottom box module where secondary windings are wound over the primary winding to obtain required ratio of the CT. The top tank primary design is also called "inverted CT design". This design has primary winding on the top head of the CT and the secondary windings are wound over the primary winding to obtain desired ratio of the CT. The series–parallel primary arrangements are also provided to derive different CT ratio. For this design the primary winding is spread in a uniform and symmetrical way around the cores, avoiding local saturation and reducing the leakage flux.

Advantages of top tank primary design CT

- Because of single HV insulated tube instead of two primary HV conductors as in the hairpin design, the volume of the insulator and the volume of oil is reduced.

- Internal ground conductor has sufficient cross-section for handling the fault current.

- Tank head design prevents projection or splitting of parts in case of faults.

  So considering the above factors, the tank head design CT is preferable to normal hairpin CT.

10. Why is combined CT and PT design preferable nowadays?

Answer

Advantages of combined instrument transformer

Because of combined unit the equipment achieves the following advantages.

- Low weight and minimum oil volume.
- Use of oil-paper insulation reduces the space for insulation.
- Use of bar type primary conductor permits higher short-circuit current and avoids the large voltage-drop across the winding.
- Only one foundation is required in switchyard, which in turn reduces the installation cost.
- Perfect transient performance and use of suitable corrosion-resistant material increases the life span.
- Because of uniformly distributed secondary windings it enhances guarantee, accuracy for both nominal rating and higher rating values.

  Disadvantage

- The one and only disadvantage of this system is that, for the case of failure of equipment, both CT and PT circuit associated with the system are hampered. So maximum Indian utilities do not prefer the use of combined CT for higher rating value. Moreover the Indian manufacturers have not yet started the manufacturing of this type of CT. However combined CT up to 36 kV range is in use in Indian scenario.

11. Can the SF_6-filled CTs be realized in practice?

Answer

Yes, SF_6-filled CTs have already been realized in practice.

Some manufacturers have filled the instrument transformer with SF_6 gas inside for better performance and to make them maintenance-free.

This type of CT enhances the life and provides reliability to the working operation. The range of these equipments is even manufactured up to 245 kV range by CGL limited in India.

The performance of this type of CT has not yet been popularly experienced by Indian utilities, for which the same has not yet been accepted openly.

12. What happens if AC voltage would be injected on secondary side of the CT being in service?

Answer

CT circuit being in service provides certain secondary voltage to drive the CT secondary current to the load. The flow of primary and secondary current on primary and secondary winding provides common flux in the core. For the case of AC voltage injection on CT secondary, the flow of current on this winding will depend upon the polarity connection of secondary emf and the value of impedance of the circuit (impedance of CT + impedance of load). Due to this voltage injection certain current will flow in the winding and accordingly the flux in the core will be developed. Now the net magnitude of flux in the core will be changed and if it becomes more than the allowable limit then current in the secondary winding will also be changed. Sometimes due to high rise of current, the secondary winding may be damaged.

13. Can a 132 kV CT be used for 33 kV system or vice-versa?

Answer

Basic constructions of CTs are all same irrespective of voltage class. But according to the voltage class, necessary electrical clearance and insulation are provided in the CTs. So use of 132 kV CT in 33 kV system is possible, but the use of 33 kV CT in 132 kV system will not be possible.

14. What are the reasons for a CT to get saturated?

Answer

Saturation of CT causes quick rise of flux in the core. The following factors are the reason for saturation of CT.

1. Use of over burden loads in CT secondary.
2. Use of CT for primary current more than the allowable limit.
3. Continuous use of metering core for protection purpose and vice-versa.
4. Continuous operation of CTs to faulty currents.
5. Use of CT to DC current components.

15. What is the meaning of "D$_1$" and "D$_{11}$" connection for delta connection of CT circuitry?

   Answer

   ### D$_{11}$ connection

   D$_{11}$ connection corresponds to the current relation between phase current in positive sequence, i.e., $[(I_r - I_y), (I_y - I_b), (I_b - I_r)]$.

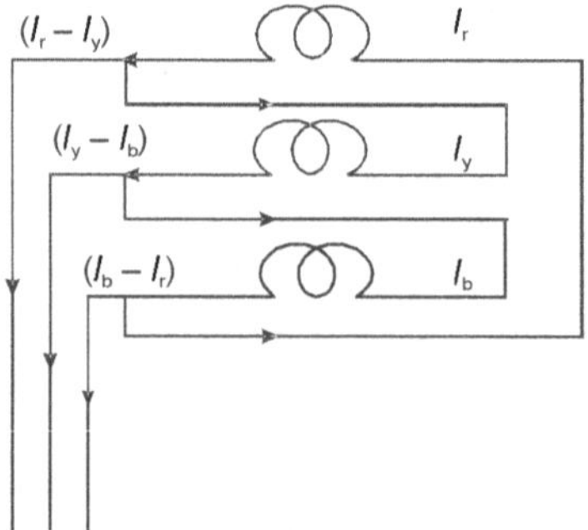

   ### D$_1$ connection

   D$_1$ connection corresponds to the current relation between phase current in negative sequence, i.e., $[(I_r - I_b), (I_y - I_r), (I_b - I_y)]$.

16. Why are CT and PT called instrument transformers?

   Answer

   The actual practical range of power system currents and voltages are in hundreds and thousands. These values cannot be taken directly for controlling or measuring the magnitude for monitoring purposes. So, to obtain the replica of the same in reduced manner within allowable limit of safe operating level, converters are needed. These converters are called CT and PT. When current conversion is done the instrument is called CT (Current transformer). Similarly, when voltage conversion is done the instrument is called PT (Potential transformer). As these are used for controlling the equipments, relays and instruments, these CT and PT are called instrument transformers.

17. Explain the concept of ISF?

   Answer

   It is called instrument safety factor (ISF). Every CT has certain limit of allowable primary current to flow in it up to which the instrument can work safely. This factor is the ratio of instrument limit primary current to the rated primary current. General standard is taken as the value less than or equal to 5. The choice and decision depends upon the purchaser and manufacturer. But general recommendation should be as low as practicable.

18. What are the standards used for CT?

   Answer

   i. BS 7626 (British Standard)

   ii. IEC 60044 (International Electro-technical Commission)

  iii. IS 2705: 1992 (Indian Standard)

  iv. IS 4201: 1983 (Indian Standard)

   v. IEEE C 57.13, C 37.110 (International Electrical and Electronics)

  vi. ANSI C57.13-19923, C53.13-19923 (American standard)

 vii. AS 1675-1986 (Australian Standard)

viii. CAN3 C13/ 1983 (Canadian Standard)

19. Why is CT error more for ring CTs as compared to conventional CTs?

Answer

The error in CT ratio develops due to the use of low quality magnetic material for the construction of CT. From the injected primary current to the CT, the amount of magnetizing current plays the role for the determination of CT error. The magnetizing current required for the ring-type CT to develop the required magnetizing effect is proportionately more than the conventional CTs. Moreover due to economical concept, the quality of magnetic material used for ring CTs is poor as compared to conventional CTs.

20. What are the factors to be considered for the selection of CT ?

Answer

For selection of CT to be used in the circuit the following factors are to be considered.

    i. Ratio

   ii. Number of areas

  iii. Rated burden for each area

  iv. Accuracy class

   v. Safety factor

  vi. Rated short-time current

 vii. Voltage clear

viii. Insulation level

  ix. Physical dimension

21. What are the types of materials used for core design instrument transformer?

Answer

Main aim for design of the core is to reduce the CT ratio error for better CT accuracy factor and low saturation factor. The materials generally used are as follows:

  i. CRGO silicon steel

 ii. Micro (μ) materials

iii. Alloy of Ni–Fe

22. What is the difference between CTs and power transformer?

    Answer

    1. CTs are current transformer and source is current but for power transformer, source is voltage.
    2. Operation of CT needs a short-circuit or least impedance secondary but for power transformer it is to be load impedance.
    3. Power rating/burden is small (15 to 200 VA) for current transformer, but for power transformer it is more.
    4. Exciting current is constant for power transformer and changes for current transformer.
    5. Windings are connected across the load for CT but for power transfomer it is in series.

23. What happens to the performance of the CT if it gets saturated?

    Answer

    When core of the CT is saturated, the magnetic nature of the same becomes different and a large proportion of primary current is required to magnetize this type of saturated core. So CTR (current transformation ratio) becomes inaccurate with the reduction of secondary current to the load. Finally it concludes with the performance of the CT and the CTR increases proportionally during the normal operation of the CT.

24. Explain the limitations of CT errors as per different standards?

    Answer

    CT is basically used for two different purposes like metering and protection of the instruments. The following tables with data are considered for the accuracy factor for metering core and protection core as per the standard IEC 60044.

Accuracy factor for metering core of CT

Accuracy class	± % Current ratio error at % of rated current					± Phase angle displacement error in minutes at % of rated current				
	1	5	20	100	120	1	5	20	100	120
0.1	–	0.4	0.2	0.1	0.1	–	15	8	5	5
0.2	–	0.75	0.35	0.2	0.2	–	30	15	10	10
0.5	–	1.5	0.75	0.5	0.5	–	90	45	30	30
1.0	–	3.0		1.0	1.0	–	–	–	–	–
0.2s	0.75	0.35	0.2	0.2	0.2	30	15	10	10	10
0.5s	1.5	0.75	0.5	0.5	0.5	90	45	30	30	30

Accuracies for higher class metering core as per IEC 60044-1

Class	± % Current ratio error at % of rated current	
	50	**120**
3	3	3
5	5	5

Accuracies for protection core as per IEC 60044

Accuracy class	Current error at rated primary current (%)	Phase displacement at rated primary current (minutes)	Composite error at rated Acc. primary current (%)
5P	± 1	± 60	5
10P	± 3	–	10
15P	± 5	–	15

25. What are the reasons of saturation of any core of the CT?

Answer

The prime reason for saturation of the CT core is the opening of CT secondary. The other reasons are also described as follows:

i. Use of low quality magnetic material

ii. Continuous use of high burden load across the CT.

iii. Use of low ratio CTs against high primary current

26. Why should the unused tapings of a particular core be kept open except the required two terminals?

Answer

For a secondary control CT, the secondary winding is taken in tapings. For obtaining the required ratio, the corresponding terminals are connected to the load. As part of the winding is connected to load the rest of the same winding should be kept open, otherwise the ratio would be different.

27. Explain the basic conditions for the selection of protection class CTs.

Answer

Protection class CTs are required to perform satisfactorily for both in steady state and transient condition for appropriate working of the relays during fault condition or abnormal rise of current in the system. The performance of the core is studied by the value of knee-point voltage (V_k), excitation current (I_e), secondary lead resistance values of the CT. So for the selection of protection class CTs these factors are to be considered.

28. State the reasons for the development of errors in CTs?

Answer

Every CT is equivalently considered as the combination of electric and magnetic circuit. These properties have certain limitations, i.e., the magnetic mmf (magneto motive force) required to produce flux and magnetizing current. This current flows and makes reduced current available for the transformation. So as per the availability of the magnetizing current the ratio error is developed. The quality of magnetic material is the prime reason for determination of CT ratio error. The other factors are like continuous use of high burden load across the CT and the use of low ratio CTs against high primary current.

29. What is the meaning of 20 VA, 5P20 for a CT?

Answer

Meaning of 20 VA is the burden of the CT and 5P20 is its accuracy class. It is stated that 5% of composite error is developed for the CT up to 20 times of rated current with maximum allowable limit of secondary burden of 20 VA. This type of CT would be best suitable for protection relays.

30. Why lead burden becomes more for a 5 A CT than for an 1A CT.

Answer

Practically burden is calculated on the basis of current flowing through the load and corresponding resistance of the load. For 1 A CT, the burden of connecting lead would be less as compared to 5 A CT.

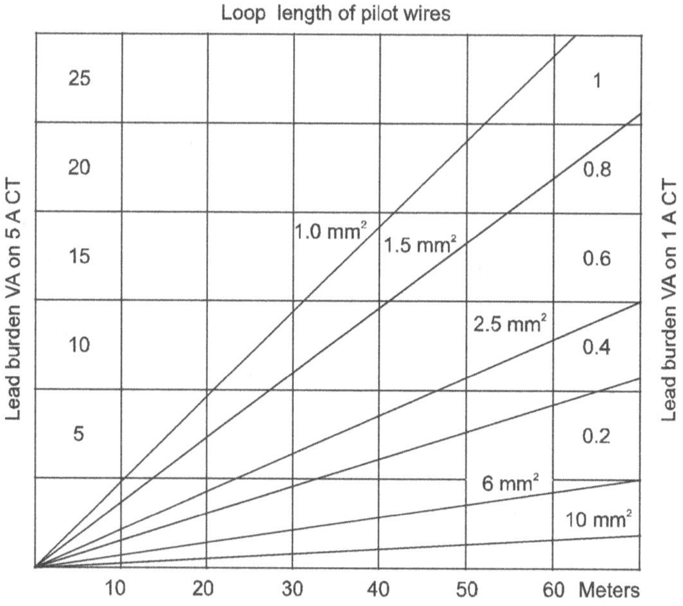

Curve for burden in VA for 1 A and 5 A CT

Example for calculating the burden using the following value.

i.   Lead resistance of 0.1Ω

ii.   1 A CT

iii.   5 A CT

For 1 A CT,

$$\begin{aligned} \text{Burden} &= V \times I \\ &= I \times R \times I \\ &= 1 \times 0.1 \times 1 \\ &= 0.1 \text{ VA} \end{aligned}$$

For 5 A CT,

$$\begin{aligned} \text{Burden} &= 5 \times 0.1 \times 5 \\ &= 2.5\,\text{VA} \end{aligned}$$

So,

$$\text{Burden} \propto I^2$$

The curve is indicative regarding the burden value plotting for both 1 A and 5 A. The common comparison of the values are providing the concept that burden at higher ampere rating CT is more than for lower rating CT.

31. What is ALF?

Answer

It is termed as accuracy limit factor and is defined as the certain standard multiplying factor of rated primary current up to which the declared limits of composite error in CT is maintained. The factors are generally of 5, 10, 15, 20 and 30 and expressed with accuracy class of CTs like the term 5P10, 10P20, etc. This term 5P10 corresponds to the accuracy class of protection core and indicates that the 5% composite error in CT is limited up to 10 times the primary current flown in the CT. Similarly the expression 10P20 indicates 10% of composite error in CT up to 20 times flow of primary current.

32. What is composite error?

Answer

During actual operation of CT, the current waveform does not become purely sinusoidal, due to non-linearity nature of exciting impedance and loading impedance. Harmonics in the system also play the role to distort the original waveform. So the current error and phase angle error contribute an error in composite form, called composite error. The composite error is generally expressed as a percentage of RMS (root mean square) values of primary current.

$$E_c = 100 / I_1 \times \sqrt{1/T \int_0^T (K_n \times i_1 - i_2)^2 \, dt}$$

where E_c = Composite error

K_n = Nominal or rated transformation ratio

I_1 = R.M.S. value of primary current

i_1 = Instantaneous value of primary current

i_2 = Instantaneous value of secondary current

T = Duration of one cycle

5P20 as composite error is indicated as 5P ($\pm$ 5% error) is maintained for 20 times normal current application.

33. What are the type tests conducted for CT?

Answer

(a) Short-time current test

(b) Temperature-rise test

(c) Lightning impulse test

(d) Switching input test

(e) Wet test

(f) Error-determination test

34. What are the routine tests conducted for CT?

Answer

(a) Polarity test

(b) Power frequency with start test for primary

(c) Partial-discharge measurement

(d) Power frequency withstand test for secondary

(e) Ratio error test

(f) Inter-turn over-voltage test.

35. Classify the outdoor CTs by its construction on the basis of primary?

Answer

It is basically of two types:

1. Live tank CT is also called inverted primary CT in which primary winding is directly insulated by oil chamber on top tank.

2. Dead tank CT also called pendulum CT and are of two types.

   i.   Hairpin primary

   ii. Eye bolt primary

For such CT the primary winding is run to bottom tank and insulated by the tank at the bottom.

36. Why nitrogen cushion is provided in oil tank CTs?

Answer

Nitrogen gas is inert in nature and does not disintegrate to form any hydrocarbon elements for deterioration of transformer oil. So this gas is used as cushion for oil bath. Moreover the pressure of nitrogen also prevents the contamination from outside atmosphere and provide hermetical sealing.

37. What are the disadvantages of primary control CTs?

Answer

Primary control CT has the following disadvantages.

1. S.C forces increase because of more primary turns.
2. Primary control can only be achieved if multi-ratio is selected in geometric proportion like 1 : 2 : 4.
3. Charge of CT ratio need line shutdown and opening of line connectors.
4. In case of multi-core CTs independent selection of ratio is not possible.

38. What are the advantages of primary control CTs?

Answer

It has the following advantages.

1. Because of fixed secondary input, the connection to external system is easy .
2. Higher ampere turn increases the accuracy with reduced core size
3. Less secondary winding helps to simplify the design of winding.

39. What is called extended primary current?

Answer

Extended primary current is defined as that current over and above which the CT performs satisfactory operation with allowable accuracy limit and continuous thermal stability. It is to be taken from 120% to 250% of normal rating.

40. What is called rated short-time current in CT?

Answer

It is the maximum thermal current which must be withstood by the CT for the period of specified declared time up to the permitted temperature. The time is taken as one second and temperature is 250°C.

41. How to decide the short-time current of CT if not specified for testing?

Answer

Name plate detail is always referred for testing of CT for various applications. The application for short-time current on various conditions like time factor, voltage level, temperature, etc. are also referred from the name plate values. But for the case of non-availability of the same, the calculation of short-time current is made from the fault level in MVA with rated voltage level from the breaker capacity associated in the scheme. It is essential that CT should withstand for both the thermal effects of short-time current and forces due to dynamic current.

42. Explain the significance of knee-point voltage for CT performance?

Answer

The concept of knee-point voltage (V_k) provides an idea regarding the saturation limit of the CT. This is the point on the magnetization curve over and above which 10% rise in voltage causes 50% rise of excitation current. For the unit protections like differential, directional and distance category, the determination of this point is important and is to be chosen in such a way that the operable value should be less than the minimum knee-point voltage. These are the primary protections and need to function satisfactorily during the time of fault current rise in the CT to be cleared by the relays in the secondary.

By keeping primary winding open, sinusoidal voltage at rated frequency is applied to the secondary terminals. The supply voltage and corresponding exciting currents are noted down to obtain the curve for excitation voltage and excitation current. During this voltage application, the point just above which 10% rise in voltage causes 50% rise of excitation current, is called knee-point voltage. The typical curve for the determination of knee-point voltage is shown in the following Figure.

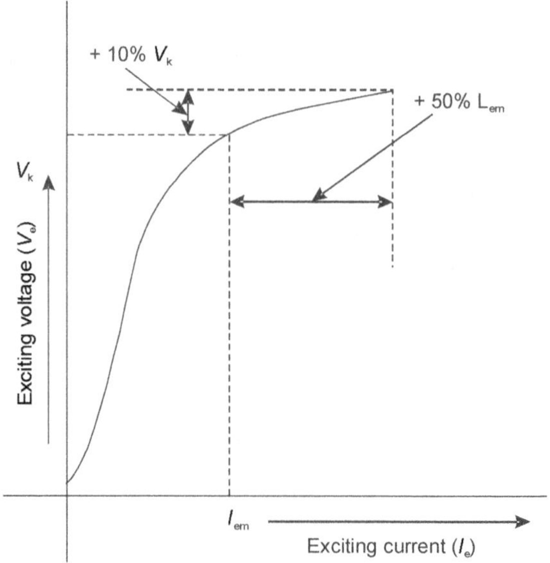

Curve for knee-point voltage determination

Minimum knee-point voltage   It is specified by the formula $V_k = k\,I_s(R_{CT} + R_L)$

where,

k =   A parameter that depends upon system fault level and characteristics of the relay.

I_s =   Secondary reflected current

R_{CT} =   CT secondary resistance at 75°C.

R_L =   Resistance of secondary circuit with lead

V_k =   Minimum knee-point voltage

43. Explain the behaviour of CT for excessive burden on secondary?

Answer

When the CT is connected to excessive burden the secondary voltage that develops across the CT may not be sufficient to drive the required current for correct operation of CT.

Example

Consider CTR = 400/1, 5P10, 10 VA, R_{CT}= 5Ω,

R_{lead} = 2Ω, Relay = 5Ω

Let fault current = 20 kA

During fault occurrance, the refeleted fault current on CT secondary passes through the relay for its actuation.

$$\text{Actual voltage} = I_{sc} \times (R_{CT} + R_L + R_r)$$
$$= 20 \times 10^3 /400 \times (5 + 2 + 5)$$
$$= 50 \times 12 = 600 \text{ volt}$$
$$\text{Design voltage} = \text{Burden} \times \text{ALF}/ I_s$$
$$= 10 \times 10/1$$
$$= 100 \text{ volt}$$

Design voltage is less than actual voltage. So it cannot drive the required current during fault condition. So the operational part of relay will suffer.

44. What are the values of voltages required to decide the insulation level in CT?

Answer

All the CTs generally withstand the continuous operational voltage and to a certain limit for over-voltages also. According to the values of these over-voltages, the insulation label of CT is decided. The insulation level includes the following factors.

1. One minute withstand voltage

2. Lightning impulse withstand voltage

3. Switching impulse withstand voltage

4. Highest system voltage

45. What are the possible reasons for CT failure in practice?

Answer

CTs are operated under high voltage and open to climatic variations. In present days, the variation of electrical parameters and electrical network traffic is very clumsy. At the same time the environment is also not that suitable and favourable to the CT operation in climatic condition. Some of the reasons for the malfunctioning of CTs are listed as follows.

1. Entry of moisture into the insulation of CT.

2. Abnormal change of atmospheric conditions.

3. Change of electrical parameters due to the conditions like

    i. Short-circuiting of primary

    ii. Opening of CT secondary

    iii. Transient conditions

4. Dielectric failure due to pre-mature ageing.

5. Opening of a delta point.

6. Other dielectric failures due to improper wrapping of paper.

7. Improper flux distribution due to abnormal core designs.

46. Explain the factors for the choice of 1 A and 5 A CTs in practice?

Answer

Choice of CT secondary depends mainly upon the secondary burden of the CT and its performance during working conditions. Following are the selection practice for 1 A and 5 A CT secondary core.

Comparison for burden selection between 1 A and 5 A CT

1 Ampere	5 Ampere
Preferred when CTs are used with outdoor type having lead burden more.	Preferred for indoor application, where lead burden is significant.
Primary ratings are moderate like (200/1), (400/1), (800/1).	Primary rating is very high like (2000/5),(10000/5), etc.
If fine turn adjustment is needed on secondary, then 1 ampere is preferred.	Secondary turn is not the prime factor.
For the case of secondary open, the peak voltage is high, but it is used for external installation. So 1 ampere is preferable.	For the case of secondary open, the peak voltage is not high compared to 1 A. So 5 A rating is preferable.

47. What is the difference between class 'P' and class 'PS' CT? Explain the significance of their uses.

Answer

Class 'P' type core in CT technology is used for protection purpose. Such core is designated by the terms like 5P10, 10P20, etc. The meaning of the same (say 5P10) is that this core is accurate up to 10 times of rated current and provides only 5% error at 10 times current. Such type of core is generally recommended for O/L, E/F relays, like back-up relays.

"PS" class core is also used for protection purpose, but it is used for the unit protections like differential REF and distance protections. For such core the limiting factor like study of knee-point voltage, exciting current, burden in terms of lead resistance are important to be considered within allowable value. If such values are not chosen properly then during fault condition, the unit protection may not operate properly.

48. What is the importance of knee-point voltage (V_k), magnetizing current (I_{mag}), internal resistance of CT (R_{CT}) and lead resistance (R_L) when used in the protection circuit and why is it necessary to check I_{mag}, V_k of the CT with the particular relay for which CT is being used?

Answer

For "PS" class protection core in the CT, the factors like knee-point voltage (V_k), permissible magnetizing current and allowable secondary burden, etc. are required to be mentioned in the circuit. This type of protection class needs to be checked for its satisfactory operation region during the time of fault current rise in the CT to clear the same by the relays used in the secondary.

Minimum knee-point voltage   It is specified by a formula $V_k = k\,I_s\,(R_{CT} + R_L)$

where,

k = A parameter that depends upon system fault level and characteristics of the relay.

I_s = Secondary reflected current

R_{CT} = CT secondary resistance at 75°C.

R_L = Resistance of secondary circuit with lead

V_k = Minimum knee-point voltage

Note

1. The knee-point voltage should not be less than the minimum knee-point voltage.

2. The value of V_k should be left to the manufacturer with assumption of R_{CT} also.

Maximum excitation current (I_{mag})   CT core needs to be identified by certain parameters for "PS" class core, magnetization nature of this core is referred with the current called maximum magnetization current at minimum level of knee-point voltage. The value should be less than the maximum magnetization current.

$$I_{mag\ (max)} = p_{mA} \text{ at } V_k / F$$

where, p_{mA} = Permissible magnetization current in mA.

F = Factor usually specified by the relay manufacturer (2 or 4 depending upon application)

Notes on "K" for knee-point voltage.

In usual practice the value of "K" is selected on the basis of fault current I_F and considered as $2I_F$.

Example   CT core for differential core with following data .

20 MVA, 132/33 kV transformer % Imp = 9%, CTR1 = 100/1 A, CTR2 = 400/1 A, Fault MVA = 2000MVA

Calculation

HT full load current = 87.5 A

LT full load current = 350 A

Corresponding fault current on the basis of fault MVA (2000 MVA) for LT side = 35 kA

Now I_F =Reflected fault current to relay = $35 \times 10^3 / 400$ = 87.5

So $K = 2\,I_F = 2 \times 87.5 = 175$

On the basis of % impedance and rating of transformer

Fault MVA = Transformer rating /(Zero source impedance) p.u impedance
= $20 \times 100 / 9$= 222.22 MVA

Corresponding fault current = 3.888 kA

I_F = Reflected fault current = 9.72

So $K = 2 \times 9.72 = 19.44 = 20$ (Say)

So K can be taken as "20" instead of "175".

49. What should be the CTR for a 16 MVA Y-Y 20/1.6 KV 3-phase transformer on 20 kV side?

Answer

The CTR selection depends upon the rating of the equipment, requirement of load and available nearest suitable CT Ratio. For this condition the full load current on 20 kV side is calculated as follows.

FL Current = 20×10^3 / Root 3 $\times$ 20 = 577.36 A

So the nearest suitable value should be considered as 600 A.

50. What is the difference between live tank CT and dead tank CT?

Answer

Considering the practical design pattern and use of insulation in the system, some comparisons can be discussed as follows:

Comparison between live tank and dead tank CT

Live tank	Dead tank
The core and secondary windings are kept in the top tank, which is live (charged with system voltage).	The core and secondary windings are kept in the bottom tank, which is earthed or dead.
The core and secondary assembly is insulated against the high voltage in this live tank.	In dead tank, the primary winding is brought down to the bottom tank and is insulated from earthed tank.
Insulation is on the core and secondary winding.	Insulation is on the primary winding
This transformer is compact and economical.	This transformer is bulky and costly.
Because the primary winding is the shortest and directly passes through the live tank, it offers high strength against the short-time current dynamic force.	Primary winding passes through the porcelain insulator stack and larger length of primary conductor produces maximum mechanical force during short-time dynamic current, which might damage the insulation.
Due to minimum length of primary winding, the I^2R loss is minimum and thermal stability would be better.	For this design due to more length of primary winding and due to complicacy in insulation design, the heat dissipation is critical. Heat generated is dissipated in two stages, i.e., winding to insulating material and to the oil.
Because of single HV conductor, the volume of insulation and oil is less.	Because of two primary conductors the volume of the insulator and oil is more.
Due to top tank design, the mechanical strength in normal condition is also not stable.	Due to dead tank being preponderant at the base, therefore its mechanical stability is more.
Heat generated between terminal connectors and primary terminal joints, affects the primary winding, as it is mounted very near to the terminals (High voltage part).	Primary winding is lengthy in comparison and takes sufficient path for heat dissipation. So, heat generation in joining terminals is less in comparison.
For the case of insulation failure it is observed that top chamber and insulator are scattered and damage the surrounding equipments.	The main insulation is at the bottom part, the failure causes less damage due to contented bottom tank.

51. For a multi-tapping secondary control CT, any of the two terminals are taken to the circuit with other terminals being open. But it is cautioned that CT secondary terminals must not be kept open. Explain the reason, why the spare terminals of the same tapping secondary are to be kept open?

Answer

For a multi-tapping secondary CT, practically the secondary side is not getting open in normal use. Out of the available terminals, any two of the tappings are generally taken to the circuit connection. The other terminals of the same used core are to be kept open to avoid the wrong CTR to the circuit. If the other available tappings are to be shorted then due to different resistance on the CT secondary, the CT ratio would be different. So only for a multi-tapping secondary control CT, any of the two terminals are taken to the circuit with other terminals being open.

52. What is hairpin-type CT? Is it same as pendulum CT?

Answer

In hairpin-type design, the primary winding is taken up to the bottom box, where secondary windings are wound over the primary winding to obtain required ratio of the CT. The major insulation is provided on the primary winding and housed properly in the bottom tank, which is earthed or dead. Because housing of the conductor is connected to the earth, such type of CT is called dead tank CT. This CT also looks like pendulum and called pendulum design CT.

53. What is inverted CT design?

Answer

The inverted CT design looks like the inverted general CT for which the main tank is taken at the top and the primary winding is spread in a uniform and symmetrical way around the core in the top tank of the CT. The secondary windings are wound over the primary winding inside this box. The core and secondary winding are insulated against the high voltage and major part of the insulation is kept on the core and secondary winding. This type of CT is also called live tank CT.

54. What is hermetic sealing in CTs?

Answer

It is the type of sealing in which the insulating mediums like oil is pushed under vacuum and sealed so that there is no leakage. Such type of sealing is basically taken to the CTs of EHT class above 145 kV level. But some of the Indian manufacturers are also adopting this practice for lower range of CTs. The test for sealing is most important for CT and must be conducted as routine test before dispatch of the equipment to the site. The test is carried out by applying the minimum pressure of 0.5 kg/sq.cm. above atmosphere for more than 6 hours and no leakage should be observed during this time. By experience it is seen that leakage is one of the important reason for the failure of CT.

55. Explain the reason for the maximum failure of live tank CT on top chamber?

Answer

For live tank CTs, the primary winding is directly insulated with oil in a chamber on the top, which is directly connected to the live supply of the system. The process of heat dissipation is related to the primary joints that are insulated to the external terminals. Practically the space available for such dissipation of heat does not become suitable for few CTs and results in the failure of top chamber.

56. Why does the maximum failure of live CT occur during night time?

Answer

For live tank CT the top tank is used with oil insulation and the primary winding is connected directly to the external terminal joints. The volumetric change in oil results due to the temperature of day and night. During night, the condensation of water vapour occurs at low temperature, which gets absorbed by the paper and oil insulation. The live tank CTs are found to be more prone to such failure as the top part of insulation is in contact due to absorption of this water drops. But in practice nitrogen cushion is provided with bellow arrangement in the system that avoids the possible reason of CT failure due to water ingress or condensation of it.

57. What is the importance of creepage distance in CT?

Answer

The creepage factor is important for the use of insulator in the equipments. CTs are generally used with porcelain insulators for which creepage distance is required to be provided. The selection values of creepage distance are generally considered as of 16, 20, 25 and 31 mm/kV.

58. What are the advantages of keeping exciting current low?

Answer

Exciting current can be made low if proper quality of core is to be used in the CT construction. Following advantages are obtained due to low excitation current in CT.

1. Flux density in the CT core becomes low
2. Ampere turns become high
3. Heating effect reduces
4. Loss in the CT becomes less
5. Accuracy of CT increases
6. Thermal stability also increases

59. Why is 1 A secondary rating preferred to 5 A rating?

Answer

The choice of secondary current rating is determined mostly by the secondary winding burden. Sometimes the user also decides the rating of CT as per the convenient practices on the system. From the standard ratings of 1 A and 5 A, the rating of 1 A is preferable because of following reasons.

1. The burden of secondary circuit becomes more for larger current.

   Example Suppose R_L = Lead resistance = 2Ω

   R_r = Relay or circuit resistance = $0.5\ \Omega$

   R_{Total} = $(5 + 0.5 + 2) = 7.5\ \Omega$

   For 5 A CT secondary the burden requirement becomes

   $$= V \times I = I \times R \times I$$

   $$= 5 \times 2.5 \times 5 = 62.5\ \text{VA}$$

   For 1 A CT the burden becomes

   $$= 1 \times 2.5 = 2.5\ \text{VA}$$

   Because of more burden requirement the CT would be expensive and large.

2. Controlling of current on secondary becomes easy for lower magnitude.

3. Secondary equipments which are in connection can be chosen.

   But sometimes for high primary current rating like 2000 A, 3000 A, etc., the possible design of 1 A secondary do not become feasible due to requirement of more number of winding turns. For such case 5 A CT secondary becomes suitable.

60. For study of pollution level, creepage distance plays an important role. Provide the table for reference.

   Answer

Creepage distance as per protection level

Protection level	Creepage distance
Light	16 mm/kV
Medium	20 mm/kV
Heavy	25 mm/kV
Very heavy	31 mm/kV

61. Suggest differential circuit for the following transformer.

   7.5 MVA, 33/11 kV Dy11, CTR1 = 150/1, CTR2 = 450/1. Also suggest whether auxiliary CT is required or not, if required then what would be the ratio of the ACT?

   Answer

   For Dy11 type winding, the connection of the differential circuit should be of Yd1 type.

   For this case line CTR1/CTR2 = V_2 / V_1 = 1/3

Due to proper matching of the line CTR, ACT use is not required as per the consideration of voltage ratio above.

But for consideration of relaying current, the full load current calculation has to be done.

Full load current = 130.725 = 131 A (HT side) and 393 A on 11 kV side

Now the secondary current on both sides = 0.8733 A.

But due to the use of delta connection (11 kV CT secondary), the current availability on the relay terminals would be $\sqrt{3}$ times of 0.8733 = 1.5125 A. This current flow of 1.5125 A will cause imbalance in the circuit. Moreover this current magnitude is even more than the normal rate of 1 A current for relay circuit. So to match the current and to keep the magnitude well within the limit of relay current, the ACT should be used on 11 kV secondary side of ratio 1/0.577.

62. Provide details of insulation level in terms of system voltage for instrument transformer?

Answer

The following table can be referred for insulation level study.

Basic insulation voltage level

Nominal system voltage kV RMS	Highest system voltage kV RMS	Power frequency withstand voltage kV RMS	Lightning impulse withstand voltage kV Peak		Nominal system voltage kV (RMS)	Highest system voltage kV (RMS)	Power frequency withstand voltage kV RMS	Lightning Impulse withstand voltage kV Peak
			List 1	List 2	220	245	360	850
Up to 0.6	0.66	3	–	–			395	950
3.3	3.6	10	20	40			460	1050
6.6	7.2	20	40	60	400	420	950 *	1175
11	12	28	60	75			1050 *	1300
33	36	70	145	170			1050*	1425
66	72.5	140	325	325	525	524	1050*	1425
110	123	185	450				1175*	1550
		230	550					
132	145	230	550					
		275	650					

* Switching impulse withstand voltage in kV (Peak)

63. What is the requirement of auxiliary CT in the static differential relay?

Answer

Regarding the requirement of auxiliary CT in the differential circuit, the following factors are to be considered.

Factors for the use of auxiliary CT

1. For matching of the CT secondary current to the relay
2. To restrict the CT secondary current within the limit of the rating of the relay
3. For matching of the impedance on either side of the differential circuit

For the case of static differential relay, use of ACT (auxiliary CT) is equally important as that of other differential relay. ACT is basically used to match the current on either side of the circuit within the limit of the relay rating current. But nowadays the static relays/ numerical relays have the facility of choosing ACT ratio by the program available in the relay. For such type of relays, ACT is not required.

64. Most of the power transformers are of Y connected, why not Δ connected. How the use of ACT and differential circuit problems are solved in Δ connection?

Answer

Power transformers are generally of higher rating units and the cost of these transformers is also high. So to compromise with the cost factor, the Y connected windings are preferable to Δ-connection. Moreover it becomes difficult to provide necessary earth fault protection to the Δ-connection winding. For bigger size transformer, even single unit with taping facility is chosen (auto transformers) to economize the cost of the winding and transformer.

For the use of ACT and differential circuit for Δ-connection winding, the following method is adopted.

Selection of transformer differential connection

Transformer winding connection	Secondary CT differential connection	Transformer winding connection	Secondary CT differential connection	Remark
Dy1	Y1d → Yd11	Y d1	D1y → Dy11	D1 connection corresponds $(I_R - I_B)$
Dy11	Y11d → Yd1	Y d11	D11y → Dy1	D11 connection corresponds $(I_R - I_y)$

65. What are the classes of CTs and how are they classified?

Answer

CTs are classified into different categories but for electrical circuit purpose CT can be classified into the following three types: 1) Metering class 2) Protection class 3) Special purpose protection class.

1. *Metering class*   Metering core accuracy class is usually mentioned as 0.1, 0.2, 0.5, 1, 3, 5, 0.2s, and 0.5s

   Lower the limit of error, more the accuracy and vice-versa.

2. *Protection class*   Protection class of accuracy is generally expressed as 5P, 10P, 15P, etc., The expression is always followed by ALF (accuracy limit factor) and expressed as 5P10, 10P20, etc.

   5P10   Stands for the accuracy class with 5% composite error for 10 times primary current in CT.

3. *Special purpose protection class*   Such type of CT core is best suitable for the relaying circuit to the balance protection schemes like differential circuit, distance relays, etc. The knee-point voltage (V_k), excitation current and secondary resistance play an important role to decide the performance of the CT.

66. What happens if a high burden CT is connected with low burden loads?

Answer

Every CT functions satisfactorily if the burden of the CT (VA rating) matches to the loads connected on the secondary side and minimum error results for the load with 50 to 75% of rated burden. But for the case of low burden the CT error becomes more. Initial charging of CT requires certain magnetizing current for the excitation of CT and this current causes the error in CT.

The phase angle error of the CT also behaves in the same manner as that of the ratio error. The details are shown in the Figures below.

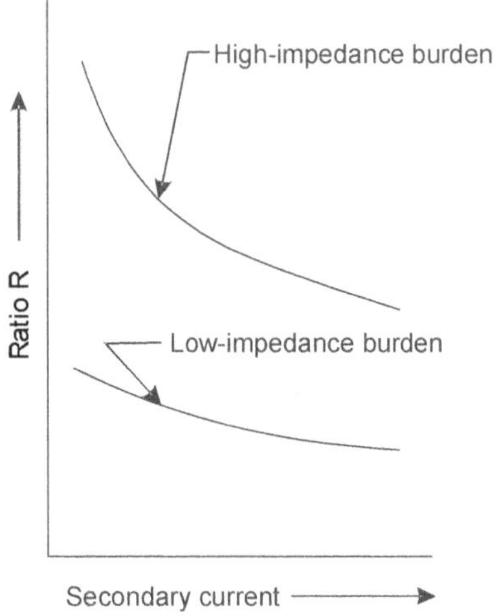

Variation of ratio *R* with secondary winding current

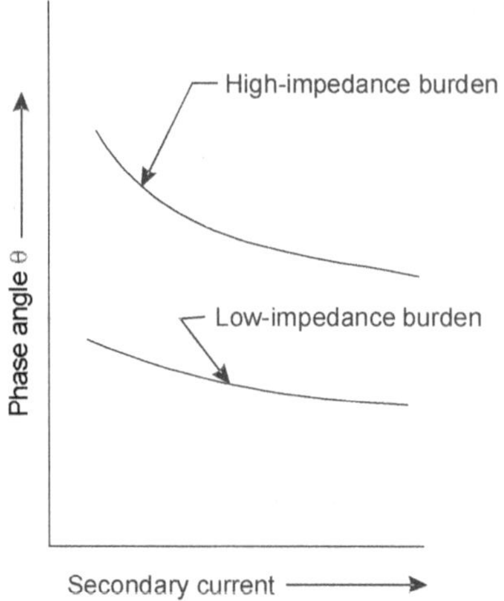

Variation of phase angle with secondary winding current

67. What is the measuring accuracy class (Cl 0.2, 0.5 etc.) for the measuring devices?

Answer

Every measuring device in electrical system if connected to the secondary side of the instrument transformers, then the accuracy of these devices depends upon the sensitivity of the instruments. The sensitivity of every device is declared with its accuracy class (0.2, 0.5, etc). The accuracy class gives an idea regarding the limit of error of the instrument. The lower the accuracy class, the limit of error is less and accuracy is more. For the accuracy class of 0.2, the limit of error is of maximum 0.2% for the allowable load on the system.

68. Why PS class core is best suitable for the application to differential relay.

Answer

PS class core is basically used for special purpose protection of the equipments. Such type of CT core is best suitable for the relaying circuit to the balance protection schemes like differential circuit, distance relays, etc. The knee-point voltage (V_k), excitation current and secondary resistance play an important role to decide the performance of the CT. The differential relay compares the currents on both sides of the system and needs to be stable for the condition of fault in the system. So the factors like knee-point voltage (V_k), excitation current and secondary resistance are to be proper for the relay application. So PS class core is best suitable for the application to differential relay.

69. What will happen to a line feeder CT with open-circuited secondary winding during idle charge of the line?

Answer

In an open-circuited condition, the flow of active current is very minimum and neglected in practice, but certain amount of reactive current flows due to capacitive effect, depending upon the length of the line in the system. This current at HV value becomes prominent to cause the rise of voltage on open CT secondary terminals and if value becomes more than that of the allowable insulation level of the secondary terminals then CT may fail. But by field experiences the following have been observed.

1. Abnormal sound of the CTs in the system

2. Failure of CT secondary insulation

3. Saturation of CT core

4. Change in CTR after rectification

Physical bursting of CTs may not result under such condition due to its design factor. But the abnormality as explained above may occur in the CT.

Note   For reference it has also been found that during maintenance of CT like change of CTR on secondary terminals, wiring modification, etc. the magnitude of CT secondary voltage is observed within the range of 100 V to 400 V even for no voltage excitation to the CT under shutdown condition. Such situation of open-circuit CT secondary voltage results due to line induction current on the primary side of the CT.

70. What are the precautions taken to save the insulations of CTs during high voltage switching?

Answer

The CTs for HV and EHT system suffer with the problems of high frequencies due to switching operation to high voltage system. During switching operation the range of frequency goes to 0.5 to 3MHz resulting in HV surges destroying the condition of insulation many a time. The use of oil-paper insulation with screens in layer manner helps to cause the distribution of this voltage within the capacitances that developed in the CT and accordingly save the insulation of the CT.

71. Explain the criticality of high frequency voltage on CT?

Answer

The development of HF voltage waveforms occurs during switching operation of the system, i.e., sudden energization of the equipment. It has been experienced that the voltage rises to the range of 1000 kV with frequency range of 100 Hz to 20 MHz for the case of 400 kV system. Such sudden rise of voltage develops unusual stresses in CT core, insulation and windings. HV/EHV CTs are accordingly designed to take this unequal rise of voltage stress in the insulation. Sometimes the failure of CT occurs when the value rises beyond the limit or remains over and above the allowable time.

72. Explain the behaviour of magnetic core of CT for the transient condition of flux?

Answer

During transient condition, the development of magnetic flux on the core of CT becomes erratic. The nature of flux becomes distorted and fluctuates due to the effects of harmonics component on it. The availability of DC components on the flux causes saturation of the core and results in abnormal current on the secondary side of CT and for the performance of the relays on the network. It is clear that, there is a large transient flux swing in the magnetic core of the CT. With this large flux swing during transient period the magnetic core of the CT of conventional design gets saturated causing undesirable effects on the performance of the protective relays connected to the secondary of the CT.

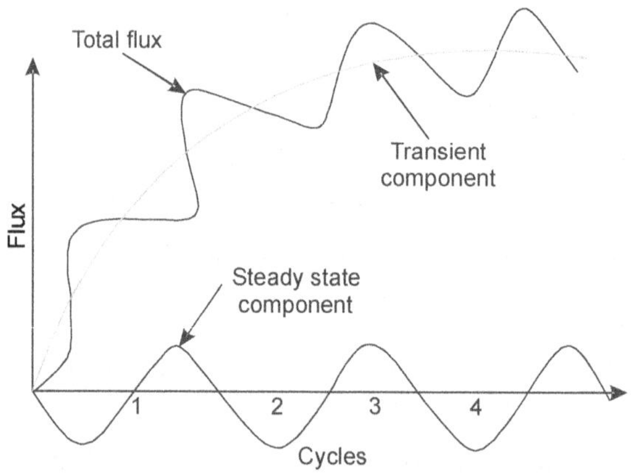

Current transformer core fluxes during transient condition

In CTs of conventional design, saturation of the core due to transient DC component of the fault current is possible within a few milli-seconds, after which their secondary current is fully distorted, resulting in inaccurate measurement of the fault current by the relay. In order to prevent an adverse effect on the performance of the relays, the CT cores must be greatly enlarged or air gaps should be provided in the cores.

73. What is the effect of magnetic core for sudden opening of CT secondary side?

Answer

Under normal condition, the flow of primary current is much more than the amount of exciting current.

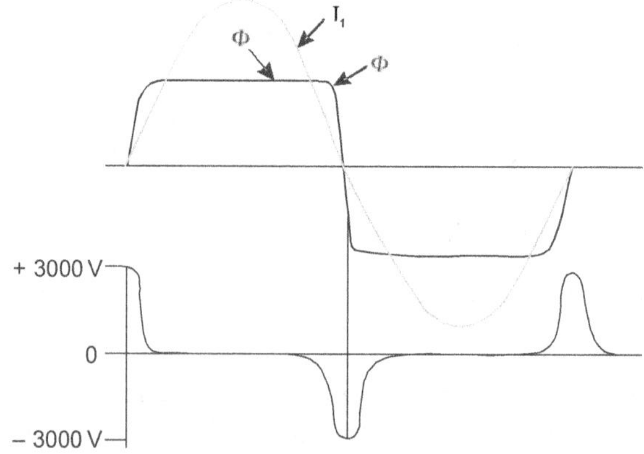

Primary current, flux and seondary voltage when a CT is open-circuit

Because of secondary side connection to the load, the amount of magnetic flux on the core becomes normal during normal loading condition. But due to sudden opening of CT, the secondary current becomes zero and the availability of primary current gets totally engaged for the magnetization of the core, resulting in saturation of the core. The behaviour of magnetic flux also becomes a flat-topped wave with steep sides . At each zero-current region, the rate of change of flux becomes extremely high for which the induced voltage peaks to several thousand volts in an open-secondary circuit.

74. Explain the CT performance during fault condition if so driven with DC offset value of current?

Answer

The DC offset value in the current waveform does not contribute to the transformation principle from primary current to secondary. So the current value on the secondary side does not become as the replica of primary side. Because of this behaviour the relays on CT secondary does not work properly during the fault condition with availability of DC offset waveform. Such behaviour of DC effect in CT results for the cause of saturation of the CT core.

75. What are those factors to decide the transient performance of current transformer?

Answer

Transient behaviour study is applicable to EHV class CTs. The important factors are the design of core size.

   i. Core size and type

  ii. Air gap

 iii. Stacking factor

 iv. Burden of the core

  v. Short-circuit current accuracy

 vi. Accuracy of core

76. Explain the behaviour of degraded insulation in CT?

Answer

For conventional CTs, the paper, film, oil are the insulating materials used between primary and secondary terminals. For any problem or degradation of insulating material the test results are changed from the normal value.

1. *Increase of tan delta value*   Value of tan delta deviates with more than 0.10% from the previous results.

2. *Increase of gas evolution*   The objectionable gases like CH_4, C_2H_2, etc. evolve as per the deterioration of insulations of the CT.

3. *Rise of capacitance value*   Capacitance value of puncturing of the material. The allowable range is <1% variation per year.

4. *Reduction of IR value* The insulation resistance of the materials so used also reduces with drop in PI (polarization index) value.

5. *Rise of PD* The formation of void, gap, etc. is the phenomena for the degradation of insulation. So the rate of partial discharge increases.

77. Explain the significance of X_m for CT equivalent circuit during saturation condition?

Answer

X_m is the magnetizing reactance that develops across the circuit and is responsible for resulting excitation current in the magnetic core of the CT. In normal condition this value remains high and results in reduced magnetizing current. But during magnetic saturation this X_m value reduces and demands more excitation current to flow through it. The error in CT ratio also increases due to bypass current flow to this circuit. The following figure is given for the explanation of the significance of X_m for the CT.

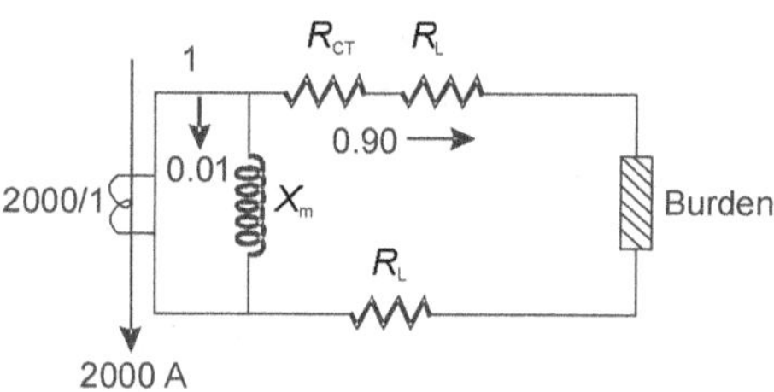

Normal current transformer

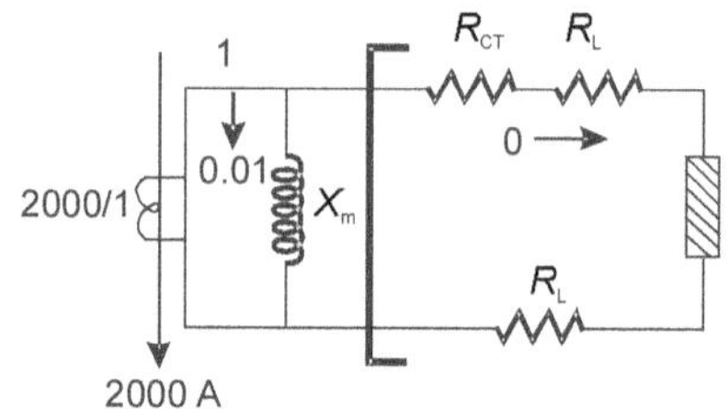

Saturated current transformer

78. Explain the behaviour of magnetic circuit of CT due to short-circuited primary condition.

Answer

For the short-circuited condition of primary side, the primary current increases several times even 50 times of the rated value, but remains not over than one second. Because of this effect the CT may enter into the saturation region. But this effect of magnetic saturation does not affect the circuit and this condition remains for a few second and gets cleared due to relaying effect on the scheme. If in case the relaying scheme fails to operate then the thermal and mechanical stresses affect the physical condition of CTs.

79. Explain the CT performance for the case of excessive primary current?

Answer

For the case of excessive primary current, the actual voltage required would be more than that of the design voltage. So the current may not be sufficient to drive during the fault condition. The performance would not be proper for the clearance of the fault.

80. Why X/R ratio is important for CT saturation study?

Answer

Value of X/R is decided as per the type of CT used and the type of circuit in the connection to the system. The availability of DC component contributes to the degree of offset and the nature of core saturation on CT. Consider a line connected directly to generation unit, where CT has been connected on the line. For such line, X/R is more and chance of saturation of CT core is also more as compared to the connection for a long line.

81. Why tan delta measurement is important for high voltage CT? Suggest the methods for tan delta measurement?

Answer

General conventional CT has oil and paper insulation between HV and LV terminals. The condition monitoring of this insulation provides the healthiness of CTs. The different online and offline methods of condition monitoring are generally used for the evaluation of the insulation. But tan delta measurement is important among them. The testing method is conducted under high voltage application of 10 to 12 kV AC supply to study the insulation.

Following are the few methods for testing of the insulation under tan delta. The connection and testing practices depends upon the type of the construction of CT.

1. *CT with test tap*—UST mode (Ungrounded specimen test)

2. *CT without test tap*—GST mode (Grounded specimen test)

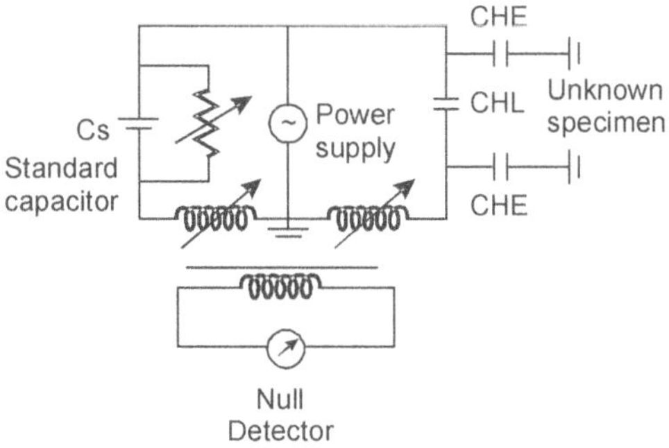

UST mode

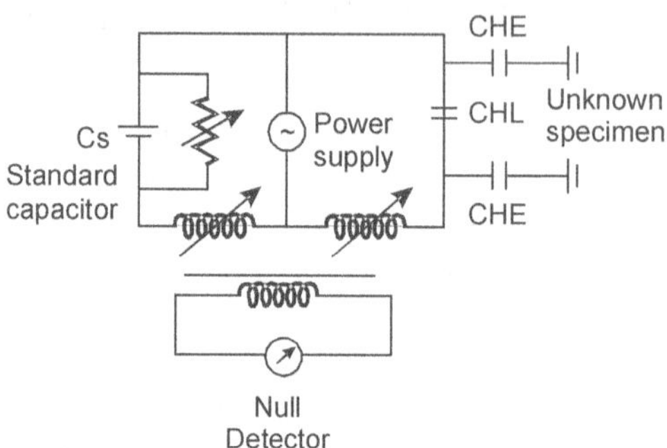

GST mode

82. Why CTs are connected between bus coupler breaker and transfer bus isolator for high voltage system?

Answer

The involvement of bus coupler breaker comes into circuit for the case of coupling the two buses in the network. These two buses in the network may be main bus and transfer bus or bus 1 and bus 2. When it is established for the case of main bus and transfer bus the following explanation holds well.

The bus coupler breaker comes into action either to provide parallel path to the system or separate individual path to maintain the faulty breaker in the network. During the current flow through the bus coupler breaker circuit, the positioning of the CTs should be available towards the load side, due to the energisation of CTs at the near end of load. So the CTs are connected between bus coupler breaker and transfer bus isolator. But it may be noted here that the placement of CTs can be done also in between bus coupler breaker and main isolator if it is found convenient in the network.

83. When a fault occurs between a feeding station and receiving station having a single circuit transmission line, how will the DP relay at the receiving end operate as the current in its CT would be zero or less during fault?

Answer

For the case of any fault in the transmission line, the respective voltage and current parameter in the line changes according to the intensity of fault. The magnitude change of these parameters recalculates the fault impedance on the line. If the calculated impedance becomes less than the set value then fault is activated by the DP relay.

For the described situation, the occurrence of fault causes the dip of voltage and rise of current on the respective faulty phase simultaneously. So the CT and PT/CVT present at both end of the transmission line starts to actuate to provide the necessary current and voltage replica to the DP relay. According to the comparison of impedance, i.e., the actual faulty impedance and the set impedance of the DP relay, the fault is initiated by the DP relay.

Moreover during fault condition, the other healthy parameters on the line are disturbed and provide the unbalanced parameter for detection of the fault, say, for example, if a fault results in R phase then the voltage of R–N gets reduced and the unbalance in voltage to the system due to this reduction takes part in fault impedance calculation. Similarly for the case of current magnitude, the value on the faulty R phase becomes more.

So the concept of zero current or less during the fault condition does not occur. But the flow of current gets disturbed at both ends. As source is connected to one end of line, during the fault condition, the source would drive the current in the forwarding direction and in a perfect manner but the current disturbance at the receiving end depends upon the unbalanced condition of voltage at that end. So the actuation and correct measurement of DP relay at the receiving end does not become perfect. However the relay actuates to the fault.

84. Explain the advantages of optical sensor CT as compared to conventional CT.

Answer

1. Optical sensor CT does not have any physical conducting wires.
2. They do not have any magnetic core.
3. There is no insulation used.
4. Digital signals are only accessed for comparison of values.
5. Limitation of CT ratio is not there.
6. Possibility of CT saturation or failure of insulation is also not there, because no mechanism saturation limit is used in this CT .
7. Changing of CTR is not required.
8. No physical hardware or copper lead is required for secondary connection.
9. Because of digital signal, the sensing of relay becomes accurate.
10. The control is microprocessor-based so retrieving of data, logical control, etc. can be done faster.

85. Why are optical sensor CTs not yet been accepted popularly in practice?

Answer

The performance study of optical sensor CTs are under development. The language for commercial use is yet to be framed, like standard code of practices, definitions, etc. As the principle for measurement of secondary current is completely different from that of conventional magnetic core CTs, the acceptance of this optical sensor CTs are to be proved in market. But the results have successfully passed the observation in laboratory and research field. After a complete and confirmed study, the same will be accepted by the users. Some of the Indian utilities have already started using this concept.

86. What is optical CT and explain the technological concept of this CT?

Answer

In the conventional CT the use of iron core, copper winding, insulation, etc., bring disadvantages for correct operation and performance of the equipment. But the non-conventional practice like use of optic fibre for instrument transformer technology can solve the possible problems as described in conventional CTs.

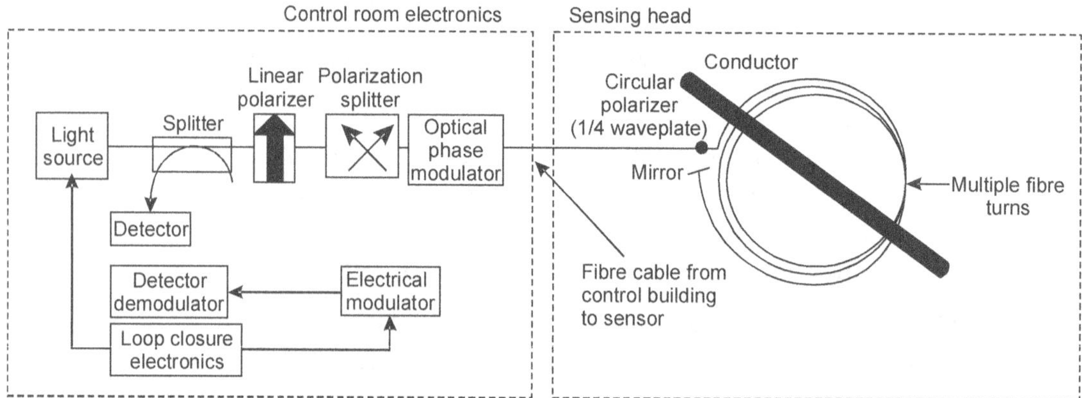

Fibre optic current sensing principle

The optic fibre is encircled on the currentcarrying conductor with one end connected to optical phase modulator and other end to a mirror.

Light from control room is sent through a wave guide polarizer and shifted to the optic fibre wound on the conductor. Finally, the light source is reflected by the mirror connected at the end. The magnetic effect and the light rays by the principle of Faraday's laws, develop the phase shifting of two sources. These effects are proportional to the current on the conductor. So the light source is calibrated for measurement of the current. Generally for such condition 5 A secondary rating is selected.

87. Explain the concept of non-conventional CTs, based on optical fibre technology?

Answer

The conventional instrument transformers work on the principle of electromagnetic induction, using a magnetic core. There are now available several new methods of transforming the measured quantity using optical methods. These concepts are based upon the following effects.

1.   Optical instrument transducer.
2.   Hybrid transducers.
3.   "All optical" transducer.
4.   Zero flux CT.
5.   Hybrid magnetic optical sensor.
6.   Rogowski coil.

88. What is Hall effect?

Answer

A semiconductor chip is placed in the gap of a magnet that develops certain voltage to the circuit due to the Hall effect on it. The current-carrying conductor is allowed to pass through the air core of the ring magnet. Now the electromagnetic effect due to this current results in certain sensing current on the semiconductor chip and it is calibrated for the measurement of the same on the secondary side.

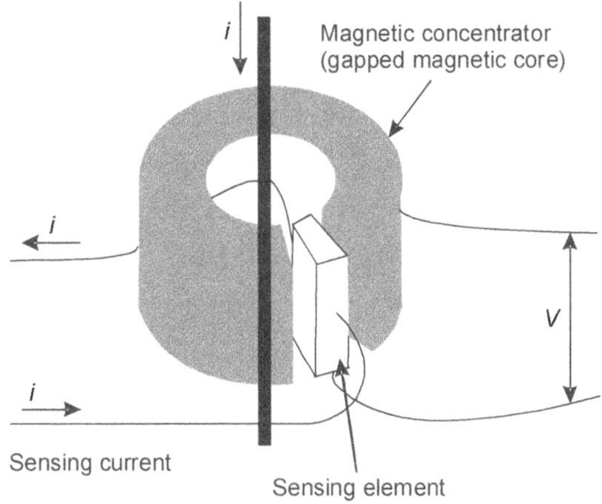

Hall effect current

89. How is the accuracy class of CT selected for the following equipments or relays?

   i. Back-up O/C relay

   ii. Energy meter

   iii. Ammeter

   iv. Differential relay

   v. Distance protection relay

   vi. Back-up E/F relay

Answer

Selection of accuracy class of CT and application of the same for different equipments or relays depends upon the characteristics of the equipments/relays. For metering purpose, accuracy class is considered as primary factor and for protection purpose, knee-point voltage, burden excitation current and accuracy limit are important.

   i. *Back-up O/C relay, E/F Relay*   Over-current relay needs to be provided with the core for which accuracy limit factor (ALF) is important. The core (CT) should be capable of providing limiting error for over-load current also. So, the accuracy class like 5P10, 5P20 type core is preferable.

ii. *Energy meter, ammeter*   Metering instruments need to be provided with most precision accuracy class CT. These instruments are used with the calculation of energy consumptions by the consumers. So accuracy class of 0.2s, 0.5s are better suitable for such instruments.

iii. *Differential relay, distance protection*   The differential relay compares the currents on both side of the system and needs to be stable for the condition of fault in the system. So the factors like knee-point voltage (V_k) excitation current and secondary resistance are to be proper for the relay application. So PS class core is best suitable for application to differential relay. Similarly for distance protection relay, the requirement of currents is needed to be stable for the fault condition. So PS class is suitable.

90. Explain the principle of Rogowski coils.

Answer

Rogowski coil is an air-cored coil, wound upon the current carrying conductor and generally connected to the secondary circuit through an amplifier circuit to measure the amplified current for the application of the scheme.

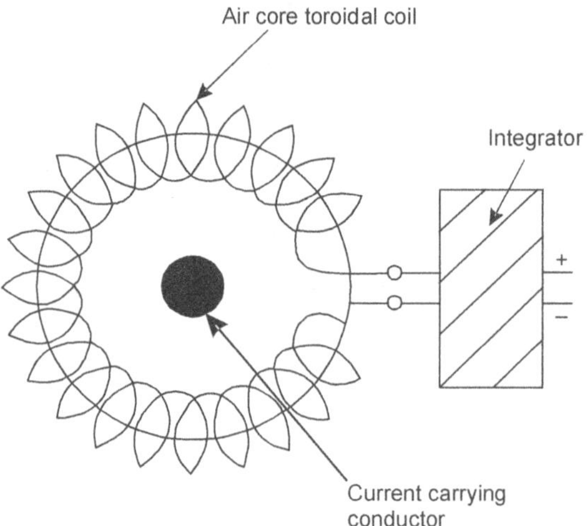

Rogowski coils

91. Explain and draw the connections of CT cores for O/L relays and E/F relays?

Answer

O/L relays are used to trace the current flow on each phase and actuates for the rise of current above certain limit as per the setting of the relay. E/F relay is used in the residual path of the phase circuit. For the case of unbalanced current flow on this circuit, the relay actuates for isolation of the fault.

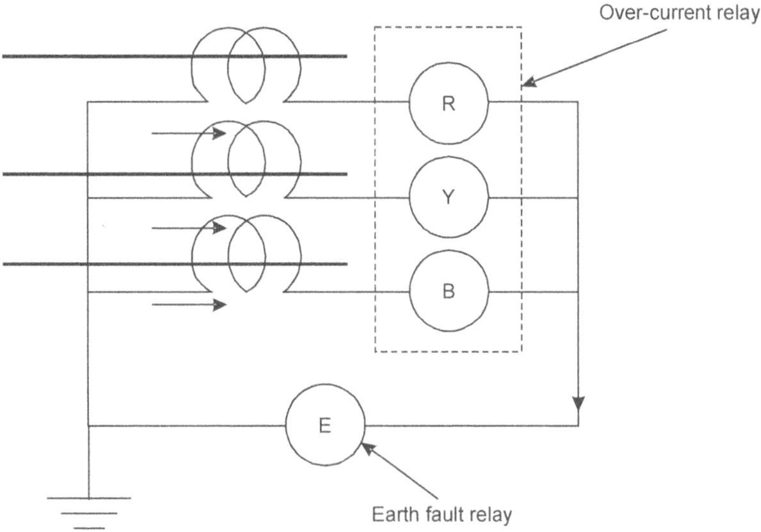

O/L and E/F relay

3

STUDIES ON CAPACITIVE VOLTAGE TRANSFORMER

INTRODUCTION

Voltage transformer is classified into two types namely electromagnetic voltage transformer or potential transformer (PT) and capacitive voltage transformer (CVT). Electromagnetic voltage transformer is a conventional type transformer that contains mainly of primary and secondary winding. But capacitive voltage transformer is different in construction and contains a capacitive voltage divider (CVD) and a electromagnetic voltage transformer (EMV). The CVTs are used for different purposes like measurement, protection and also for carrier communication.

CVT is used with its primary side connected directly either between two phases or one phase and earth, depending upon the requirement and application. But the general practice of use is between phase and earth. Secondary winding is connected across the rated burden of loads. According to the application and suitability of CVT, the number of cores is chosen in the system. These cores are generally used for metering or protection purpose in the secondary circuit. So it is categorized as

1. Measuring voltage transformer (Metering core)
2. Protection voltage transformer (Protection core)

BASIC CONSTRUCTION

The construction of CVT depends upon the arrangement of capacitor unit and electromagnetic unit on the equipment (Figure 3.1). Basically CVT comprises of two important units—CDU (capacitor divider unit) and EMU (electromagnetic unit).

CAPACITOR DIVIDER UNIT (CDU)

Capacitor divider unit consists of a capacitor stack inside a porcelain unit filled with oil. The stack contains mainly of capacitor elements connected in series. The total number of capacitors are grouped under two main units as HV capacitor (C_1) and intermediate voltage capacitor (C_2).

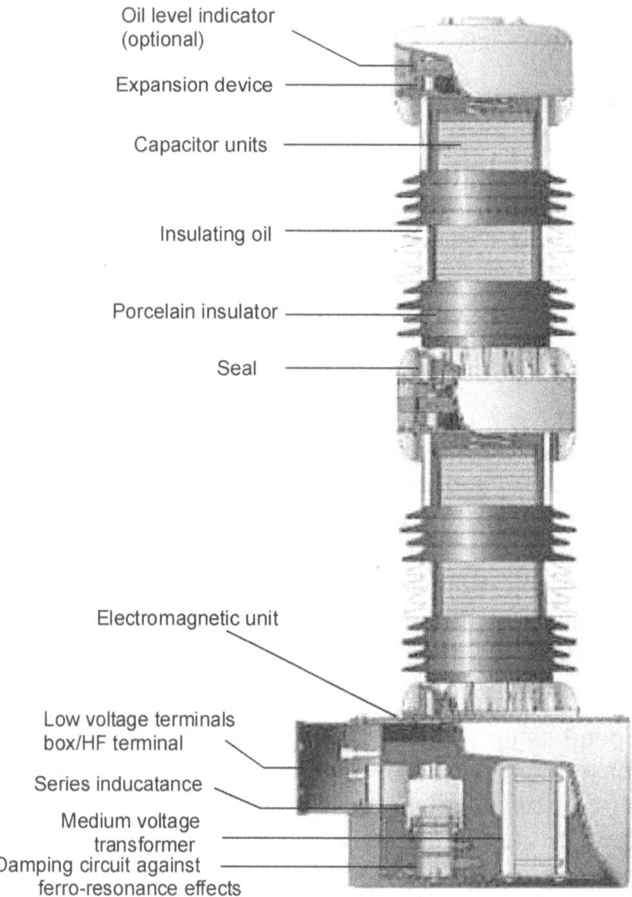

Figure 3.1 **Cross section of a typical CVT**

Individual capacitor element consists of aluminum foils as electrodes and certain insulating material between them as the dielectric medium. For connection of these elements in series, tinned copper taps are inserted over the aluminum foils. Each capacitor unit is independently hermetically sealed. In some CVT, stainless steel bellow is also provided inside each unit to compensate volumetric changes in oil due to the variation of ambient temperature. In between the capacitor units, intermediate voltage tap is taken out for connection to the EMU. Similarly the end point of capacitors is connected to the line conductor stud connection as the top end and NHF (high-frequency carrier terminal) as the bottom end. The bottom end is connected to earth through the NHF link. In some units a protective gap is also provided across NHF and earth, along with cable link across the gap.

As a regular practice the capacitor unit C_1 contains maximum number of capacitor elements as compared to the capacitor elements in C_2. Typical number of elements used in 420 kV-class CVT is $C_1 \approx 280$ nos. and $C_2 \approx 20$ nos.

ELECTROMAGNETIC UNIT

The electromagnetic unit (EMU) comprises the following electrical components that are shown in Figure 3.2.

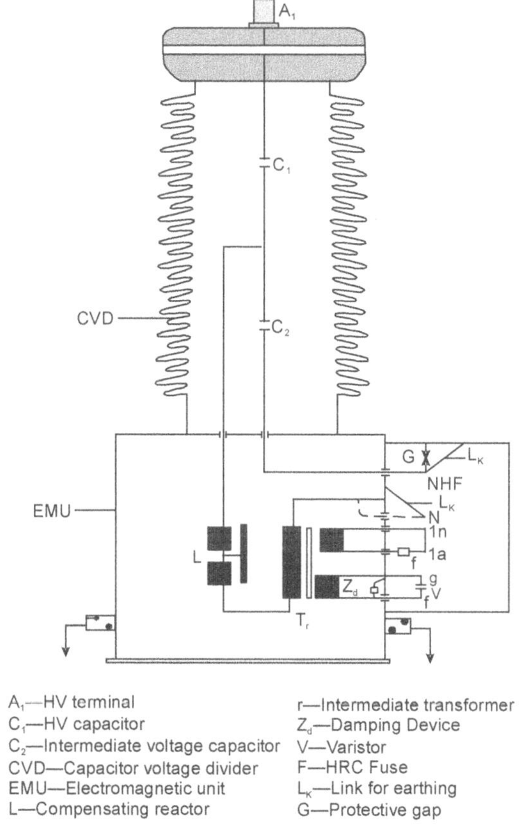

A₁—HV terminal
C₁—HV capacitor
C₂—Intermediate voltage capacitor
CVD—Capacitor voltage divider
EMU—Electromagnetic unit
L—Compensating reactor
r—Intermediate transformer
Zd—Damping Device
V—Varistor
F—HRC Fuse
Lₖ—Link for earthing
G—Protective gap

Figure 3.2   CVT with electrical circuit

1. Compensating reactor (L)
2. Intermediate transformer unit (T_r)
3. Damping device (Z_d)
4. Varistor (V)
5. Protective gap across NHF terminal
6. Miniature circuit breaker (MCB) or HRC fuse

7. Earth link
8. Secondary terminal box
9. Carrier protection device
    i.   Drain coil
    ii. Lightning arrestor
    iii. Earth switch
10. Other accessories
    i.   Oil level indicator
    ii. Earthing terminal
    iii. Oil sampling valve
    iv. Oil filling vent
    v.   Name plate
    vi. Lifting lugs

The compensating reactor is connected in the circuit from the potential divider terminal in series to the primary electromagnetic winding of the intermediate transformer. These units are housed inside the oil-filled hermetically sealed EMU tank. On the secondary side of the intermediate transformer, the secondary voltage coils are arranged according to the requirement of number of cores. Sometimes even three cores are provided on the secondary circuit (2 nos. of core for protection purpose and 1 no. for metering purpose). The availability of secondary side cores depends upon the use and choice of customer.

Secondary Terminal Box

This box is the external terminal box, mounted along with the EMU and contains the terminal elements. The secondary cables from different secondary coils are made available at terminal connectors through VT fuses or MCB poles for the connection of external circuit. The protective device circuit is also available in this box. This circuit contains Z_d (damping device) and V (varistor) across the auxiliary secondary winding. The link arrangement for both capacitor unit and intermediate transformer units are also provided inside this box. The capacitor unit link is marked with NHF and intermediate transformer unit link is marked with "N".

In some CVT, the earth link box is also separately available instead of getting housed in the secondary box. The protective air gap adjustment across the NHF point and earth is also another protective device in CVT.

Carrier Protection Device

The "HF" terminal of the CVT is used for carrier purposes. But for the protection of the carrier circuit, three-element carrier protection device is used in the system. This device contains drain coil, lightning arrestor and an earth switch. The link between HF terminal and earth is disconnected and terminal is taken through the carrier protection device as shown in Figure 3.3. For carrier purpose, the circuit needs a modular coupler unit. This modular coupler unit forms a programmable high-pass filter circuit, which contains a tunable drain coil, the coupling capacitor, one tunable series capacitance and a shunt inductance. Use of transformer with taps provide potential separation between line side and equipment side.

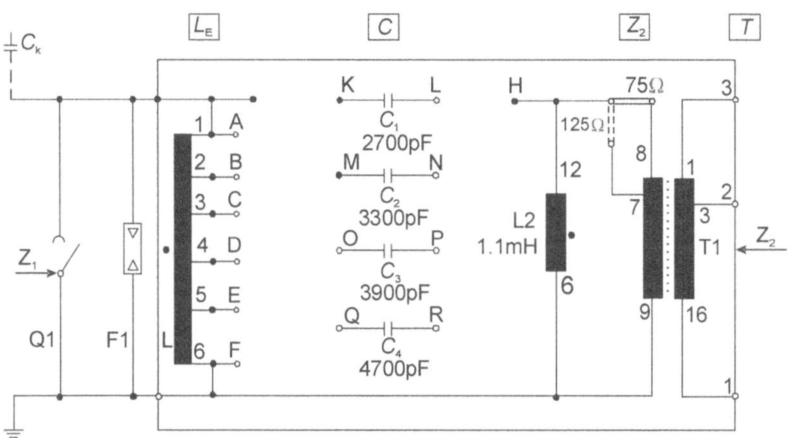

Standard setting guide

Z_2 nom = 75Ω /125Ω

| CK (pF) | Z_1 nom = 240Ω | | | | | | | | | | | | Z_1 nom = 320Ω | | | | | | | | | | | |
	f_1 (kHz)	LE				C						T	f_1 (kHz)	LE			C							T
1500–2199	232	AB	DE	EF		HK	LP	GQ				3	180	BC	EF		HL	LP	KN	NQ	MR	GQ		2
2200–2699	158	AB	EF			HK	LN	NR	GM	MQ		3	128	EF			HM	NR	GQ					2
2700–3299	132	BC	EF			HM	NP	PR	GQ	OQ		3	102	CD	DE		GM	LN	NR	HK	KQ			2
3300–3899	115	AB	BC	CD	DE	GQ	LP	PR	HK	KO		3	90	AB	BC	DE	GQ	NP	PR	HM	MQ			2
3900–4699	96	AB	BC	CD		GM	HN					3	78	AB	DE		GQ	LN	NP	PR	HK	KM	MQ	2
4700–5599	80	CD	DE			HK	KO	LN	PR	GM	MQ	3	76	BC	CD		GK	KM	NQ	PQ	HL	LR		3
5600–6799	70	AB	BC	DE		GK	KM	NO	PQ	LR	RH	3	70	AB	BC	DE	GK	KM	NQ	HL	LR			3
6800–7499	60	AB	DE			GK	KO	PQ	HL	LR		3	58	AB	DE		GM	MO	PQ	HN	NR			3
7500–	52	BC	DE			GK	KM	HL	LN			3	52	AB	BC		GQ	QN	KM	HL	LR			3

Figure 3.3 PLC coupling circuit

Function of drain coil This coil is connected in parallel to the circuit and diverts the power frequency current flowing through the coupling capacitor to earth.

This coil has following specification.

i. Power frequency impedance $\leq 1.5\ \Omega$
ii. Short-time current (0.2 s, 50 Hz) = 50 A
iii. Continuous current (50 Hz) = 1.5 A

Functions of LA   The lightning arrestor (LA) is also connected in parallel to the circuit and drain coil. The function of lightning arrestor is to limit the voltage peaks to a harmless value for the circuit that comes from the line.

It has following specifications.

i. Rated voltage = 660 V
ii. Maximum impulse spark over voltage, peak value 1.2/50 μs wave = 3300 V
iii. Rated discharge current peak value 8/20 μs wave = 5 kA

Functions of earth switch   Earth switch is also connected in parallel to the circuit. The basic function of E/S is to provide safety path of current flow during the time of servicing/checking/commissioning of modular coupler unit in the scheme.

Note   The element carrier protection device is the optional and extra unit of CVT. The normal function of CVT is independent of the availability of this unit.

WORKING PRINCIPLE OF CVT

It is already discussed under the basic construction of CVT that it consists of a capacitor potential divider in conjunction with an electromagnetic intermediate transformer. This CVT is connected across the phase conductor and earth. So the voltage (line voltage/$\sqrt{3}$) becomes available across the primary side of CVT.

Now due to capacitive voltage divider unit and use of C_1 (HV capacitor) on the upper stack, considerable amount of voltage drops across the capacitance C_1. After drop of the voltage, a suitable and required quantity in the range of $10\,\mathrm{kV}/\sqrt{3}$ to $20\,\mathrm{kV}/\sqrt{3}$ becomes available across the EMU. This range of voltage ($10\,\mathrm{kV}/\sqrt{3}$ to $20\,\mathrm{kV}/\sqrt{3}$) is selected according to the design and different voltage class of the transformers. Electromagnetic unit intermediate voltage transformer with compensating reactor in series with it, steps down the available voltage ($10\,\mathrm{kV}/\sqrt{3}$ to $20\,\mathrm{kV}/\sqrt{3}$) to the desired secondary voltage of $110\,\mathrm{V}/\sqrt{3}$ across the different secondary windings. Finally different secondary windings are made available at the secondary terminal box for connection of the secondary circuit.

Function of Intermediate Voltage Capacitor (C_2)

This capacitor unit becomes a part of the circuit parameter of the carrier communication circuit. The lower end of the capacitor unit is connected to the NHF terminal and during the use of carrier communication circuit, this terminal is connected to the circuit. The capacitor unit C_2 is also called coupling capacitor and according to the magnitude of this capacitor, the other parameters and corresponding connections are decided for the modular coupler unit of the circuit. Table 3.1 provides the PLCC technical data with their nominal values.

Table 3.1   PLCC technical data (nominal value)

Data	Value
DC supply required for carrier set	48 V
AC current required for single carrier	1 A
Characteristic impendance of coaxial cable	75/125 Ω
Capacitance of CC/CVT	4400 pF (220 kV) 5575/6000 pF (132 kV)
Inductance of wave trap	0.5 mH (132 kV) 1 mH (220 kV)
Line current capacity of wave trap	630/800/1200 A
Nominal Tx at coaxial termination (By SLM)[#]	+ 4 to +12 dBM
Nominal Tx at coaxial termination (By SLM)[#]	–26 to 0 dBM

[#]   **Measurement procedure** Tune SLM to Tx/Rx frequency + Pilot frequency of that carrier type, for which selection is done.

Local loop   Using dummy load in place of hybrid print, local Tx/Rx can be looped back for ensuring healthiness of local carrier set. It is the method of choosing local testing of PLCC.

Function of HV Capacitor (C_1)

This capacitor unit is the main component of CVT and present on upper stack of the CVT. Maximum number of capacitor elements being in series form this unit. Technically it can be defined as the capacitance connected between HV terminal and intermediate voltage terminal. During working operation of CVT, maximum voltage drop occurs across this unit and only the required suitable voltage is made available across the intermediate transformer unit for further step-down to the secondary circuit.

Capacitor divider ratio   From the basic circuit of CVT (Figure 3.4) it is found that capacitor C_1 and C_2 are regarded as the capacitor units being in series and connected across the source voltage (V_s).

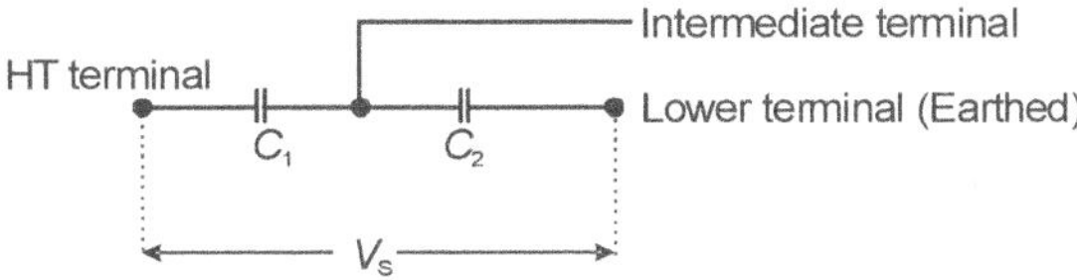

Figure 3.4   Typical capacitor circuit of CVT

Capacitor divider ratio is defined as the ratio of sum of the total capacitors to the HV capacitor value.

$$\text{Capacitor divider ratio} = \frac{(C_1 + C_2)}{C_1} = \frac{V_s}{V_2}$$

From the voltage divider rule

$$V_2 = \text{Voltage across } C_2 = \frac{(V_s \times C_1)}{C_1 + C_2}$$

So
$$\frac{(C_1 + C_2)}{C_1} = \frac{V_s}{V_2}$$

Note   During the calculation of divider ratio the stray capacitance values are generally negligible.

Functions of Compensating Reactor (L)

CVT is basically a capacitor potential divider. During working conditions, when source voltage results in a current flow through this capacitor unit, the necessary voltage drop becomes capacitive in nature and the available voltage for working of intermediate transformer needs to be compensated.

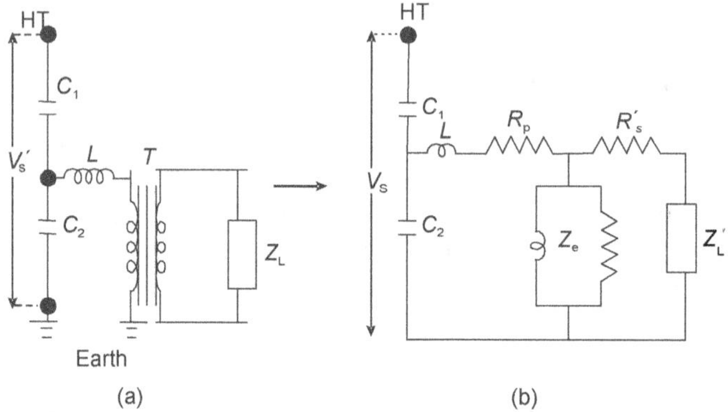

Figure 3.5   **Electrical equivalent circuit of CVT**

where,

$$Z_e = \quad \text{Exciting circuit impedance}$$
$$R_p = \quad \text{Resistance of primary side}$$
$$Z_L' = \quad \text{Load impedance referred to the primary and}$$
$$R_S' = \quad \text{Secondary resistance referred to the primary}$$

Now by the use of inductance in series to the primary circuit of intermediate transformer the compensation to the circuit can be achieved. The value of inductance (L) is so adjusted that the circuit resonates for the supply to the standard frequency source (Figure 3.5a and b).

Now neglecting the effect of exciting impedance and considering burden of the circuit being resistive in nature the circuit can be modified as follows (Figure 3.6).

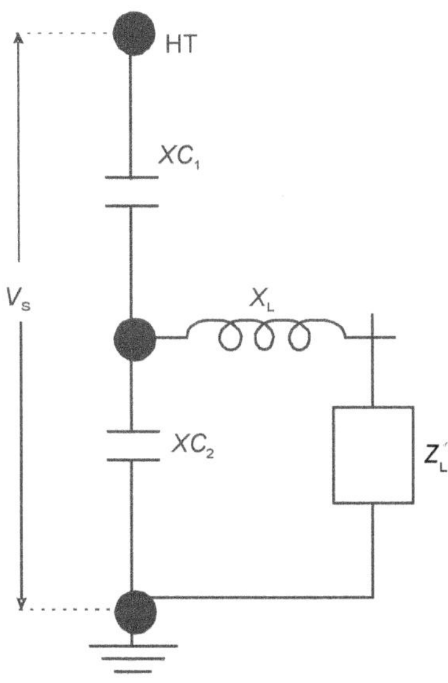

Figure 3.6   **Reduced electrical circuit of CVT**

For proper adjustment of circuit and to develop resonance circuit

$$X_L = (X_{C_1} \times X_{C_2}) / (X_{C_1} + X_{C_2})$$

But from the practical study of the circuit, it is seen that X_{C_2} is very small compared to X_{C_1}. So the above equation can be written as

$$X_L = X_{C_1}$$
$$\omega L = 1 / \omega C_1 \text{ or } L = 1 / \omega^2 C_1$$

Functions of Intermediate Transformer

The working principle of this transformer is similar to the potential transformer with electromagnetic winding in the system. As per the use of primary and secondary turns in the transformer, the rated voltage on secondary windings are made available. The standard voltage across the secondary winding is taken as $(110 V / \sqrt{3})$. According to the suitability and requirement of the system, the numbers of secondary windings are arranged upon the common core of the transformer. These secondary windings are used either for the metering scheme or protection scheme.

Functions of Damping Device (Z_d) and Varistor (V)

The overall circuit of CVT comprises all the circuit parameters (R, L and C). Availability of these parameters in a circuit results in the resonance condition for certain value of system frequency. The varistor (V) and damping device (Z_d) are connected across an auxiliary secondary winding. The varistor is mounted inside the CVT secondary box and helps to suppress the over-voltages due to transients on the system. The damping device (Z_d) is mounted inside the EMU tank and damps down the transient oscillations in the CVT.

Notes on ferro-resonance The circuit parameters (R, L and C) of CVT in the form of potential divider capacitance, compensated inductance, intermediate transformer and burden impedance result in the resonance circuit. If this circuit is subjected to a voltage impulse due to switching ON/OFF the supply voltage to the CVT, then some degree of oscillation results in the system. Now the frequency of the circuit is considered for the development of oscillation, and if it is slightly less than one-third of the system frequency, it becomes possible for energy to be absorbed from the system to result in oscillation. Depending on the value of components, oscillations at the fundamental frequency or at other subharmonic or multiples of supply frequency are possible but the third subharmonic is one of the most disturbing frequency for oscillation. The transient phenomenon becomes prominent for the case of this oscillation. Some of the typical pattern voltage waveforms on ferro-resonance effect are shown in the Figures 3.7a and b.

The ferro-resonance phenomenon in CVT is developed due to the following situations.

1. Secondary short circuit fault on the secondary cable and opening of the same by fuse protection or by MCB.
2. Saturation of the auxiliary voltage transformer on the secondary circuit of CVT by any temporary over-voltage condition. Subsequently when over-voltage goes away, the normal burden conditions are removed.
3. Sudden application of more burden and removal of it.

To reduce this ferro-resonance effect, the suppression circuit can be used in the system. This circuit comprises varistor and damping device as already discussed before.

Notes on transient response to CVT Every CVT is a resonance circuit. The behaviour of this circuit for the voltage application to the CVT depends upon the magnitude and nature of the electrical parameters of the CVT. Compensating inductance is used to match the capacitance of the potential divider unit and tuning of CVT is done at fundamental frequency of 50 Hz. In theory, these parameters (L and C) are regarded as the ideal loss-less components and accordingly design factor is considered for tuning by considering the suitable value of L and C.

However, the compensation could not be completed because of loss components in the circuit. In practice, any sudden fault in the system and clearance of the same results in subharmonic oscillation in the CVT. When a short-circuit occurs on the HV side of the CVT, the secondary voltage does not become zero immediately but it oscillates around zero axis for a few milli seconds before dying down to zero. These oscillations occur due to energy stored in capacitance and inductance of CVT. This transient response of CVT is much different from that of an electromagnetic VT. The transient phenomenon becomes prominent for the case of transmission line with compensating reactor in the system. Shunt reactors are generally used for the compensation of the line capacitance effect on the transmission line. With this line, when circuit breaker on either side gets open, an oscillation between inductance, shunt reactor and line capacitance results. The frequency of this oscillation ranges from 60 to 80% of normal frequency and depends upon the degree of compensation. But sometimes, it is seen that the frequency becomes subharmonic in nature and causes more oscillation. The transient response of the system becomes prominent, and takes time to settle down completely.

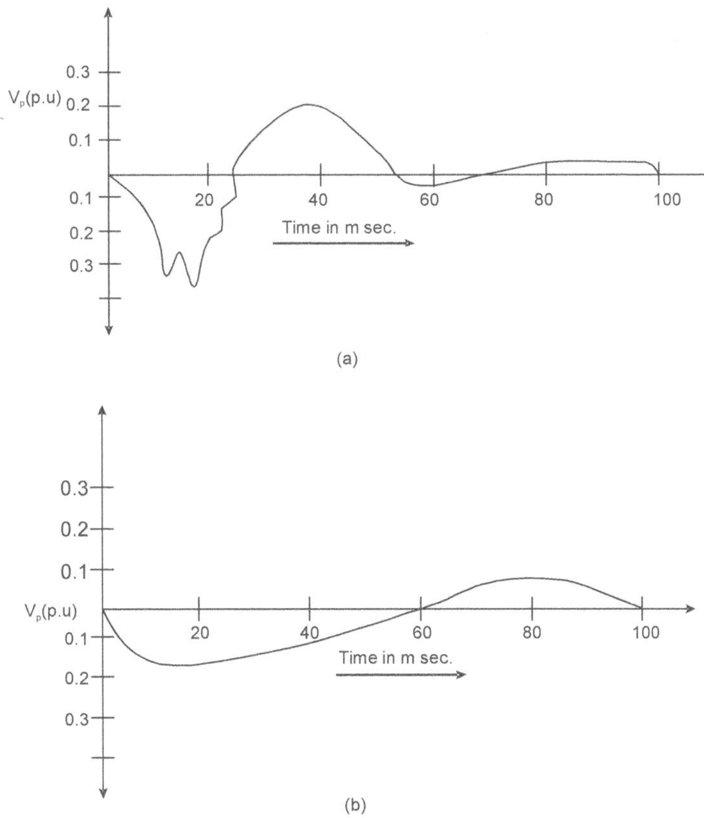

Figure 3.7   Typical pattern voltage waveforms on ferro-resonance effect

To reduce this effect and to quickly dissipate the energy in the elements (L, C), damping circuits are generally connected across the burden of a CVT. This circuit contains L_d and C_d in tuned manner as shown in Figure 3.8. Some of the typical patterns of the secondary voltages of a CVT due to the faults close to the CVT terminals are shown in Figure 3.9a and b.

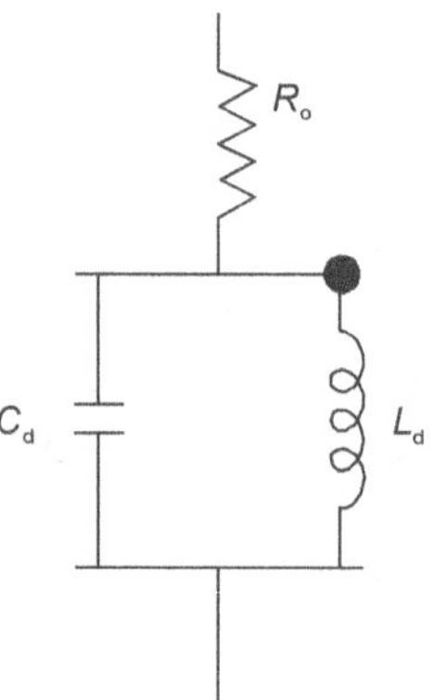

Figure 3.8   A typical damping circuit with parameters

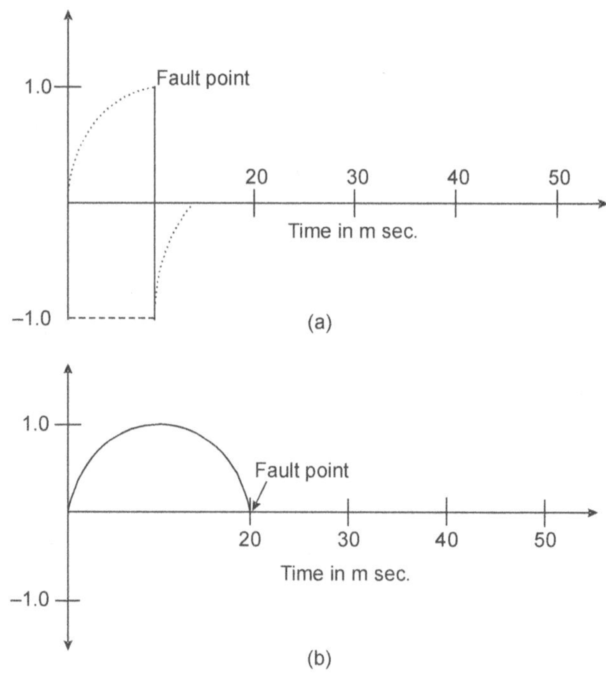

Figure 3.9 (a) Transient response of CVT for fault at voltage maximum and (b) Transient response of CVT for fault at voltage zero

NOTES ON ERRORS IN CVT

In practice, under working conditions of the CVT, the voltage on the primary side does not respond as the ideal replica to the secondary side. This is due to the electrical parameters used in the CVT. All the elements like capacitance divider unit, compensating inductance, electromagnetic transformer, etc. have their own limitations of idealism towards the working principle. This results in errors in CVT.

These errors are grouped as follows:

1. Voltage ratio error
2. Phase displacement error

Voltage Ratio Error

By the application of supply voltage to the primary side of CVT, when corresponding secondary voltage is measured, the value of such voltage does not become equal to the rated value. This deviation and comparison of the actual ratio to the nominal transformation ratio is termed as voltage ratio error and explained as follows.

$$\% \text{Voltage error} = \frac{(K_n V_S - V_p)100}{V_p}$$

$$= [(K_n / K) - 1]100$$

where,

K_n = Rated transformation ratio (rated PTR)
K = Actual transformation ratio (obtained PTR during measurement) = V_p / V_s
V_p = Actual primary voltage
V_S = Actual secondary voltage.

Phase Angle Error

For ideal CVT the difference in phase angle between the primary and secondary vector is considered to be zero. But for actual condition, this phase angle difference instead of zero deviates to certain value and this deviation is termed as phase angle error.

This error is considered to be +ve, when secondary voltage vector leads the primary voltage and –ve, if it lags and usually expressed in minutes. This definition is valid for the response to sinusoidal supply only.

NOTES ON CONNECTION PRINCIPLE

Polarity and Connection

The primary and secondary side of CVT are identified by polarity marking using symbols like *A* and *N* for primary and *a* and *n* for secondary. But usually top primary terminal that is connected to the line conductor, is regarded as terminal *A* even if it is not marked. Now the secondary terminal marking is done, according to the primary terminal marking. The polarity marks have the same significance as for current transformers. When primary current enters the *A* terminal, the secondary current leaves the *a* terminal. It is shown in Figure 3.10.

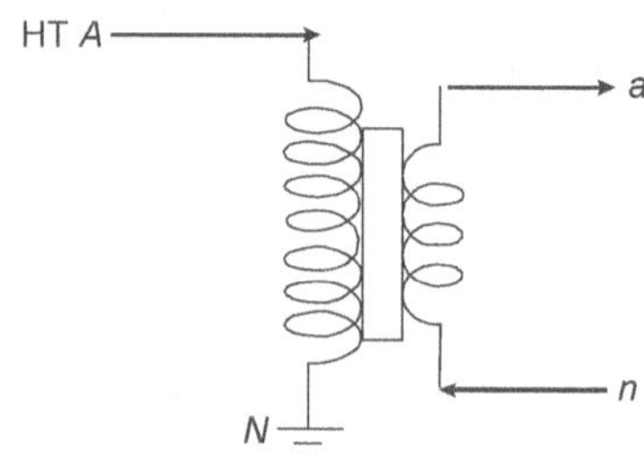

Figure 3.10   Polarity and terminal marking of CVT

Concept on CVT Circuit Connection

The secondary windings of different phases in a three phase system are connected either in star (Y) or in delta (Δ) connection. The type of connection is considered as per the requirement of the secondary burdens to the circuit.

Star connection circuit (Y connection)   For this connection, any one of the similar polarity of each phase on secondary winding is connected together to form the common point. Generally the terminal marked with *n* of each phase is connected together and earthed. The other terminals say *a*, *b*, *c* of respective phases are carried to the loading circuit (Figure 3.11).

Voltages on these windings are vectorially related and expressed in sequential components.

$$V_R = V_{R1} + V_{R2} + V_{R0}$$
$$V_Y = V_{Y1} + V_{Y2} + V_{Y0} = a^2 V_{R1} + a V_{R2} + V_{R0}$$
$$V_B = V_{B1} + V_{B2} + V_{B0} = a V_{R1} + a^2 V_{R2} + V_{R0}$$
$$\text{So, } V_R + V_Y + V_B = 3V_{R0} = 3V_{Y0} = 3V_{B0}$$

where,

1, 2, 0 designate +ve, −ve and zero sequence components and a and a^2 are the operators.

Note 1   Voltage vectors are in +ve sequence only with voltage on R phase taken as reference.

Note 2   Star connection is the regular practice for different secondary load connections like distance relays, metering circuit and directional O/C relays, etc.

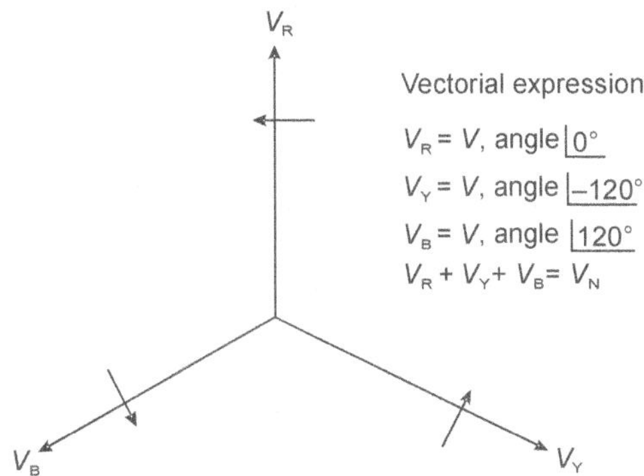

Figure 3.11   Vector representation of star connection

Delta connections circuit (***Δ-connections***)   For delta connection circuit the pattern of connection can be made in two possible ways. D_{11} connection and D_1 connection.

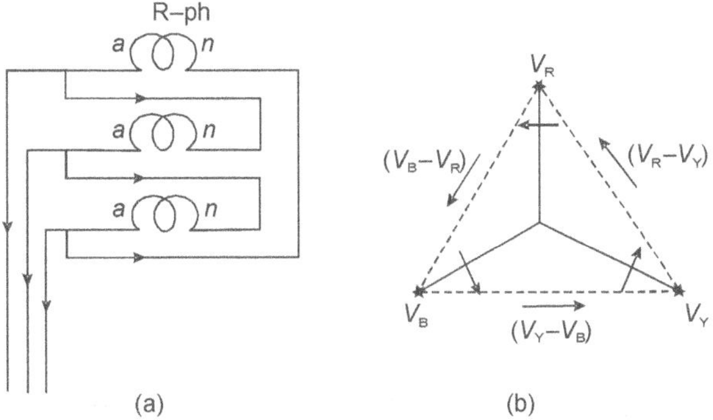

Figure 3.12   (a) Delta connection winding in D_{11} pattern (b) Vector diagram of D_{11} connection

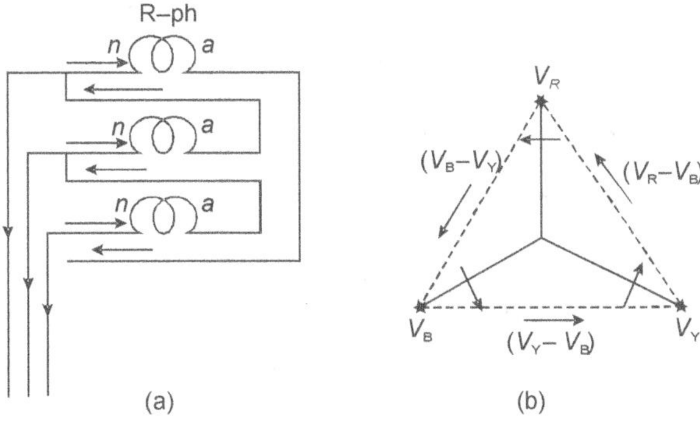

Figure 3.13   (a) Delta connection winding in D_1 pattern (b) Vector diagram of D_1 connection

a) D_{11} *connection*   For this type of connection *a* terminal of one phase is connected to the *n* terminal of next phase in regular sequence (R, Y and B). So, the current terminal maintains the sequence $[(V_R - V_Y), (V_Y - V_B), (V_B - V_R)]$ (Figure 3.12a and b).

b) D_1 *connection*   For this type of connection *a* terminal of one phase is connected to the *n* terminal of next phase in opposite sequence (R, B and Y). So, the current terminal maintains the sequence $[(V_R - V_B), (V_Y - V_R), (V_B - V_Y)$ (Figure 3.13a and 3.13b).

Note   But it is found that the connection and terminal extension from delta connection of CVT circuit is generally not used in practice.

Open delta connection   For this type of connection, the secondary circuit is connected in delta manner either in D_1 or D_{11} pattern as discussed earlier. But the final connection is kept open to form this open delta connection (Figure 3.14). For this connection the voltage available across the broken terminals is as follows:

$$V_{mn} = V_{an} + V_{bn} + V_{cn} = 3V_{a0} = 3V_{b0} = 3V_{c0}.$$

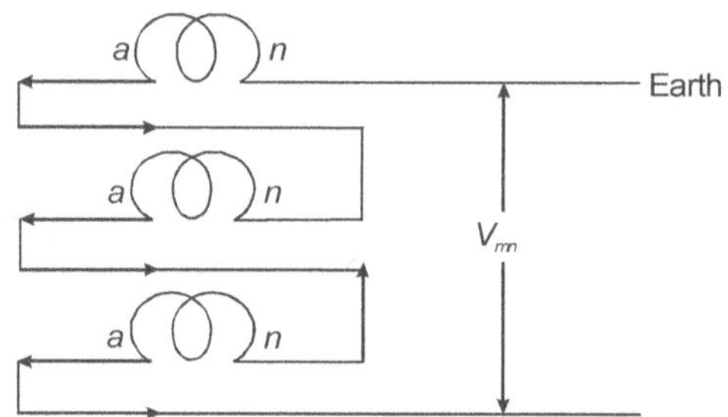

Figure 3.14   **Open delta connection**

This voltage is called polarizing voltage and becomes three times the zero sequence component of voltage of any phase.

Note   This broken delta voltage is needed for directional ground fault relay. So during working conditions, when any ground fault results in the system, the voltage on the respective phase is reduced, which causes an unbalanced voltage across the broken delta terminal and results in the operation of ground fault relay.

BASIC INSTALLATION PRACTICES

Transportation and Shipping

The basic structure of CVT is similar to the electromagnetic PT and dead tank CT. The electromagnetic unit is of greater dimension and filled with insulating oil which is the base unit of CVT. Similarly upper part of the CVT contains capacitance divider unit inside the oil-filled porcelain insulator.

Transportation of CVT is to be done with proper care, and supports are to be provided during transportation to protect the porcelain insulator of the CVT. The CVTs are generally packed in wooden crates and transported in horizontal position. But vertical position transport is always safe for CVT to avoid sliding of capacitor stacks, spilling of oil and disturbance of other internal parts.

Note   For higher class CVT, the base unit and copper capacitor units are sometimes transported separately.

Unloading, Handling Practice

The unloading and handling procedures are generally provided by the manufacturer and the instruction manual is sent with the CVT package also. During unloading and handling the CVT, these instructions are to be followed strictly. But in general, the unloading should be done vertically by the suitable hoist or crane. The string is to be tied and fastened through the lifting lugs, encircling the lifting ropes (Figure 3.15).

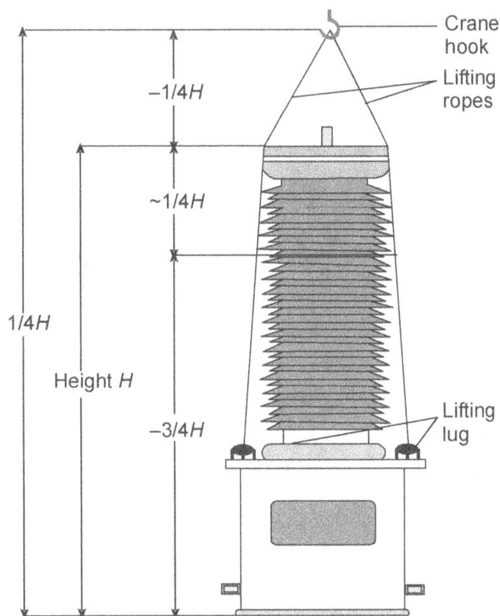

Figure 3.15   Unloading and handling of CVT

Store the unit in vertical position on a hard, dry and even surface to avoid wear and tear due to external climatic conditions.

Installation Practice

1. By the use of suitable hoist or crane, the CVT is to be handled vertically and lifted to the available height for the installation on the platform.

2. Mounting holes on the base unit are matched according to the foundation drawing and with suitable nut and bolt, the same is fastened properly.

3. Upper class CVT should installed with caution and care. Certain precautions to be followed during installation are:

   - Ensure that the capacitor units of same serial number are coupled.
   - Use the hardware provided by the manufacturer for the assembly of upper capacitor unit with the base unit.
   - Make sure that the studs are tightened to the torque as per instruction for coupling the capacitor units.
   - Proper care is advised during erection of capacitor units—upper and lower capacitor units have to be shorted and connected to earth to avoid charging current in the system.
   - Avoid tampering of other factory set screws.

4. **Earthing**   CVT is connected between the line conductor and earth. So proper earthing to the CVT is highly important and fixing of substation earthing at the necessary earth pad of the CVT has to be ensured. Earthing of the CVT is a part of the working terminal of electrical circuit. So the lower end (N) of the high voltage winding of the intermediate voltage transformer when brought inside the terminal box has to be connected firmly through earth link or directly to the earthing screw. Similarly, the LV terminal of the capacitor unit (NHF) is also connected firmly through earth link.

Pre-commissioning test   Before installation to the platform, the necessary pre-commissioning tests/checks are done to the CVT for confirmation of the status, PTR, etc.

   - By primary injection into the CVT from the available low-voltage supply (230 V to 1000 V) the secondary voltage is measured for the confirmation of voltage ratio.
   - Ensure the polarity of CVT cores.
   - Measure the resistance value of each secondary winding.
   - If possible and available, before assembly of the capacitor unit and electromagnetic unit, the "tan δ" and capacitance value should be noted for record.
   - Maintain the record of the values obtained in pre-commissioning test for future record in the prescribed format.

Pre-commissioning checks   Before energization of the CVT, the following points are to be checked in a systematic manner (Table 3.2).

Table 3.2   Pre-commissioning check of CVT

Points to be checked	Check	Remark
Tightness of the bolts and screws		
i. Upon base structure	OK/Not OK	
ii. Between capacitor unit assemblies		
iii. Between capacitor unit and base unit		

Table 3.2   (Continued)

Points to be checked	Check	Remark
Tightness of primary connection	OK/Not OK	1[*]

Tightness of secondary connection
  i. Core no. 1
  ii. Core no. 2
  iii. Core no. 3
	OK/Not OK	

Open of spare (unused) secondary core	OK/Not OK	
No secondary winding is short	OK/Not OK	
Correct and suitable secondary terminal in use	OK/Not OK	
Firm earthing of base structure	OK/Not OK	
NHF terminal is firmly earthed through earth link or earth screw, when not used for carrier communication	OK/Not OK	
N terminal is firmly earthed through earth link or earth screw	OK/Not OK	
Star connected terminals are earthed at CVT secondary box	OK/Not OK	
Connected burdens of each core is within the limit of rated value	OK/Not OK	
Polarities of secondary terminals are as per the schematic diagram	OK/Not OK	
Visual inspection of varistor	OK/Not OK	

Check oil condition in EMU
  i. Oil level
  ii. Oil colour
  iii. Oil leakage
	OK/Not OK	

Check oil/bellow condition in CDU	OK/Not OK	
Correct rating of fuse or MCB	OK/Not OK	
Secondary box cover tightly closed	OK/Not OK	
Noting down of Sl. no., reference no., etc.	OK/Not OK	

[*] To avoid corrosion, do not connect aluminum terminals with copper cables/tubes directly. Use weather-resistant protective coating and suitable washers and wire ribbon, etc.

STANDARDS AND PRACTICES

Standards and practices are the common limitation terms to which the different members, associated with the electrical utilities, refer to maintain the performance of the equipments.

Manufacturers, purchasers, customers, users and others dealing with the electrical equipment have to know about these standards and practices of the equipments. According to the standard values, the equipments are generally manufactured. But sometimes according to the requirement and suitability, the manufacturer and purchaser make the agreement and compromise with some limits mentioned in the standard.

In India, different institutions like CBIP, CPRI, IRDA, NPL, ERTL, IDEMI, IEEMA, SEBs, PSUs work on different R&D fields of electrical equipments to device the limitation of the equipments and accordingly frame the standard of the electrical/mechanical parameters of the equipment. For CVT, our national standard is IS 3156(Pt-4)1992.

The standard of CVT basically covers the information on the following factors.

   i.   Ratio
  ii.   Accuracy class
 iii.   Rated burden
 iv.   Basic insulation level and voltage ratio
  v.   Service condition
 vi.   Testing of CVT
vii.   Marking of CVT

RATIO OF CVT

Ratio of CVT is mentioned as the rated primary voltage to the rated secondary voltage. Rated primary voltage is the nominal system voltage that should appear across the primary side during the working of the CVT in the system. Similarly the value of the secondary voltage which shall appear across the secondary terminals on which its performance is based can be called rated secondary voltage.

Use of CVT in field becomes economical above the range of 66 kV class. So in Indian practice, the CVT above this class of voltage is used, and connected individually across the line conductor and ground, avail the phase voltage on the primary side. For 3 phase system, three single phase CVTs are connected across each phase in star formation. The primary and secondary rated voltages are mentioned as according to the normal standard of the nominal voltage values. In CVT from the working principle and basic construction, it is found that certain intermediate voltage is made available across the intermediate transformer primary, which steps down to the rated secondary voltage across the secondary winding. So this magnitude of voltage is to be considered and taken as the standard also. Following Table 3.3 can be used as such standard.

Example

A CVT marked with $(220\,\mathrm{kV}/\sqrt{3})/(20\,\mathrm{kV}/\sqrt{3})/(110\,\mathrm{V}/\sqrt{3}-110\,\mathrm{V}/\sqrt{3}-110\,\mathrm{V}/\sqrt{3})$ ratio indicates that rated primary voltage is $220\,\mathrm{kV}/\sqrt{3}$. Intermediate voltage is $20\,\mathrm{kV}/\sqrt{3}$ and has three secondary winding cores of rating $110\,\mathrm{V}/\sqrt{3}$.

Table 3.3   Standard rating voltage of CVT

Primary rated voltage (kV)	Intermediate rated voltage (kV)	Secondary rated voltage (V)
$66\,\text{kV}\,/\,\sqrt{3}$		
$110\,\text{kV}\,/\,\sqrt{3}$		
$132\,\text{kV}\,/\,\sqrt{3}$	$(10\text{ kV to }20\text{ kV})/\sqrt{3}$	$110/\sqrt{3}$
$220\,\text{kV}\,/\,\sqrt{3}$		
$400\,\text{kV}\,/\,\sqrt{3}$		
$525\,\text{kV}\,/\,\sqrt{3}$		

ACCURACY CLASS

According to the application of the secondary winding and the type of core used in practice, the accuracy class of the same is decided. Basically CVTs are used for two different purposes (metering, protection). Metering class CVT core needs better accuracy as compared to the protection class. But protection class needs robust performance even during saturation region.

Class of accuracy for metering core   Standard accuracy classes for measuring voltage transformers shall be (0.1, 0.2, 0.5, 1.0, and 3.0). In practice, all the mentioned accuracy class CVT has certain limits of errors. These limits are considered as the standard of the accuracy class. IS 3156(pt-2)1992 is the Indian standard for this metering core CVT. The importance and significance of the core is judged on the basis of accuracy class. Lower the limit of error causes more accuracy and vice-versa.

Class of accuracy for protection core   Standard accuracy class of protection core is mentioned as (3P, 6P). For this type of protection core, accuracy limit of error is not considered that important as compared to the measuring core VT. But nature of magnetization and corresponding field of electrical characteristics are to be designed perfectly to meet the purpose of protection. The limit of voltage error and phase displacement is mentioned in the following Table 3.4.

Class of accuracy for residual voltage obtained from CVTs   Sometimes the secondary winding of voltage transformer is directly used for the development of residual voltage to the ground fault relays. Open delta or broken delta connection is developed among the winding of the secondary circuit from all the individual CVTs used in 3 phase system. The accuracy class of such interconnected system shall be "6P" as mentioned earlier.

Note 1. For special purpose use of residual voltage winding, the accuracy class can be agreed between manufacturer and purchaser.

   2. For auxiliary voltage winding, used for damping purpose, accuracy class factor is not necessary.

Table 3.4 Errors in CVT

Metering core			Protection core		
Accuracy class	**± % Voltage ratio error**	**± Phase angle displacement error in minutes**	**Accuracy class**	**± % Voltage ratio error**	**± Phase angle displacement error in minutes**
0.1	0.1	5	3P	3	120
0.2	0.2	10	6P	6	240
0.5	0.5	20			
1.0	1.0	40			
3.0	3.0	-			

Note: Errors at any voltage between 80 to 120% of rated voltage, with burdens between 25 to 100% of rated burden at p.f 0. 8 (lag).

Note 1: Errors at 5% rated voltage and voltage multiplied by voltage factor (1.2, 1.5 or with burdens between 25 to 100% of rated burden at p.f 0.8 (lag).

Note 2: Errors at 2% rated voltage shall be twice as high as given in the table with similar burdens to Note 1.

Class of accuracy for residual voltage obtained from three-phase transformer

The residual voltage, that developed by the open delta connection of the individual windings from each phase of the individual CVT as described in the previous section is normally used in regular practice. But specifically three phase transformer being the single unit can also be used to avail the residual voltage from the open delta of the winding. The typical kind of winding is availed as the tertiary winding in the transformer.

The class of accuracy of such transformer is different as that of the residual voltage described earlier. The standard accuracy classes of these transformers are expressed as (5PR, 10PR).

Limits of the voltage errors and phase displacement errors for residual voltage is shown in Table 3.5.

Table 3.5 Limits of errors for residual voltage transformer

Accuracy class	%Voltage error	Phase angle displacement error in minutes
5PR	± 5.0	± 200
10PR	± 10.0	± 600

RATED BURDEN

The equipments, loads, circuits, etc. connected on the secondary winding of the CVT are called "Burden of CVT" at the rated secondary voltage with some particular power factor up to specified accuracy limit.

For CVT, the secondary windings are used for different applications and rated with certain voltage as its output. Each secondary winding has its own limit of accommodating the external circuit connection and the capacity is declared in terms of burden and rated with VA. So, the rated output (VA) is regarded as the apparent power, which the CVT can deliver to the secondary circuit at its rated voltage, by maintaining its accuracy to certain limit. Beyond its rated output, if the equipments are connected on the secondary side, then the accuracy is lost proportionately.

The standard output in VA at power factor of 0.8 (lag) is given as

(10, 15, 25, 30, 50, 75, 100, 150, 200, 300, 400 and 500 VA)

i.   The underlined values are preferred.
ii. The other standards are also taken as required by the customers.

Burdens of standard instruments   Some standard instruments have certain burdens that are connected to the metering core. The burden of these instruments have been expressed in the following Table 3.6.

Table 3.6   Burdens of standard instruments

Loads	Burdens (VA)
Voltmeter	1.0
Pressure Coil of Watt/VAR meter	1.5
Volt. Coil of Energy meter	2.5
Volt. Coil of P.F meter	2.5
Volt. Coil of TV meter	5.0
Leads between CVT and meters	2.0*

* Burden calculation of lead as assumed above of value 2 VA is only taken as reference. But for lead length above 300 m, the burden of the lead becomes more in comparison.

Note 1   The value of the burdens expressed in the above table is considered on the basis of maximum limit. But for recent development, the use of numerical relays, instruments, etc. provide reduced burden to the system.

Note 2   The error of the secondary winding is associated with the length of cable in connection. The voltage drop caused in the load is linked with the burden of the core.

Example

CVT has following particulars

Burden   = 200 VA
Lead resistance = (2 × 0.5) Ω
Accuracy class = 0.5

Rated voltage $= 110\text{V}/\sqrt{3}$

Considering the full burden being connected to the CVT secondary circuit, the voltage drop becomes $= (200 \times$ lead resistance$)/ (110\text{V}/\sqrt{3}) = 3.149\,\text{V}$.

Now 3.149-V drop is remarkably high and 4.96% of the rated value. If the accuracy class to be considered of 0.5, and the allowable limit is accounted, then for –0.5% error, the voltage drop would be still more and actual error for the calculation will be more in comparison.

So, the lead resistance voltage drop is the important factor for CVT error calculation.

To avoid the above problem and to limit the error, the lead length should be taken as minimum as possible.

Rated voltage factor   It is the ratio of maximum operating voltage to the normal voltage of the equipment. The factor is decided on the basis of earthing condition to the system. The primary winding of the CVT is connected to the earth for the formation of star point and according to the category of earth connection the voltage factor is decided. The time duration of the factor is also mentioned to decide the intensity of the voltage factor.

Note 1   Normal standard is 1.2 (continuous) and 1.5 for 30 seconds.

Note 2   For other factor refer clause 6.6.2 of IS 3156 (Pt 1) : 1992

BASIC INSULATION LEVEL (BIL)

Voltage is one of the electrical parameter that decides the withstanding limit of the electrical equipment.

Table 3.7   **Rate of insulation level for HSV from 0.66 kV to 245 kV**

Nominal system voltage kV (RMS)	Highest system voltage kV (RMS)	Power frequency withstand voltage kV (RMS)	Lightning impulse withstand voltage kV Peak	
			List 1	List 2
Up to 0.6	0.66	3	–	–
3.3	3.6	10	20	40
6.6	7.2	20	40	60
11	12	28	60	75
33	36	70	145	170
66	72.5	140	325	325
110	123	185	450	
		230	550	
132	145	230	550	
		275	650	

So while declaring the ratings of the equipment, different range of voltages are mentioned in the rating plate. Some of the voltages like power frequency withstanding voltage, lightning impulse voltage, etc. are provided to study the insulation level of the equipment. The values of voltages for different withstanding capacity are described in Tables 3.7 and 3.8.

Table 3.8   Rate of insulation level for HSV from 420 kV TO 765 kV

Nominal system voltage kV (RMS)	Highest system voltage kV (RMS)	Power frequency withstand voltage kV (RMS)	Lightning impulse withstand voltage kV (Peak)
220	245	360	850
		395	950
		460	1050
400	420	950 *	1175
		1050 *	1300
		1050*	1425
525	524	1050*	1425
		1175*	1550

*   Switching impulse withstand voltage in kV (Peak)

SERVICE CONDITION

Some standards are followed to put the CVTs in service condition. These standards include ambient condition, atmospheric weather condition, earthing to the system, etc.

Ambient Temperature Condition

1. Maximum ambient temperature $\leq 45°C$
2. Maximum daily average ambient temperature $\leq 35°C$
3. Minimum ambient temperature $\geq 5°C$

Note   The values mentioned are as per the Indian standard conditions.

Altitude

Sometimes few users insist upon the manufacturers to declare altitude factors for the installation of CVT. But this standard is not so important for the installation of CVT in the system. However for standard practice, this factor can be chosen up to 1000 m above mean sea level.

System Earthing

For both protection and safety of the system, earthing to the equipment is considered as one of the most important factors. The following points are to be followed as the normal standard and practice regarding the earthing of the equipments.

1. The structure, framework upon which equipment is installed should be earthed with two different terminals. The equipment base, marked with earth point has to be connected to solid earth point.
2. The earthing terminals should be of required size and protected against corrosion.
3. The earthing of secondary star terminal should be done at one point only and preferably it is to be done at switchyard instead of at control or relay panel.
4. The star terminal of different core available in CVT should be separately earthed.

TESTING OF CVT

It is discussed in detail under the section Testing Procedures.

MARKING OF CVT

Marking on the CVT contains the following details that are mentioned clearly on the name plate/rating plate.

i.   Identification No. (Sl.No, Designation, Type, etc.)

ii. Ratio of primary and secondary voltage with number of cores.

   Example $[220/\sqrt{3}[110/\sqrt{3} - 110/\sqrt{3}]$ This indicates that it is a multi-core CVT, earth reference with two secondary cores.

iii. Rated working frequency

iv. Normal system voltage and HSV (Highest system voltage)

v.   Rated insulation level (mentioned with power frequency withstand voltage and lightning impulse voltage)

vi. Rated voltage factor (Voltage factor for continuous and for 30 sec. to be mentioned)

vii. Reference to the standard

viii. Name of the manufacturer and detail address

ix. Equivalent capacitance

x.   Capacitance of top unit (C_1)

xi. Other details

- Weight of capacitance unit oil
- Weight of EMU oil
- Weight of core + Winding

- Total weight of CVT
- Reference drawing, PO., etc.
- Other caution remarks
    1. When NHF terminal is not used for HF transmission, it must be connected to earth.
    2. Under no condition, the burden values mentioned should exceed the rating
    3. Short-circuit the capacitor unit during installation and erection.

xii. Detailed core identification with its rating
    1. Metering core (Rated burden, accuracy class, voltage rating)
    2. Protection core (Rated burden, accuracy class, voltage rating)
    3. Protection core (Residual value) (Rated burden, accuracy class, voltage rating)

xiii. Connection diagram and detail of index used

The typical values in the form of name plate detail for 132 kV and 220 kV CVTs have been mentioned in Tables 3.9, 3.10, 3.11 and 3.12.

Table 3.9 Typical name plate details of 132 kV CVT

Particulars	Rating/value	Particulars	Rating/value
Make		Sl. No.	02314
Type		Rated voltage	$132 kV/\sqrt{3}$ kV
Highest system voltage	145 kV	Rated insulation level	275/650 kV
Total weight	420 ± 10% kg	Rated frequency	50 Hz
CAP. Oil	25 ± 10% kg	Standard	IS 3156
EMU Oil	85 ± 10% kg	HV (Pry) Capacitance	(6511 + 10% – 5% pF)
Intermediate V (sec) capacitance	(35418 + 10% –5% pF)	Equiv. Cap (C_n) for PLCC	(5575 + 10% – 5% pF)
Nominal intermediate voltage	13 kV	Total SIM. burden/class	150 VA/0.5
Total thermal burden	300 VA	Month/Year of manufacturing	–
Voltage factor	1.2 Continuous/ 1.5 for 30 sec.	1 ph solidly earth connection	

Table 3.10 Specification of secondary core of 132 kV CVT

Rated secondary voltage	Terminal marking	Rated burden (VA)	Accuracy class
$110/\sqrt{3}$ V	1a-1n	100	0.5
$110/\sqrt{3}$ V	2a-2n	100	3P

Table 3.11 Typical name plate details of 220 kV CVT

Particulars	Rating/value	Particulars	Rating/value
Make		Sl. No.	2204147
Type		Rated voltage	$220/\sqrt{3}$ kV
Highest system voltage	245 kV	Rated insulation level	245/460/1050 kV
Total creepage	6125 min. (mm)	Rated frequency	50 Hz
Weight of oil	140 kg	Standard	IEC : 60186/IS : 3156
Total weight	750 kg	HV (Pry) capacitance	4840 pF
Intermediate V (sec.) capacitance	48400 pF	Equipment capacity (C_n) for PLCC	(4400 + 10% −5% pF)
Nominal intermediate voltage	$20/\sqrt{3}$ kV	Temp. category	−5 to 55° C
Total thermal burden	750 VA	Class of insulation	A
1 ph solidly earth connection		Suitable for hot line washing	
Month/year of manufacturing	−	Voltage divider ratio	$220/\sqrt{3}$ kV/ $20/\sqrt{3}$ kV
Voltage factor	1.2 continuous /1.5 for 30 sec	G.A Drg. No.	−

Table 3.12 Specification of secondary core of 22 kV CVT

Rated secondary voltage	Terminal marking	Rated burden VA	Accuracy class
$110/\sqrt{3}$ V	1a–1n	150	0.5
$110/\sqrt{3}$ V	2a–2n	150	3P
$110/\sqrt{3}$ V	3a–3n	50	3P

GTP AND CLASSIFICATION

General technical particular (GTP) of any equipment provides the detailed technical declaration of the manufacturer regarding the item/set/equipment. It is similar to the rating plate of the equipment, but some extra data are also mentioned in GTP. During the time of common inspection, the customer if so desired can insist upon the manufacturer to show the declared values in the GTP by testing the equipment, according to the standards mentioned thereof. GTP of CVT includes the electrical and some mechanical data about the equipment also. For reference, GTP of a 220 kV CVT is mentioned with all data in Table 3.13.

Specification of any equipment is generally considered as an agreement between the customer and the manufacturer with the reference of some normal standards, rules, etc. Sometimes during the preparation of the specification, the purchaser/customer declares some values according to the requirement and limitation of his own accord. These values may be different from that of normal standards and practices.

GTP of a Particular 220 kV CVT

Table 3.13   GTP of a typical 220 kV CVT

Particulars	Technical values
Manufacturer name and address	–
Type	–
Type of installation	Outdoor/pedestal
Rated primary voltage	$220/\sqrt{3}$ kV
Type of voltage transformer	Capacitive
Rated frequency	50 Hz
No. of phases	Single
HSV	$245/\sqrt{3}$ kV
No. of secondaries	Three
Rated output VA	200/200/100 VA
Accuracy class	0.5/3P/3P
Standards applicable	IEC60186/IS3156
Rated secondary voltage	$110/\sqrt{3}$ - $110/\sqrt{3}$ - $110/\sqrt{3}$
Rated voltage factor	1.2 continuous, 1.5–30 sec
Rated thermal burden	750 VA

(Contd.)

Table 3.13 (Continued)

Particulars	Technical values
Rated simultaneous burden	200 VA/0.5, 400 VA/3P
Equivalent capacitance	4400 pF +10% to –5%
High-voltage capacitance	4840 pF +10% to –5%
Natural frequency of coupling (self-tuning frequency)	≥ 700 kHz
Bandwidth	40 to 500 kHz
Temperature rise above ambient at 50°C	IS 3156/IEC186
One minute power frequency test voltage of secondary winding	3 kV
One minute power frequency test voltage of HF terminal	4 kV for enclosed type
One minute power frequency test voltage of CVT (dry and wet kV RMS)	460 kV
1.2/50 μs impulse withstand voltage of CVT (kV peak)	< 1050 kV
Corona inception voltage	< 225 kV
Corona extinction voltage	< 156 kV
Radio interference voltage at 1 MHz at $1.1\,\mu\mathbf{m}\,\sqrt{3}\,(\mu\mathrm{V})$	< 1000
Capacitance test coefficient	0.07%
Rated intermediate voltage	$20/\sqrt{3}$ kV
tan δ value of capacitance unit	≤ 0.0005
Provision for accepting the change in volume	Metal bellows
Value of stray capacitance and stray conductance in carrier frequency range (40 to 500 kHz)	300 + 0.05 Cn pF and 50 μs
Minimum creepage distance	6125 mm
Facility of tan δ point measurement	Yes
Quantity of oil (kg)	140
Total weight (kg)	750
Facility of oil sample collection	Yes
Treatment of external ferrous surface	Epoxy painted
CVT hermetically sealed	Yes

Specification Inquiry of CVT

While placing the order of the CVT, different technical information has to be provided. These specifications, GTP, etc. can also be called as inquiry of CVT. Such inquiry may not be same for all utilities. The purchaser may inquire according to the requirement of its own references. But some of the following points are the guidelines to inquire about the CVT.

1. Type of CVT   Outdoor/pedestal single unit/separator unit
2. Rated system voltage parameters Normal system voltage/highest system voltage
3. Type of insulation and BIL as in GTP
4. Rated transformation ratio
5. Voltage factor and duration
6. Reference standards
7. Rated core detail
    i. No. of cores
    ii. Type of cores
    iii. Burden, accuracy class, rating of the cores
8. Service conditions if any speciality is there like the use of CVT in altitude over 1000 m and other environmental factors
9. Facilities to be extended like
    i. Monitoring status
    ii. Collection of oil sample
    iii. Measurement of tan δ value
    iv. Coupling capacitance for carrier communication
    v. Other protection circuits for CVT
    vi. Other particulars as per suitability and requirement of the user.
10. Frequency other than 50 Hz.

MAINTENANCE PRACTICES

CVT does not require any special maintenance schedule. Periodically it requires some scheduled checking and some testing practice as per demand. Following maintenance schedule may be followed for CVT, as described in Table 3.14.

Table 3.14 Maintenance schedule of CVT

Periodicity	Checking/testing	Actions to be taken
Daily	i. Oil leakage ii. Abnormal noise iii. Other visual checking	Follow-up action
Weekly	i. Oil level ii. Voltage reading of secondary circuit on each core iii. Visual check of varistor (Bulging or burning)	
Monthly	i. Analysis of voltage readings on the secondary circuits ii. Analysis of zero sequence voltage, monitoring value for open delta winding iii. Terminal checking of secondary circuit at terminal box iv. Earthing of PLCC link v. Analysis of level of metal bellow	
Yearly	i. Earthing of base plate ii. Connection checking of secondary circuit at both junction box and C/R Panel iii. Measurement of earth resistance iv. Checking of voltage ratio v. Cleaning of porcelain insulator vi. Checking of primary terminals connection vii. Checking of corrosion of metal parts viii. Checking of temperature rise of emu ix. Dielectric strength of oil used x. Detail of oil leakage	
Five-yearly	i. DGA of oil used ii. tan δ and capacitance measurement iii. Checking of typical characteristics of oil sample	

TESTING PROCEDURES

Testing of the CVT is classified into three types. They are

1. Type test
2. Routine test
3. Optional test

TYPE TEST

To compare and confirm the major parameters of the electrical equipment, some tests, are required to be done by the suitable methods, available in different standards. These tests are called type test. Following are the few tests, categorized in type test.

1. Temperature-rise test
2. Lightning impulse test
3. Ferroresonance test
4. Transient response test
5. Test for accuracy

Temperature-rise Test

EMU (Electromagnetic Unit) is the unit in CVT that contains coil and winding. During working conditions this unit suffers with temperature variation. So, for temperature rise test, the variation of temperatures on this unit can be considered as the specimen for temperature rise test.

For testing of temperature rise on CVT following conditions are to be satisfied during test.

1. The supply voltage should be as per the standard, adopted for testing.
2. The secondary burden should also be connected at any power factor between 0.8 power factor lag to unit power factor.
3. Mounting of the transformer should be similar to that of the mounting to be done in real practice.
4. The ambient temperature shall not exceed 40°C.

Now the temperature rise is measured and the values are compared with the standards and limits for final conclusion.

Various methods are adopted to measure the temperature rise in the winding.

Some methods among them are

1. Variation of resistance method
2. Thermometer or thermocouple method

Lightning Impulse Test

Every CVT is rated with certain impulse voltage on its rating plate. During the lightning impulse test of the CVT, this rated impulse voltage is referred for the testing of the CVT. The values in the testing standard methods can also be compared for the finalization of the result.

For Indian standard practice, the test shall be conducted in accordance with IS 2071(part I) and IS 2071 (part II) of 1974.

Testing method During test, the following points/factors are to be followed for obtaining the correct result.

   i. Test voltage shall be applied across the primary side of the CVT. The lower/earthing terminal should be properly earthed (body, framework, other metal body part, secondary terminals are to be earthed together).
   ii. Application of number of impulses depends upon the type of CVT to be tested.
   iii. The following Table 3.15 can be used for voltage application.
   iv. The peak values and wave shape of impulse voltages are to be recorded for the comparison with the values as specified before under the section "basic insulation level".

Table 3.15 Voltage impulse application for impulse test

CVT	No. of impulses
Outdoor ≤245 kV	15 continuous full-wave impulse of each polarity without correction for atmospheric condition.
420 kV onwards	3 consecutive full-wave impulses of each polarity without correction for atmospheric condition.

Confirmation of the test CVT shall have passed the test if the following conditions are satisfied

   1. No disruptive discharges occur in the non-self-restoring insulation.
   2. No flashovers occur along the non-self-restoring with external insulation.
   3. No more than two flashovers occur across the self-restoring external insulation.
   4. No other evidence of failure is detected.

Ferroresonance Test

For testing the ferroresonance effect on a CVT, the following methods are to be adopted.

Method 1 Apply 120% of the rated primary voltage. During the test, secondary terminals shall be short-circuited for at least 0.1 second (This kind of shorting can be done by fuse protection or MCB arrangement on the CVT secondary circuit). This test shall be conducted for a minimum of 30 times at the 120% of rated voltage.

Result   After every sudden removal of short-circuit on the secondary terminals, the peak of the secondary voltage shall revert to a value which does not differ from its normal value by more than 10% after 10 cycles of rated frequency.

Method 2   By this method the CVT is supplied to a voltage corresponding to its rated voltage factor. Similarly during the test, the secondary terminals shall be short-circuited for the minimum time of 0.1 sec. This test shall be repeated for about 10 times at a primary voltage corresponding to voltage factor.

Result   For this test, after removal of every sudden short-circuit on the secondary terminals, the ferroresonance effect shall not remain for more than 2 seconds.

Transient Response Test

Transient response in CVT is a complex behaviour and to study such behaviour, under testing method, some of the equivalent network, equivalent source, etc. have to be considered.

Test Method   The CVT under test is to be connected with a burden of 25% to 100% of the rated value. The rated primary voltage is also to be applied to the CVT. Suddenly the high-voltage terminal and low-voltage earthed terminal are short-circuited for the study of the secondary voltage.

Result   Following a short-circuit of the supply as described above, the secondary output voltage of a CVT shall decay, within one cycle or rated frequency, to a value of less than 10% of the rated secondary voltage.

Note   The test shall be made at the option of the manufacturer, either 10 times at random or twice at the peak of the primary voltage and twice at the zero passage of primary voltage. In the latter case, the phase angle of the primary voltage shall not differ by more than ± 20 electrical degrees from the peak and zero passage.

Test for Accuracy

Every CVT is used mainly for two different secondary purposes like metering scheme and protection scheme.

For the measurement of errors and characteristics of these cores, some of the following points are to be considered during the type test of CVT.

1. Test shall be made at the rated frequency.
2. Temperature for testing should be at room temperature, also for both extreme temperature as decided by the manufacturer and user.
3. Equivalent circuit of the CVT can also be used for comparison.

4. Calculation on the basis of temperature coefficient for the actual capacitive divider to be considered.

5. The accuracy result should confirm the limits of errors as discussed before.

Note For the type test with accuracy class measurement, the user and manufacturer must have to decide upon the standard to which both should agree during the time of testing.

ROUTINE TEST

The regular and normal tests, that are conducted to study the performance of equipment are called "routine test". Following are the few test methods grouped under the routine test of CVT.

1. Terminal and polarity marking
2. Power frequency test for CVT
3. Power frequency test for low-voltage terminal
4. Power frequency test for electromagnetic unit
5. Tests for accuracy.

Terminal and Polarity Marking

Marking on the primary terminals are not provided in CVT. Usually the top conductor of CVT is taken as the HT terminal; and bottom earth is considered as the earth terminal. But secondary windings are clearly marked with its terminals. The circuit diagram of CVT is also drawn on the rating plate. For configuration of the marking, the polarity test can be done as follows (Figure 3.16).

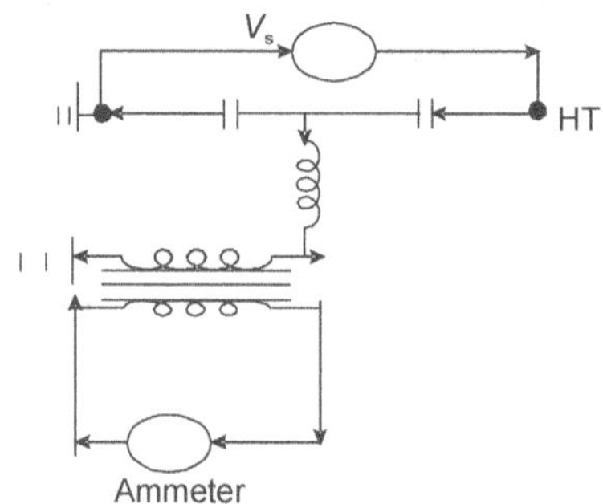

Figure 3.16 **Polarity test of CVT**

1. Connect the supply (AC voltage) to the primary side of CVT.
2. Use an ammeter on the secondary winding.

3. Now the combined current through the supply wire, entering into the CVT and current through the secondary wire, entering into the ammeter are to be measured. The net current should be additive if the polarity matches properly.

Power Frequency Test for CVT

The application of supply voltage to the capacitive voltage divider should be according to the values mentioned in BIL table. The testing method has to be agreed by the user and the manufacturer. For Indian practice the normal standard used for testing of the power frequency for CVD is IS 9349:1971.

Power Frequency Test for Low-voltage Terminal

Low-voltage terminal of CVT is also to be tested for power frequency withstand. But the magnitude of voltage application is less and restricted to 10 kV (RMS) value for 1 minute withstand, between the low-voltage terminal and earth terminal. But for special conditions like use of the terminal for carrier frequency device, the voltage application could be 4 kV (RMS).

But for these cases, the user and manufacturer has to agree upon the method of test.

Power Frequency Test for Electromagnetic Unit

The electromagnetic unit has to be tested separately. Application of the voltage is to be decided according to the rating of the electromagnetic unit, and its corresponding BIL. The terminals chosen for voltage application are the intermediate voltage terminal and earth of the EMU.

Sometimes application of the voltage from the secondary terminals can also be considered for the testing. But this time the reduced voltage as per the calculation of ratio between rated BIL to the voltage ratio shall be applied.

The frequency of exciting voltage may be increased above the rated frequency. If the frequency exceeds twice the rated frequency, the duration of the test may be reduced as given below.

Note 1 Duration of test frequency with a minimum of 15 seconds = (Twice the rated frequency × 60 sec)/ Test frequency.

Note 2 For all the power frequency tests, the protective device connected to the CVT has to be removed during the test.

Tests for Accuracy

Methods used for the routine test for accuracy are similar to the methods to be used in type test also. For routine test, certain reference temperature has to be chosen at a standard frequency. The actual value of test frequency and test temperature shall be part of the test report.

In normal practice, for routine test 100 per cent of rated voltage is applied with burden being 25% to 100% of rated value across the secondary. The permissible value for ratio error and phase displacement can be referred to obtain the accuracy of the CVT. It is always advisable to test the total CVT instead of taking its equivalent circuit. But for some conditions as obtained from type test of accuracy, the equivalent circuit can be used for testing purpose.

Note   Routine test is a normal standard test practice. Temperature standard may be chosen as the room temperature, voltage standard as the 100% rated nominal voltage, burden standard as the 25% to 100% of rated burden.

OPTIONAL TEST

There are no specific tests to be followed for this optional test. But following few tests may be chosen as the special test for CVT:

- Chopped lightning impulse rest.
- Short-circuit withstand capability test.

These tests are critical in nature and need high cost involvement for conducting the test. Moreover, the manufacturer and purchaser have to agree to conduct the same under specified limited conditions. For further detailed test the IS 3156 can be referred.

Note 1   These tests are generally conducted by the mutual understanding between manufacturer and purchaser.

Note 2   But in case of lot purchase of the CVT, any one/two CVTs may be selected randomly and tested for the special test as described above.

Note 3   Testing of CVT under type test and special tests should be conducted under the following conditions.

- Agreements shall be made between the purchaser and manufacturer regarding the points below
- Reference standards to be followed during test.
- Testing methods.

CASE STUDIES

CASE STUDIES ON CVT CIRCUITRY

Case Study 1

One of the 220/132 kV grid substation was found with following voltage across the open delta point of secondary core, during regular energization of three-phase system.

Analysis   The individual voltage output of each phase winding was checked and found in rated magnitude. The detailed vector diagram has been drawn as per the measurements conducted during analysis (Figure 3.17).

Secondary voltage in volt						
rn	yn	bn	ry	yb	br	Open delta
64.85	65.10	65.28	65.12	65.12	112.10	131.13

From the readings and vector analysis of the measured currents, it was concluded that Y ph CVT has some polarity problem. Then the polarity of Y ph CVT was checked and was found with reverse connection internally. The same was rectified.

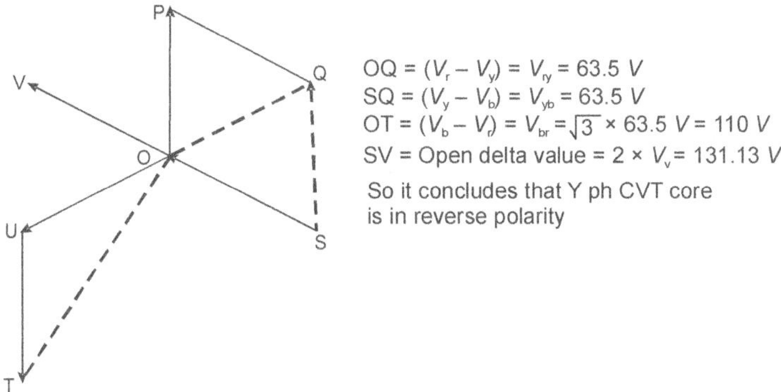

Figure 3.17   Vector diagram for case study 1

Case Study 2

In one of the grid substation abnormal tripping was observed from the distance protection relay.

Analysis   The detailed checking was done for the CVT circuit and CVT secondary circuit. From the physical connection of the circuit no abnormality was detected. But from the measurement of terminal voltages it was concluded that the polarity of R ph CVT has been altered. The detailed vector analysis and measurement is shown in Figure 3.18.

Secondary voltage in volt					
rn	yn	bn	ry	yb	br
63.45	64.12	63.56	64.07	112.10	64.34

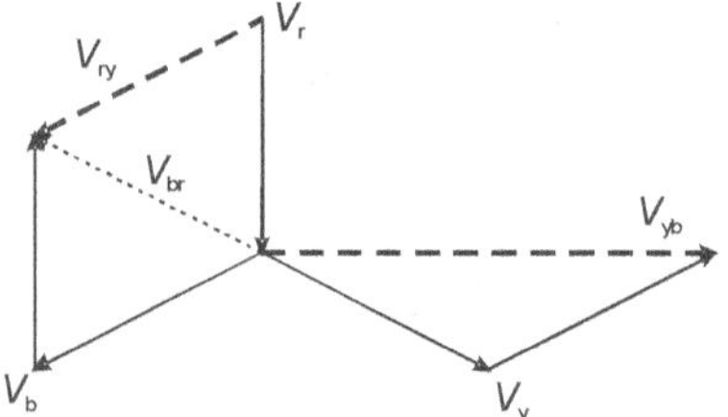

Figure 3.18 Vector diagram for case study 2

Because of this reverse polarity connection of R ph CVT, the relay was tripping abnormally. The polarity was changed and the problem was rectified.

Case Study 3

In one of the 220/132 kV substations, the secondary voltages for different CVTs were measured for routine check-up and the following were found on the metering core.

Month	Phase	220 kV Fdr1	220 kV Fdr2	220 kV Fdr3
February	Rph	63.2	63.8	**65.2**
	Yph	63.8	**60.1**	63.8
	Bph	63.4	63.4	63.3
March	Rph	63.3	63.7	**65.4**
	Yph	63.7	**59.8**	63.6
	Bph	63.6	63.3	63.4
April	Rph	63.0	63.6	**65.8**
	Yph	63.5	**59.7**	63.7
	Bph	63.2	63.2	63.3
May	Rph	62.9	63.0	**65.0**
	Yph	63.3	**58.9**	63.4
	Bph	63.1	63.1	63.3

Investigation   After observing the abnormality from February month, the monitoring was attempted for other cores of all the three feeders from March month onwards. The readings of other cores were also found with similar abnormality as of the metering core. From these observations of the readings it could be apprehended regarding the problems in CDU (Capacitance divider unit).

Analysis   CVT contains two capacitor blocks as top unit and bottom unit. Top unit consists of capacitor stacks of more number of elements and lower unit has less number of elements.

For the typical 220 kV CVT

C_1 = 4840 pf (Top unit),

$C_2 = 48400$ (Bottom unit)

Voltage divider ratio $= (220\,\text{kV}/\sqrt{3}) / (20\,\text{kV}/\sqrt{3}) = 11$

Assume number of elements in $C_1 = 200$

Number of elements in $C_2 = 20$

Healthy voltage divider ratio $= (C_1 + C_2)/C_1 = 11$

Case 'A' suppose one element is faulty

New $C_1 = 4840 \times 200/199 = 4864.32$ pF

Voltage divider ratio $= (4864.32 + 48400)/4864.32 = 10.95$

Now, new secondary voltage $= (110/\sqrt{3}) \times 11/10.95 = 63.8$

For '2' elements faulty

New secondary voltage $= 63.52 \times 11/10.89 = 64.08$ V

Similarly other calculations can be done.

Note   For the faulty capacitor elements on the top unit the secondary voltage magnitude increases. Symptom of rise in secondary voltage of top unit compared to secondary voltage in other phase indicates about the problem in top capacitor unit.

Case 'B' suppose one element is faulty in bottom unit

New $C_2 = 48400 \times 20/19 = 50947.37$ pF

New voltage ratio $= (4840 + 50947.37)/4840 = 11.53$

Now, new secondary voltage $63.5 \times 11/11.53 = 60.58$

For '2' elements faulty

New secondary voltage $= 63.5 \times 11/12.11 = 57.7$ volt

Note   For the faulty capacitor elements on the bottom unit, the secondary voltage magnitude decreases. Symptom of fall in secondary voltage of bottom unit, compared to secondary voltage in other phase indicates about the problem in bottom capacitor unit.

1. From the recorded values of secondary voltages it is apprehended that for feeder 2, "Y ph" CVT might have suffered with problems on capacitor elements of bottom unit.
2. For fedeer 3; it is for problem with capacitor elements of top unit.

Remark   The said CVTs were measured for capacitance value and the faulty units were replaced by the good ones.

CASE STUDIES RELATED TO OTHER FIELD

Case study 1

During energization of a long transmission line, the CVT on both ends were damaged. 'Y' phase CVT became faulty at one end and 'B' phase CVT at other end.

Post-incident investigation   The faulty CVTs were found with no physical damage. The secondary protection circuit was found burnt on either end of the line. For the performance study of the CVT, low-voltage injection was carried and found with no voltage on any of the secondary windings. The defective CVTs were replaced by good ones for energization of the line.

Analysis   This line is approximately 220 km long. The electrical circuit of the line has certain capacitance and inductance in the system. During opening of circuit breaker, these parameters develop an oscillation in the system due to the resonance effect of the line. The resonance circuit is formed with the circuit parameters (R, L, C components of CVT and transmission line). On the day of the above incident it was reported with the following operations. The electrical circuit representation is shown in Figure 3.19 for study.

1. The line was first charged at one end to extend the supply to the other end.
2. Then the same circuit was planned to be closed at the other end for power flow in the circuit.

But during the closing command from breaker switch, the charged line was tripped due to wrong operation of the switch and immediately it was closed followed with tripping action.

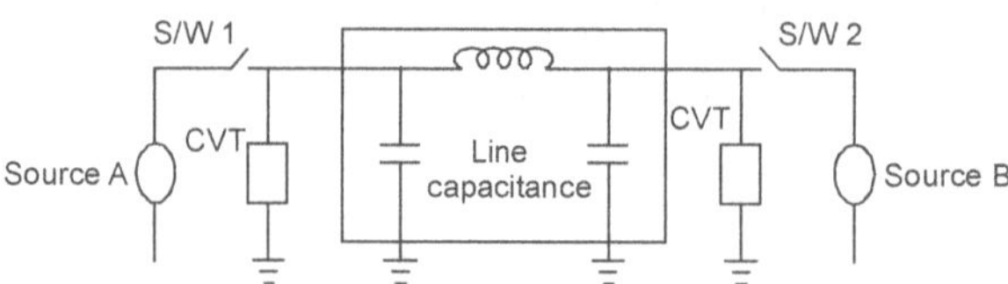

Figure 3.19   Electrical circuit of the long line for case study 1

For the initial charging of the long transmission line from one end, the voltage could be more than the rated value at the other end due to long line capacitance effect (Ferranti effect). Here for charging at 'B' end, there resulted a voltage rise at open terminal of source 'A'. But due to wrong operation of the switch (1), at end 'A' the charged line got de-energized suddenly and before complete die-down of the oscillation and voltage spike the closing command was followed quickly for energization of the line again. This sudden de-energization and quick energization of a line with resonating effect, developed voltage surge and finally the damage of the CVT at both end terminals. Though all the CVT sets at both end suffered with this effect but due to unequal voltage spike the Y ph CVT at one end and B ph CVT at other end got damaged.

Case Study 2

In one of the radial charged line one of the remote CVT failed due to bursting of PT at one end.

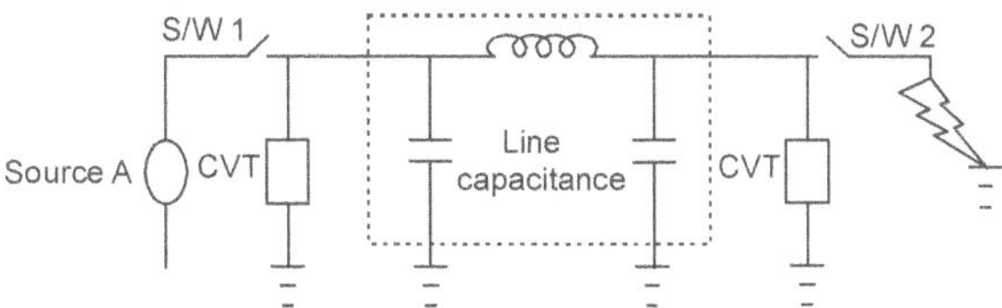

Figure 3.20   Electrical representation of case study 2

Post-incident investigation   Because of some problem, the 'R' ph Bus PT got burst at the load end station. Accordingly the radial line was tripped immediately on distance protection relay at the station (A). From the usual practice of trial charge the same line was again charged with fault being in the system. This time immediate tripping was resulted by the actuation of DP relay at station (A). Refer Figure 3.20 for explanation.

The faulty PT was isolated from the system for charging of the radial line. But this time the secondary voltage from 'R' ph CVT was not obtained to the circuit at station (A). Physically the secondary circuit was checked of this CVT and found with no voltage even at the CVT secondary box terminals. This behaviour confirmed the failure of CVT.

Analysis   The primary elements of CVT (L,C), along with line inductance and capacitance, give rise to ferroresonance circuit. In this incidence, due to first occurrence of fault at the remote end PT, there resulted the tripping of breaker at source end 'A'. But because of ignorance of the fault on the remote end, when the second attempt was made for charging of the line, due to the short-circuit of primary side voltage, the severe short-circuit current might have resulted in the damage of the corresponding 'R' ph CVT at source end 'A'. The CVT at remote end (load end) did not fail due to driving source energization at end 'A' and quick response of DP relay at this end.

Case Study 3

During normal service condition one of the CVT failed at one of 220/132 KV grid substation on a 220 kV feeder.

Post-incidence investigation   The faulty CVT was not affected physically. Only secondary windings were found with no voltage at the terminals. This situation was observed from indicating instruments at control room. So suspecting the cable breakage terminal open, the wires, terminals, etc., were checked with continuity tester and found with no abnormality. Then the affected CVT was injected with LT supply voltage and the voltages at secondary terminals were measured and no voltage was found. The failure of the CVT was confirmed after injection test only.

Analysis   The failure of CVT during normal service condition is rare in comparison to the problems resulting due to faults in the switching of the system voltage. But sometimes due to frequency variation in the system, if it becomes and gets tuned for subharmonic frequency, then ferroresonance effect may result in the oscillation in the system, if regular frequency rides over then abnormal variation in system voltage may damage the CVT secondary winding. But in practical situations this practice is rare.

COMPARATIVE STUDY WITH THRESHOLD VALUE

Comparison of Broken Delta Voltage with Reference

Broken delta or open delta voltage is the vector sum of the voltages of all the three phases. In healthy/normal condition, this value becomes zero and no current flows in the circuit. When the ground fault occurs, the voltage that appears across the broken delta burden corresponds to 3 times the zero-phase-sequence component of any one of the three phases to ground voltage at the potential device location and stated as "3V0", but the actual magnitude depends upon the following factors.

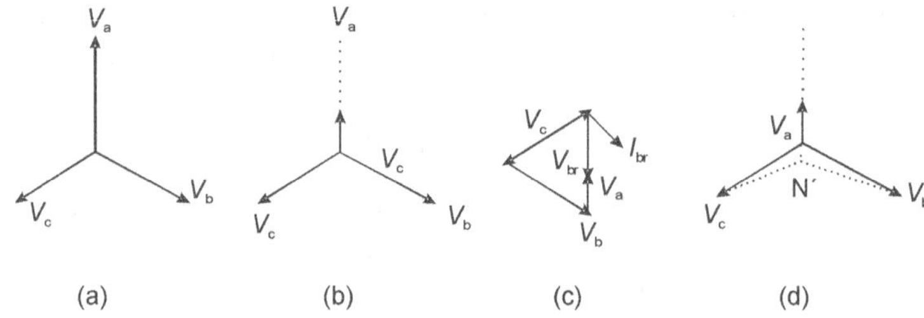

Figure 3.21 (a) Normal voltage condition (b) Fault in 'a' phase and no neutral shift (c) V_{br}—Broken delta voltage with fault I_{br}—broken delta current with fault (d) Partial neutral shift

1. Neutral ground of the system.
2. Fault zone with respect to the potential device location.
3. Intensity of fault like solid ground fault or impedance ground fault, etc.

During the time of ground fault, the possibility of neutral shifting may occur in any direction. So the voltages of the unfaulted phase windings are not nearly as variable as the broken-delta burden voltage. The winding voltages of the unfaulted phases vary from approximately rated voltage to $\sqrt{3}$ times rated, while the broken-delta burden voltage varies less than rated value to approximately 3 times rated. So, the broken delta voltage differs with respect to the reference value of zero voltage. Some of the typical explanations are shown in Figure 3.21a, b, c, d.

Comparison of Unit Capacitance in CVT

Every CVT contains two capacitor units (top unit and bottom unit). Top unit contains capacitance of lower value and bottom unit of higher value. These comparisons of values depend upon the design of the CVT intermediate voltage terminal. For economical and safe design of the intermediate voltage across the electromagnetic transformer, the capacitive voltage drop is allowed more across the top unit and the maximum numbers of capacitance element stacks are connected in series on the top unit.

In general practice, the number of capacitor elements vary from 200–400 on the top unit. The bottom unit contains the elements from 15–30 in number. The usual value of intermediate voltage varies from $(10\,\text{kV to }20\,\text{kV})/\sqrt{3}$.

The voltage divider ratio plays the role for the choice of capacitor elements to be made available in different units. It is considered as the ratio capacitors in the units as follows.

$$\text{CVDR} = (C_1 + C_2)/C_1$$

Example 1

For 245 kV CT

If $C_1 = 4840$ pF

Intermediate voltage to be chosen $= 20\ \text{kV}/\sqrt{3}$

Then C_2 can be calculated accordingly from the CVDR as follows:

CVDR $=$ Primary voltage/intermediate voltage $= (220/\sqrt{3})\,/\,(20/\sqrt{3}) = 11$

So $(C_1 + C_2)/C_1 = 11$

$C_2 = 10\ C_1 = 10 \times 4840$ pF

$C_2 = 48400$ pF

Example 2

For a typical 420 kV CT

If $C_1 = 4650$ pF

Intermediate voltage to be chosen $= 20$ kV

Then C_2, can be calculated accordingly from CVDR as follows:

CVDR $=$ Primary voltage/intermediate voltage

$= (400\,/\,\sqrt{3})\,/\,(20\,/\,\sqrt{3}) = 20$

So $(C_1 + C_2)/C_1 = 20$

$C_2 = 19\ C_1 = 88{,}350$ pF

Comparison of Ferroresonance Effect with Reference

Ferro-resonance is applied to the circuit that contains capacitors and inductors. It is a non-linear resonance phenomenon that affects the power networks which are made up of large number of saturated inductances (power transformers) as well as capacitors (long transmission line, CVTs, etc.). Ferroresonance occurs due to the following changes in the network circuit.

    i. Change of inductance in the network
    ii. Change of capacitance in the network
    iii. Change of supply frequency
    iv. Change of supply voltage
    v. Combination of the above factors

Basically intensity of ferroresonance effect depends upon the system parameters of the line and the CVT in the system. These parameters are considered as the reference values for the ferro-resonance effect. Comparison of these values with the actual parameters are regarded as the effect of ferroresonance in the circuit.

Effects of ferroresonance in the system   Following symptoms/effects may be accompanied in the system network for the occurrence of ferro-resonance.

    i. Displacement of neutral point voltage
    ii. Distortion of voltage and current waveform
    iii. Heating of transformer unit
    iv. Noise in transformers and reactors
    v. Damages in electrical equipments (Due to thermal effect or insulation breakdown)
    vi. Unbalanced voltage magnitude in the secondary circuit.

CONCLUSION

This topic, related to the study on capacitive voltage transformer covers the detailed information regarding the basic construction, working principle, maintenance practice, etc. of the CVT.

The failure rate of CVTs is more as compared to the electromagnetic PT during the initial days of early services. The basic causes of failure of CVTs are as follows:

1. Switching operation of long transmission line.
2. Ferro-resonance effects on the CVT.
3. Wrong design of electromagnetic network.
4. Others (Short-circuit faults in line; lightning effects on the line, etc.).

However, maximum care has been adopted for the modification and development in the design and manufacturing of CVTs. Particularly the use of suitable damping network, spark gap system, fuse protections, etc. are the advance systems, which is used in CVT to avoid the failure rate.

Because of multi-purpose uses like providing the signal for measuring, protection circuit and coupling capacitance to the tele-communication circuit, CVT has gained popularity for the use in EHT system network.

APPENDIX

APPENDIX 6

COMPARISON OF CVT WITH ELECTROMAGNETIC VT

Table 1

Electromagnetic VT	Capacitive voltage transformer
It contains primary and secondary windings, according to the requirement of voltage transformation by the equipment. VT works on the principle of mutual induction, according to the number of turns on the primary and secondary side, the voltage is transferred.	It contains capacitor units across which maximum voltage gets dropped on the primary side and availability of intermediate voltage across the EMU (electromagnetic unit), causes the voltage transformation by the principle of mutual induction.
For the voltage class of 66 kV and above, VT is not economical as compared to CVT.	For the voltage class of 66 kV and above, CVT are less costly as compared to VT.
Carrier communication circuit connection does not become possible for VT circuit. Because this circuit does not have coupling capacitance in this system.	Along with measuring, protection system, the CVT can be used for carrier communication circuit also. The unit capacitance of the CVT, provides coupling capacitance for the system.
This is accurate in performance and provides less error for the circuit connection.	Because of transient behaviour of the circuit parameters, the CVT provides certain error for the circuit performance.
The Ferroresonance effect is absent for VT.	Ferro-resonance is the common effect for CVT, because CVT contains R, L, C parameter.
The failure rate is less whereas only insulation failure results for VTs.	The failure rate is more in comparison to VT, because the effects like voltage transients, ferroresonance, and persistence of high voltage in the CVT, etc. are the usual phenomenon for CVT. Sometimes due to absence of protection circuit, effects become prominent and results in the failure of CVT.

(Contd.)

Table 1 (Continued)

Electromagnetic VT	Capacitive voltage transformer
Basically the VT rarely fails, but for the case of insulation failure if it fails, then the VT is splitted into pieces and scattered severely and causes damage to the adjacent equipments.	In general, failure of CVT does not cause any physical devastation. In most of the incidents, the secondary circuit of CVT fails, because of transient oscillation of voltage on the circuit parameters. But failure of insulation also results in splitting of CVTs into pieces and scattering them.
Bus PT connection in a system, providing common voltage to all the protection circuits for all the transmission lines from the system becomes economical in comparison to the CVT connection in each and individual lines.	CVT connection on each and individual lines provide separate protection to the respective transmission line. So, protection scheme is effective for the use of CVT in the circuit.
The status of incoming voltage cannot be known when using bus PT, So, synchronization of the incoming and running voltage does not become possible.	The status of the incoming voltage, can be known when using line CVT and by comparison with the existing bus voltage, the synchronization becomes possible.
The availability of voltage supply to the protection scheme is effective and relay memory action is active when using bus PT. So, for any energization of a transmission line with fault in the system, the protection system being active, actuates properly and clears the fault immediately.	The availability of voltage supply to the protection becomes effective after energization of the said transmission line when using the line CVT. So, till the energization of the line, the relay memory action is inactive. The line with a fault in the system being in momentary activeness, provides wrong protection action for the system.

APPENDIX 7

LIST OF STANDARDS

Standards	Title
IS 335:1993	Insulating oil used for transformer
IS 2017:1974(Part-I)	Methods of HV testing (General definition and test requirement)
IS 2017:1974(Part-I)	Methods of HV testing (Test Procedures)
IS 2099: 1986	Bushings for AC voltage above 1000 V.
IS 2165:1977(Part-I)	Insulation coordination (Ph to earth insulation coordination)
IS 3716:1978	Application guide for insulation coordination
IS 4146: 1983	Application guide for voltage transformer
IS 5547:1983	Application guide for CVT
IS 9676: 1980	Reference ambient temperature for electrical equipment
IS 11322:1985	Method for PD measurement in instrument transformer
12360: 1988	Voltage bands for electrical installation, including preferred voltages and frequency.
IS 3156:1992(Part-I)	Voltage transformer- specialists (general requirements)
IS 3156:1992(Part-II)	Measuring voltage transformer
IS 3156:1992(Part-III)	Protective voltage transformer
IS 3156:1992(Part-IV)	Capacitive voltage transformer
IS 9348:1979	Coupling capacitors and capacitor devices
IEC 186 (1987)	Details on voltage transformer
BS 3941:1975	Voltage transformers with amendments

QUESTIONS AND ANSWERS

1. Why does CVT become economical for the voltage range above 66 kV class?

Answer

Electromagnetic VT with voltage range above 66 kV class has primary and secondary winding and needs proper insulations to maintain required voltage class on primary winding. It is observed that the insulations required for voltage class above 66 KV becomes costly as compared to the voltage class of CVT of same rating. For the case of CVT, capacitor stacks are used on the top unit and also on the bottom unit. As because dielectric medium is used between the electrodes of each element, the same medium provides also certain insulation to the system. So, quantity of extra insulation that is required for the CVT becomes less when compared to others and 66 kV class CVT becomes economical.

Moreover, for every transmission line, PLCC is required to be provided as the communication channel. For voltages above 66 kV range, this PLCC becomes compulsory feature in the system. So by the use of the CVT, the capacitor units that are available in the equipment provide coupling capacitance $\left(C_\chi\right)$ to the carrier communication circuit. Because the capacitor units, can be used for many purposes, the CVT above 66 kV class becomes economical in comparison to the use of VT.

2. Why is intermediate inductor (choke) used in CVT?

Answer

As already discussed under the working principle, CVT contains capacitor stacks both in top and bottom unit. When supply voltage is applied the capacitor on the top unit causes voltage drop and according to the magnitude of capacitance value, the necessary voltage is made available across the EMU.

The intermediate inductance (L) plays the role of compensatory circuit for the formation of a resonance circuit. The value of inductance is so adjusted that the circuit resonates for the supply with the standard frequency source. For practical study of the circuit, the value of inductance is chosen as follows.

$$L = \frac{1}{\omega^2 C_1}$$

where,

$$C_1 = \text{Capacitance of the top unit}$$
$$\omega = 2\pi f$$
$$L = \text{Compensatory inductance}$$

3. How is the value of intermediate inductance obtained?

Answer

Intermediate inductance is used for the compensation of the circuit to adjust for the development of resonance circuit during the normal working range of supply frequency. To design/calculate the value of its inductance, the following circuits can be considered.

For proper adjustment of the circuit parameters, the capacitance and inductance of the circuit should be designed accordingly. Now, the inductive reactance of the system should be equal to the capacitive reactance.

$$X_L = (X_{C_1} \times X_{C_2})/(X_{C_1} + X_{C_2})$$

But from the practical study of the circuit it is seen that X_{C2} is very small compared to X_{C1}. So the above equation can be written as

$$X_{\mathrm{L}} = X_{C_1}$$

$$\omega L = 1 / \omega C_1 \text{ or } L = 1 / \omega^2 C_1$$

So inductive inductance (L) should be designed according to the value of capacitance C_1.

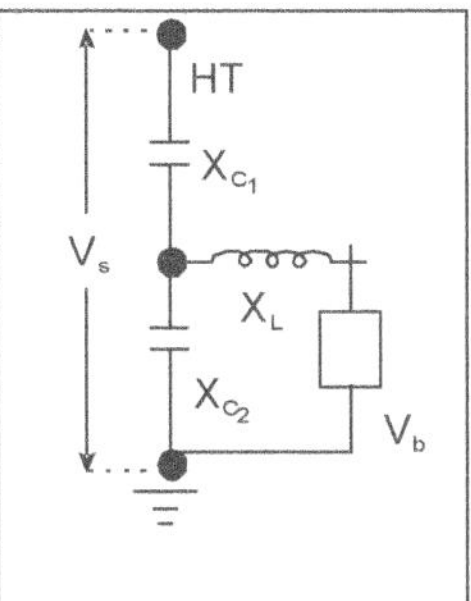

4. Why CVT is preferable to VT nowadays?

Answer

CVT is extensively used for the line protection and for carrier communication. The coupling capacitor in CVT works for the principle of carrier communication and EMU helps for the line protection. The single unit (CVT) with the provision of EMU and coupling capacitance provides this multi-application, and becomes preferable to VT.

Moreover the cost of the CVT above 66 kV class becomes economical as compared to PT. The other features have been explained in the comparison table given in the Appendix.

5. What is ferroresonance? Explain the circuit configuration for which this effect is prominent.

Answer

The circuit parameters of the CVT combined with the line parameters of the system cause the ferroresonance in the CVT. The R, L and C network of the system and abnormality in supply voltage, results in certain transient oscillations in the circuit. This effect is called ferroresonance.

Ferroresonance occurs due to the changes in the following parameters in the network.

 i. Inductance in the network
 ii. Capacitance in the network
iii. Supply voltage and frequency
iv. Combination of the above factors

Following circuit configurations in the system result in the ferroresonance.
1. CVTs connected between phase and earth of an isolated neutral system.
2. Transformers connected through long or capacitive lines.

3. Unloaded or lightly loaded power or voltage transformer, switching operation of capacitor banks, unloaded lines or insulation faults, lightning, etc. can result in the ferroresonance.

It is basically understood that for the question of sudden rise of voltage in the system, with abnormal loading pattern on the CVT causes transient oscillation that leads to ferroresonance.

6. Which method of monitoring practice is preferable for CVT?

   i. Secondary voltage measurement of each core

   ii. Residual voltage measurement of broken delta secondary

Answer

The practice with secondary voltage measurement of each core is preferable and advantageous as compared to the other system.

Because for the measurement of secondary value of each core, the monitoring of all the individual cores, becomes possible and the analysis value with respect to the commissioning result helps to know the status of the secondary core and corresponding CVT also. The detailed practice of studying the status of the CVT has been explained in case study 3, earlier.

But for the monitoring practice with the measurement of residual voltage for the broken delta secondary, the magnitude of each and individual secondary value cannot be known. For the unbalanced condition of any of the core involved for the open delta circuit formation, the residual voltage is only developed. After knowing the residual voltage value/magnitude, the affected core and status of the same cannot be known, without checking the individual core. So development of residual voltage is only the indicative regarding the discrepancy in the system. But monitoring with individual secondary value measurement is advantageous as compared to residual value measurement.

In practice, residual measurement is done to know the indication regarding the problem in the CVT. But for details and conformation, the voltage on each secondary core can be measured for proper diagnosis.

7. Why is the capacitance in the bottom unit chosen more commonly as compared to the top unit?

Answer

The function of capacitance in the CVT is to provide the capacitive reactance for the following purposes.

1. Causes necessary voltage drop to avail the required voltage across the EMU.
2. Provides the necessary parameter for resonance.
3. Provides coupling capacitance to the carrier communication circuit.

   The value of top unit is necessary to be designed/taken less as compared to the bottom unit. The maximum amount of voltage is allowed to be dropped across the top unit. For a standard 220 kV CVT system, intermediate voltage is chosen as $20\,kV/\sqrt{3}$

$$\text{CVDR} = \text{Primary voltage/Intermediate voltage} = (220 / \sqrt{3}) / (20 / \sqrt{3}) = 11$$

So $(C_1 + C_2)/C_1 = 11$

$C_2 = 10\ C_1 = 10 \times 4840$ pF

$C_2 = 48400$ pF

8. Explain the behaviour of CVT for the compensated transmission line.

Answer

Compensated transmission line consists of shunt reactors to compensate the capacitive effects of the line. Basically for the long line case, this phenomenon of shunt reactor compensation is used in the system. If the line containing shunt reactor compensation is opened at both ends, transient oscillation results between the inductance of shunt reactor and line capacitance. Practically every CVT contains capacitor dividers, compensating inductive reactance which are tuned to the system frequency of 50 Hz. When the compensated line opens at both ends, the combined effects of tuning parameters do not become same as compared to the previous. So, the magnitude of voltage for the system due to this effect does not become uniform. The abnormal and transient phenomenon results with ferro-resonance in the circuit. Sometimes the following effects are observed in the system network.

   i. The over-voltage relay actuates and results in system disturbance.
   ii. During the time of normal tripping at both ends, sometimes the CVT fails due to high voltage impulse in the system.
   iii. For normal operation, the actuation of over-voltage relay leads to confusion regarding the behaviour of EHV equipments in the system.
   iv. Voltage stress on the system equipment also results.

9. What is polymer housing insulation compared to porcelain unit insulation?

Answer

For the construction of CVT, the capacitance units are required to be provided with maximum care to avail the proper insulation in the system. The main components of such units are polypropylene film, tissue papers and other insulating materials as the stack is to the capacitor element. The oil is also provided as the insulation and cooling medium for the design of CVT. These capacitors along with the insulations are housed in the porcelain insulators. It is seen that during the time of failure of CVT, scattering of porcelain chips cause damage to the nearby equipments.

To avoid the above problem, polymer housing are used in place of porcelain unit. But this housing is 5 to 6 times costly as compared to porcelain insulators. So the use of this housing is still under trail. This type of housing is generally used for CVT class above 400 kV ranges. Basically polymer material is light in weight and provides better insulation in comparison. Moreover due to better insulation, the size of the CVT can be reduced in comparison.

10. Explain the circuitry for the protection scheme of a distance protection relay and directional E/F relay from a particular CVT secondary core?

Answer

For the DP relay protection scheme, it is needed to avail the CVT secondary voltage of standard value 110 V from the star connected voltage core. But for the directional E/F relay protection scheme, it is needed to avail the voltage from the open delta CVT secondary core. From a particular CVT secondary core, the star connection is formed at the CVT console and terminals are carried directly connected to the DP scheme. By the use of auxiliary PT of ratio 1:1, the broken delta secondary voltage is availed from the secondary side of the auxiliary PT as shown in Figure 3.22.

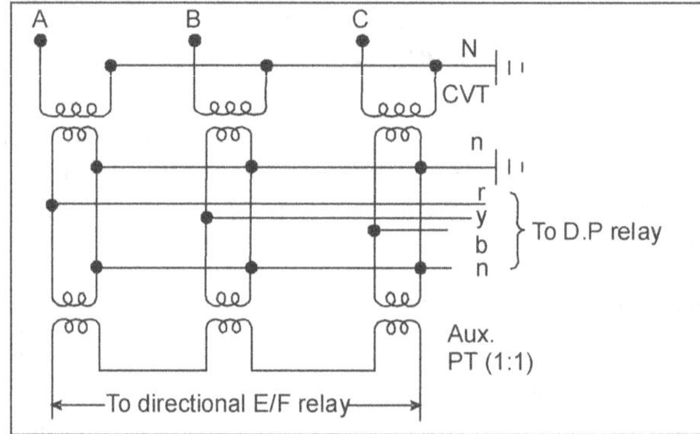

Figure 3.22 CVT secondary circuit for relay connection

11. What is the difference between PT and CVT?

Answer

Construction

1. PT is constructed with the principle of electromagnetic induction and has primary and secondary winding.
2. CVT works on the principle of capacitive voltage divider. It is coupled with capacitor stack and inductive voltage transformer and reactor in series to the circuit.

For detail comparison the following table can be referred.

Electromagnetic potential transformer	**Capacitive voltage transformer**
It contains primary and secondary windings, according to the requirement of voltage transformation by the equipment. VT works on the principle of mutual induction, according to the number of turns on the primary and secondary side, the voltage is transferred.	It contains capacitor units across which maximum voltage drop occurs on the primary side and availability of intermediate voltage across the Electromagnetic unit (EMU), causes the voltage transformation by the principle of mutual induction.
For the voltage class of 66 kV and above, VT is not economical as compared to CVT	For the voltage class of 66 kV and above CVT is less costly as compared to VT.
Carrier communication circuit connection does not become possible for VT circuit. Because this circuit does not have coupling capacitance in this system.	Along with measuring, protection system, the CVT can be used for carrier communication circuit also. The unit capacitances of the CVT, provides coupling capacitance for the system.
This is accurate in performance and provides less error for the circuit connection.	Because of transient behavior of the circuit parameters, the CVT provides certain error for the circuit performance.
The ferroresonance effect is absent for VT.	Ferroresonance is the common effect for CVT. Because CVT contains R, L and C parameters.
The failure rate is less whereas only insulation failure results for VTs	The failure rate is more in comparison to VT. Because the effects like voltage transients, ferroresonance, and persistence of high voltage in the CVT, etc. are the usual phenomenon for CVT. Sometimes due to absence of protection circuit, these effects become prominent and result in the failure of CVT.
Basically the VT rarely fails, but for the case of insulation failure if it fails, then the VT is splited into pieces and scattered severely and causes damage to the adjacent equipments.	In general, failure of CVT does not cause any physical devastation. In most of the incidents, the secondary circuit of CVT fails, because of transient oscillation of voltage on the circuit parameters. But failure of insulation also results in splitting of CVTs into pieces and scattering them.
"Bus PT" connection in a system, providing common voltage to all the protection circuits for all the transmission lines from the system becomes economical in comparison to the CVT connection in each and individual lines.	CVT connection on each and individual lines provides separate protection to the respective transmission line. So protection scheme is effective for the use of CVT in the circuit.

Electromagnetic potential transformer	Capacitive voltage transformer
The status of incoming voltage cannot be known, when using bus PT. So, synchronization of the incoming and running voltage does not become possible.	The status of the incoming voltage, can be known, when using the line CVT and by comparison with the existing bus voltage, the synchronization becomes possible.
The availability of voltage supply to the protection scheme is effective and relay memory action is active when using bus PT. So, for any energization of a transmission line with fault in the system, the protection system being active, actuates properly and clears the fault immediately.	The availability of voltage supply to the protection becomes effective after energization of the said transmission line when using the line CVT. So, till the energization of the line, the relay memory action is inactive. The line with a fault in the system being in momentary activeness, provides wrong protection action for the system.

12. What are the parameters to be considered for the selection of metering CTs/CVTs.

Answer

Measurement value always needs to be precise and accurate during normal rating of currents or voltages for any equipment. So the instrument transformers used for this purpose should be selected with low saturation limit and high accuracy factor. The available burden of the instrument transformers should also be more to meet the satisfactory loads to the circuit.

13. Why is CVT preferable to PT for EHV feeder bays?

Answer

Capacitive voltage transformer is extensively used in the line protection due to its accuracy in measurement and use as potential transformer and coupling capacitor. CVT is composed of one or more capacitive units depending on the voltage level and impregnated with high grade dried and degassed dielectric oil. The electromagnetic unit which includes the MV transformer and series inductance is located in a hermetically sealed oil filled tank with a protective device thus avoiding over-voltage and ferroresonance.

But for the PT connection physical windings are made available across the line conductor and earth. This works on the principle of mutual induction, but does not provide coupling capacitance. So the CVT is preferable to the PT.

14. Mention the accuracy class of voltage transformer?

Answer

The accuracy class of the CVT for application in the circuit is decided according to the secondary winding and the type of core used in practice. Basically CVTs are used for two different purposes (metering, protection). Metering class CVT core needs better accuracy as compared to the protection class. But protection class needs robust performance even in saturation region.

According to the requirement to the circuit the following accuracy class error table can be referred.

Metering core			Protection core		
Accuracy class	± % Voltage ratio error	± Phase angle displacement (error in minutes)	Accuracy class	± % Voltage ratio error	± Phase angle displacement (error in minutes)
0.1	0.1	5	3P	3	120
0.2	0.2	10	6P	6	240
0.5	0.5	20	**Note 1:** Errors at 5% rated voltage and voltage multiplied by voltage factor (1.2, 1.5 or with burdens between 25 to 100% of rated burden at p.f 0. 8 (lag)		
1.0	1.0	40			
3.0	3.0	-			
Note: Errors at any voltage between 80 to 120% of rated voltage, with burdens between 25 to 100 % of rated burden at pf 0. 8 (lag).			**Note 2:** Errors at 2% rated voltage shall be twice as high as given in the table with similar burdens to Note 1.		

15. Why is PT of different rating used (110V /v3 and 110V)?

Answer

The rating of the PT on secondary side depends upon the connection practice of the same on primary side. Usually primary side is connected between line conductor and earth terminal. So the primary side of such PT is declared as the magnitude of Line Voltage/v3. Now the secondary voltage is accordingly considered for the standard practice of 110V /v3. But for some special purpose like the sensitive ground fault protection (directional), the broken delta secondary voltage is required to be availed from the PT secondary. Sometimes it is needed to avail 110V supply to this protection system.

Note   In general practice the voltage range 110 V/v3 is the standard. Availability of 110 V is the special case as described above.

16. Explain the specifications to be adhered during the selection of VT (voltage transformer).

Answer

Selection of VT depends upon the application and use of instruments on the secondary side of the VT circuit. The ratings, accuracy class, burdens, etc. are some of the important factors for the selection of the VT. However the following points can be described as the specifications for the selection of VT in the circuit.

Specification of VT

1. Rating (primary and secondary side)
2. Number of VT cores
3. Burden of each core
4. Accuracy class
5. Withstand voltages
6. Frequency
7. Physical dimensions
8. Type of construction
9. Over-load limit factors
10. Others like type of insulation used, precautions, etc.

17. Explain the burden on the VTs?

Answer

In general the loads that are connected on the secondary side of the VT are called the burdens of the transformer. It is expressed in VA (volt ampere) at rated voltage and at a particular power factor. Ohmic value of the burden is calculated as follows.

$$Z_b = V_s^2 / P$$

where,

V_s = VT secondary voltage in volt

Z_b = Secondary burden in ohm.

P = Burden in VA

Standard values of burdens are 10, 25, 30, 75, 100, 150, 200, 300, 500 VA for VT and CVT.

18. Explain the basic constructional design of CVT?

Answer

The construction of CVT depends upon the arrangement of capacitor unit and EMU (electromagnetic unit) on it. It consists of two parts.

1. CDU (Capacitor divider unit)
2. EMU(Electromagnetic unit)

   CDU contains two set of capacitor stacks as HV capacitor stack (C_1) and intermediate capacitor stacks (C_2).

   EMU contains mainly of compensating reactor, intermediate transformer unit and ferroresonance suppression circuit.

The detailed Figure is shown for reference.

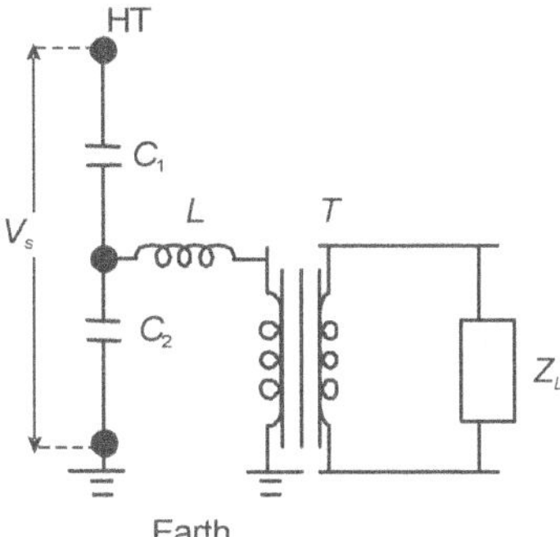

19. Explain the concept of capacitance ratio for a 245 kV CVT?

Answer

For a 220 kV CVT structure the capacitors that are used on primary and intermediate side are responsible for the voltage division on the CVT. In general the number of capacitor stack on HV capacitor (C_1) is more than the stacks on intermediate range (C_2). The standard range of values are of C_1 = 4840 pF and C_2= 48400 pF for 245 kV CVT.

So C_2/C_1 = 10.

But voltage division depends upon the ratio as $V_1/V_2 = (C_1 + C_2)/C_1 = 11$

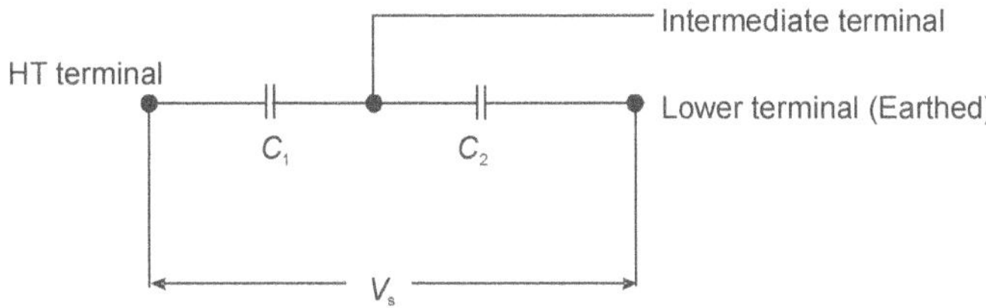

20. Describe the method of condition monitoring of CVT?

Answer

The electrical methods for CVT condition monitoring are described as follows.

1. Secondary voltage measurement technique.

2. Tan δ and capacitance measurement of stacks.

Secondary voltage measurement technique

The development of secondary voltage depends upon the condition of capacitor units used in CVT.

V_2 = Voltage across 'C_2' = $V_s \times C_1 /(C_1 + C_2)$

   Now variation of secondary voltage indicates regarding the problem in capacitor unit as per the following description.

i.   Puncture of individual capacitor of top unit (C_1), results in the rise of secondary voltage and puncture of individual capacitor of bottom unit (C_2), results in the decrease of secondary voltage.

ii. For a standard 400 kV CVT, puncture of C_1, increases voltage by 0.35 to 0.45% (0.22 to 0.28 V) and for C_2 it is decreased by 5 to 7% value (3.2 to 4 V).

Preliminary information regarding the capacitor unit could be obtained by the secondary voltage measurement method of the CVT.

Tan δ and capacitance measurement of stacks

By the method of capacitance and tan δ measurement, the status of the units can be confirmed. It is to be considered that change of capacitance value above 10% is objectionable and CVT needs to be replaced. Similarly change of tan δ value by ± 0.004 from original value should be considered for replacement.

21. Explain the use of compensating reactor in CVT and derive the relation with the capacitor units?

Answer

CVT is basically a capacitor potential divider unit and contains capacitor stacks on top and bottom units. During working condition the supply voltage gets divided across the capacitor units as per the proportion of the capacitance value. Top unit has capacitance value lower than the bottom unit. So the maximum of the voltage drop occurs across the top unit and the voltage that is made available across the electromagnetic unit (EMU) becomes capacitive in nature, which needs to be compensated for suitable use by the intermediate transformer. So certain magnitude of inductance is required to be provided in series to the circuit. This inductance is called compensating reactor.

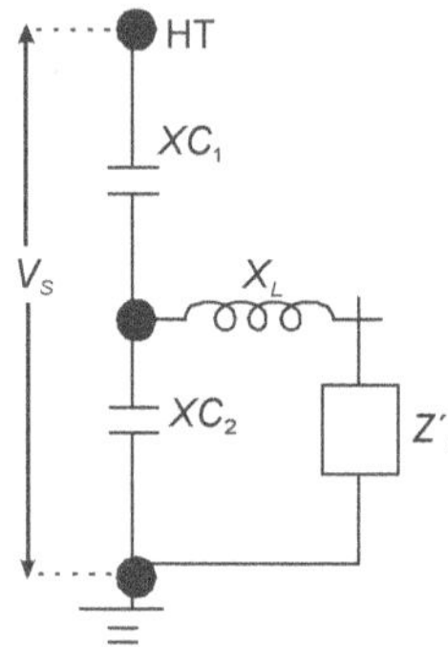

22. What are the reasons for replacement of CVT?

Answer

The reasons are many, but careful attention to the condition monitoring reduces the failure rate.

Some of the possible reasons are listed as follows:

1. Manufacturing defects
    i. Problems in aluminum foils
    ii. Abnormal soldering
    iii. Poor quality of insulating material
    iv. Pin holes in bellows
    v. Wrong alignment of capacitor and EMU
    vi. Rusting in flange, bolt
    vii. Looseness of core bolts
    viii. Failure of resonance circuit
2. Climatic condition
    i. Entry of moisture in capacitor stacks
    ii. Moisture entry due to poor gasket quality
    iii. Overheating of core and other materials.
3. Electrical condition
    i. Ferroresonance in the circuit
    ii. Abnormal rise of system voltage
    iii. Short-circuiting of secondary loads.

23. Explain the reasons for which the capacitor dielectrics get deteriorated in CVT?

Answer

For oil-immersed CVT, the dielectric medium that is sused for capacitor stacks are of oil-impregnated papers with insulating films between aluminum foils. The insulating degradation during the service of CVT takes place as a result of one or more of the following phenomenon.

1. *Chemical degradation*   The oil in due course of service span results in chemical degradation due to its compatibility properties. But such deterioration is a long-term ageing phenomenon.
2. *Over voltage*   The effect of over-voltage due to reasons like lightning surge, switching surge cause electrical stress on the insulating materials during service operation of CVT. This nature of electrical stress slowly deteriorates the insulating material and finally causes the total failure of the same.
3. *Moisture ingress*   Sometimes due to improper sealing system of CVT, moisture ingress results into the capacitor unit. The oil present inside gets deteriorated and subsequently the insulation fails to provide its action.
4. *Mechanical defects*   The mechanical defects of the insulation like presence of void, air bubble, moisture, etc., result in partial discharge and in due course of service in operation, deteriorates the insulation system and finally fails to provide required insulation.
5. *Natural ageing*   The insulation has its own life span till which its provides satisfactory insulation to the system. But with due course of operation the insulation starts

deteriorating slowly over several years. This method of deterioration can be grouped under natural ageing process. The usual span of this natural ageing is more than 20 years for CVT.

24. What is V-V connection in CVT?

Answer

For this type of connection, two windings are only connected across the phase voltages. One set of winding is connected across R-Y and other set of winding is connected across Y-B on primary side of the CVT. Similarly the secondary side windings are connected in the same fashion as that of primary side. This type of winding connection is called V-V connection. The concept of zero sequence is not produced for such type of connection.

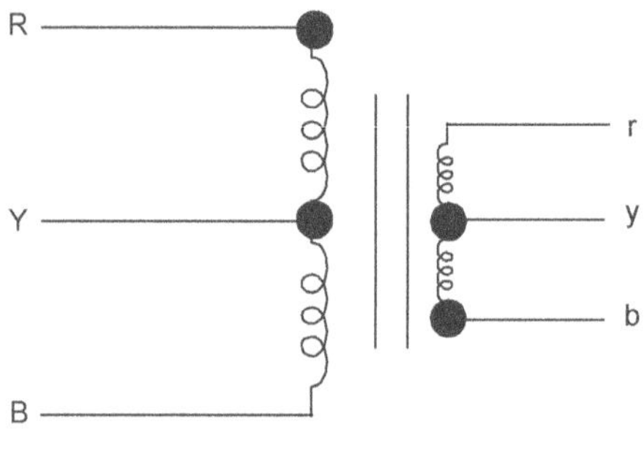

V-V connection of CVT

25. Explain the category of insulating materials used for capacitor elements?

Answer

Basically four categories of insulating materials are used for the construction of capacitor elements and these insulations are taken in combination form. The development of insulation technology is always in progression as per the demand and requirement of the customers. So only the normal uses of insulation materials are described here.

The list of the insulations can be grouped as follows.

1. Paper + mineral oil
2. Film + paper + mineral oil
3. Film + paper + synthetic oil
4. Film + mineral oil
5. Film + synthetic oil

26. Suggest few designs of magnetic core to develop the linearization of excitation curve in voltage transformer?

Answer

The excitation curve is drawn from the relationship between exciting current and induced voltage on the secondary winding under no load condition. The core is designed as per the recommended standard of secondary exciting current at the rated voltage. The linearization–excitation curve in practice is not possible to obtain. But by the following methods the design of the core can be attended to the approximation of the linearization.

1. Interleaved core design
2. Cut-core construction
3. Metered core construction
4. Combination of the above

27. What is ferroresonance in CVT?

Answer

The circuit parameters (R, L, and C) of CVT in the form of potential divider capacitance, compensated inductance, intermediate transformer and burden impedance result in the resonance circuit. If this circuit is subjected to a voltage impulse due to switch on/off of the supply voltage to the CVT, then some degree of oscillation results in the system. The frequency of the circuit is considered for the development of oscillation and if it is slightly less than one third of the system frequency, it becomes possible for energy to be absorbed from the system to result in oscillation. Depending on the value of components, oscillations at fundamental frequency or at other sub-harmonic or multiple of supply frequency are possible but the third sub-harmonic is one of the most disturbing frequencies for oscillation. The transient phenomenon becomes prominent for the case of this oscillation.

The ferroresonance phenomenon in CVT is developed due to the following situations.

1. Secondary short-circuit fault on the secondary cable and opening of the same by fuse protection or by MCB.
2. Saturation of the auxiliary voltage transformer on the secondary circuit of CVT by any temporary over-voltage condition. Subsequently when over-voltage goes away, the normal burden conditions are removed.
3. Sudden application of more burden and removal of it.

28. How is ferroresonance effect defined as per IEC 186?
    (a) After clearance of a short-circuit at the secondary terminals, the CVT being energized at 120% of rated voltage, the crest value of the secondary voltage must return to less than 10% error after 10 cycles.
    (b) After short-circuit clearance at 150% of the rated voltage the ferroresonance must be eliminated within less than 2 seconds.

29. What are the routine tests to be conducted for CVT?
    Answer

The routine test to be conducted for CVT are as follows.

   i. Polarity test
  ii. Determination error

   iii. P.D measurement

   iv. Power frequency withstand test on secondary

   v. Power frequency withstand test on primary

   vi. Ferroresonance test

   vii. Sealing test

30. What are the type tests to be conducted for CVT?

Answer

The type tests to be conducted for CVT are as follows.

   i. Temperature-rise test

   ii. SC with start test

   iii. Lighting impulse test

   iv. Switching impulse test

   v. Wet test for outdoor transformer

   vi. Error determination.

31. Can the optical sensor CVT be possible? Explain the principle behind it?

Answer

Yes, the optical sensor CVT is possible in practice. Practically for this type of CVT, the electrodes or the conductors across which the voltage is applied are set at certain distance. The optical fibres are kept under the electric field of these conductors. Now the core of the electro–optic sensor can be calibrated for the response measurement of the magnitude of the voltage that is applied across the electrodes.

32. Explain the advantages of optical CT and PT?

Answer

Optical instrument transformers do not have any magnetic material and do not carry electrical current to flow through in the transformer. So the following advantages can be availed by the use of this type of instrument transformers.

1. The effect of CT satuaration is absent due to the absence of any magnetic material.
2. The possibility of ferroresonace, effect can also be avoided for this type of transformer.
3. Insulating materials like oil, SF_6, etc. are not required.
4. It is light in weight also.
5. It is electrically isolated from the system.
6. The situation like CT opening, CVT shootings can also be avoided.
7. It has wide range of bandwidth application.
8. It provides digital communication system and easy data acquisition system

33. What might be the problem for the following situation of voltage availability on CVT secondary system?

Secondary voltage in volt					
RN	YN	BN	RY	YB	BR
64.85	65.10	65.28	65.12	65.12	112.10

Answer

CVT secondary connection of a particular core is generally connected in star mode. So the phase to neutral voltage becomes 63.5 volt and phase to phase becomes 110 volt. But for the situation described in the question the voltage value for "RY" and "YB" are found to be 65.12 volt instead of corresponding (v3 × 65.12 volt) . So it is concluded from the result that the associated "Y" phase core has some problem. Practically "Y" phase CVT polarity is in reverse mode for this situation. The vector diagram has been drawn for such situation to explain the measured voltage as described.

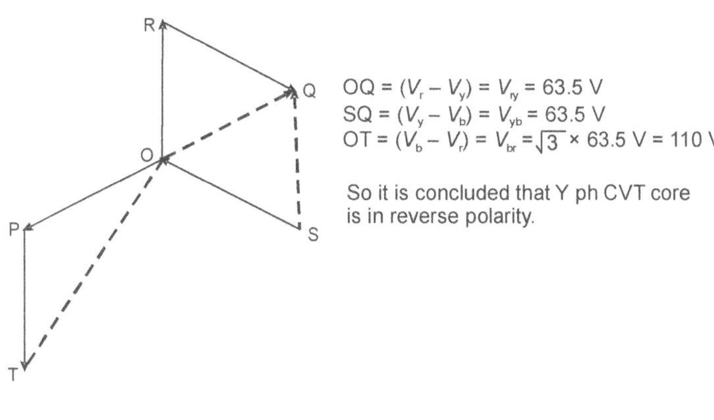

4

DATA ON ELECTRICAL SYSTEM

INTRODUCTION

Data on the electrical system are the most useful tool for electrical engineers to know the limit of their working arena. According to the limitation of data, proper planning, estimation and confirmation regarding the equipments can be achieved. Basically, the practice engineer is very cautious regarding the value of threshold on the electrical parameters, upon the field works on the equipment on which they work to obtain the efficiency of the assigned project. Many a time they get confused and are unable to make a decisive conclusion, due to the non-availability of a ready reckoner for electrical data. This chapter covers all the possible zones of electrical field topics with some reference data in a concise manner. The reference tips on the maintenance schedule, testing methods, installation practice and commissioning patterns can help the electrical engineers to acquaint themselves and develop a healthy environment by refining their practice towards the field works.

DATA ON ELECTRICAL CIRCUIT STUDY

STANDARD CONNECTING WIRE

Letters	Purpose	Letters	Purpose
A	CT secondary circuit special	K	Control (Trip, Close, etc.).
B	Bus bar protection	L	Alarm, indication, annunciation circuit
C	Protection AC circuit (CT)	M	Motor control circuit
D	Metering circuit	N	Transformer tap changer control circuit
E	PT secondary	S	Marking of secondary CT polarity
F	Fan control circuit	U	Used for spare wires
H	LT AC supply	X	Used for spare wires
J	DC main circuit		

DETAIL DEVICE IDENTIFICATION CODE

1. Master element is the initiating device, such as a control switch, voltage relay, float switch, etc. which serves either directly or through such permissive devices as protective and time-delay relays to place equipment in or out of operation

2. Time-delay starting or closing relay is a device which functions to give a desired amount of time delay before or after any point or operation in a switching sequence or protective relay system, except as specifically provided by device functions 62 and 79.

3. Checking or interlocking relay is a relay which operates in response to the position of a number of other devices, or to a number of predetermined conditions, in an equipment to allow an operating sequence to proceed or to stop or to provide a check of the position of these devices or of these conditions for any purpose.

4. Master contactor is a device, generally controlled by device No:1 or equivalent and the necessary permissive and protective devices, which serves to make and break the necessary control circuits to place an equipment into operation under the desired conditions and to take it out of operation under other or abnormal conditions.

5. Stopping device is a device whose primary function is to place any hold any equipment out of operation.

6. Starting circuit breaker is a device whose principal function is to connect a machine to its source of starting voltage.

7. Anode circuit breaker is one used in the anode circuits of a power rectifier for the primary purpose of interrupting the rectifier circuit if an arc-back should occur.

8. Control power disconnecting device is a disconnecting device such as a knife switch, circuit breaker or pull-out fuse block-used for the purpose of connecting and disconnecting respectively, the source of control power to and from the control bus or equipment.

Note   Control power is considered to include auxiliary power which supplies such apparatus as small motors and heaters.

9. Reversing device is a device which is used for the purpose of reversing a machine field or for performing any other reversing functions.

10. Unit sequence switch is a switch which is used to change the sequence in which units may be placed in and out of service in multiple unit equipments.

11. Reserved for future application

12. Over-speed device is usually a direct connected speed switch which functions on machine overspeed.

13. Synchronous-speed device is a device such as a centrifugal speed switch, a slip-frequency relay, a voltage relay, an undercurrent relay, or any type of device which operates at approximately synchronous speed of a machine.

14. Under-speed device is a device which functions when the speed of a machine falls below a predetermined value.

15. Speed or frequency marching device is a device which functions to match and hold the speed or the frequency of a machine or of a system equal to, or approximately equal to, that of another machine, source, or system.

16. Reserved for future application

17. Shunting or discharge switch is a switch which serves to open or close a shunting circuit around any place of apparatus (except a resistor), such as a machine field, a machine armature, a capacitor, or a reactor.

Note   This excludes devices which perform such shunting operators as may be necessary in the process of starting a machine by devices 6 or 42, or their equivalent, and also excludes device 73 function which serves for the switching or resistors.

18. Accelerating or decelerating device is a device which is used to close or to cause the closing of circuits which are used to increase or to decrease the speed of a machine.

19. Starting-to-running transition contactor is a device which operates to initiate or cause the automatic transfer of a machine from the starting to the running power connection.

20. Electrically operated valve is a solenoid or motor operated valve which is used in vacuum, air, gas, oil, water, or similar lines.

Note   The function of the valve may be indicated by the insertion of descriptive words such as "Brake" or "Pressure Reducing" in the function name, such as "Electrically operated brake valve".

21. Distance relay is a relay which functions when the circuit admittance, impedance, or reactance increases or decreases beyond predetermined limits.

22. Equalizer circuit breaker is a breaker which serves to control or to make and break the equalizer or the current balancing connections for a machine field, or for regulating equipment, in a multiple unit installation.

23. Temperature control device is a device which functions to raise or to lower the temperature of a machine or other apparatus, or of any medium, when its temperature falls below, or rises above, a predetermined value.

Note   An example is a thermostat which switches on a space heater in a switch gear assembly when the temperature falls to a desired value as distinguished from a device which is used to provide automatic temperature regulation between close limits and would be designated as 90T.

24. Reserved for future application

25. Synchronizing or synchronism-check device is a device which operates when two AC circuits are within the desired limits of frequency, phase angle, or voltage to permit or to cause the paralleling of these two circuits.

26. Apparatus thermal device is a device which functions when the temperature of the shunt field or the amortizes winding of a machine, or that of a load-limiting or load-shifting resistor or of a liquid or other medium, exceeds a predetermined value; or if the temperature of the protected apparatus, such as a power rectifier or of any medium decreases below a predetermined value.

27. Under-voltage relay is a relay which functions on a given value of under-voltage.

28. Reserved for future application.

29. Isolating contactor is a contactor which is used for disconnecting one circuit from another for the purposes of emergency operation, maintenance, or test.

30. Annunciation relay is a non-automatically reset device which gives a number of separate visual indications upon the functioning of protective devices, and which may also be arranged to perform a lockout function.

31. Separate excitation device is a device which connects a circuit such as the shunt field of a synchronous converter to a source of separate excitation during the starting sequence; or one which energizes the excitation and ignition circuits of a power rectifier.

32. Directional power relay is one which functions on a desired value of power flow in a given direction or upon reverse power resulting from arc-back in the anode or cathode circuits of a power rectifier.

33. Position switch is a switch which makes a breaks contact when the main device or piece of apparatus, which has no device function number, reaches a given position.

34. Motor-operated sequence switch is a multi-contact switch which fixes the operating sequence of the major devices during starting and stopping or during other sequential switch operations.

35. Brush-operating or slip-ring short circuiting device is a device for raising, lowering or shifting the brushes of a machine, or for short circuiting its slip rings, or for engaging or disengaging the contacts of a mechanical rectifier.

36. Polarity device is a device which operates, or permits the operation of another device, on a predetermined polarity only.

37. Undercurrent or under-power relay is a relay which functions when the current or power flow decreases below a predetermined value.

38. Bearing protective device is one which functions on excessive bearing temperature or on other abnormal mechanical conditions, such as undue wear, which may eventually result in excessive bearing temperature.

39. Reserved for future application.

40. Field relay is a relay that functions on a given or abnormally low value or failure of machine field current or on an excessive value of the reactive component of armature current in an AC machine indicating abnormally low-field excitation.

41. Field circuit-breaker is a device which functions to apply or to remove the field excitation of a machine.

42. Running circuit-breaker is a device which functions to apply or to remove the field excitation of a machine.

43. Manual transfer or selector device is a manually operated device which transfers the control circuits so as to modify the plan of operation of the switching equipment or of some of the devices.

44. Unit sequence starting relay is a relay which functions to start the next available unit in multiple-unit equipment on the failure or on the non-availability of the normally proceeding unit.

45. Reserved for future application.

46. Reverse-phase or phase-balance current relay is a relay which functions when the poly-phase currents are of reverse phase sequence, or when the poly-phase currents are unbalanced or contain negative phase-sequence components above a given amount.

47. Phase-sequence voltage relay is a relay which functions upon a predetermined value of poly-phase voltage in the desired phase sequence.

48. Incomplete sequence relay is a relay which returns the equipment to the normal, or off, position and locks it out if the normal starting or operating or stopping sequence is not properly completed within a predetermined time.

49. Machine or transformer thermal relay is a relay which functions when the temperature of an AC machine armature or of the armature or other load-carrying winding or element of a DC machine, or converter of power rectifier or power transformer (including a power rectifier transformer) exceeds a predetermined value.

50. Instantaneous over-current or rate of rise relay is a relay which functions instantaneously on an excessive value of current or on an excessive rate of current rise, thus indicating a fault in the apparatus or circuit being protected.

51. AC time over-current relay is a relay with either a definite or inverse time characteristic which functions when the current in an AC circuit exceeds a predetermined value.

52. AC circuit-breaker is a device which is used to close and interrupt an AC power circuit under normal conditions or to interrupt this circuit under fault or emergency conditions.

53. Exciter of DC generator relay is a relay which forces that DC machine field excitation to build up during starting or which functions when the machine voltage has built up to a given value.

54. High-speed DC circuit breaker is a circuit breaker which starts to reduce the current in the main circuit in 0.01 second or less, after the occurrence of the DC over-current or the excessive rate of current rise.

55. Power factor relay is a relay which operates when the power factor in an AC circuit becomes above or below a predetermined value.

56. Field application relay is a relay which automatically controls the application of the field excitation to an AC motor at some predetermined point in the slip cycle.

57. Short-circuiting or grounding device is a power or stored energy operated device which functions to short-circuits or to ground a circuit in response to automatic or manual means.

58. Power rectifier misfire relay is a relay which functions if one or more of the power rectifier anodes fail to fire.

59. Over-voltage relay is a relay which functions on a given value of over-voltage.

60. Voltage balance relay is a relay which operates on a given difference in voltage between two circuits.

61. Current balance relay is a relay which operates on a given difference in current input or output of two circuits.

62. Time delay stopping or opening relay is a time delay relay which serves in conjunction with the device which initiates the shutdown, stopping, or opening in an automatic sequence.

63. Liquid or gas pressure level of flow relay is a relay which operates on given values of liquid or gas pressure, level or flow, or on a given rate of change of these values.

64. Ground protective relay is a relay which functions on failure of the insulation of a machine, of a transformer, or of other apparatus to ground, or flashover of a DC machine to ground.

Note   This function is assigned only to a relay which detects the flow or current from the frame of a machine or enclosing case or structure of a piece of apparatus to ground, or detects a ground on a normally ungrounded winding or circuit. It is not applied to a device connected in the secondary circuit or secondary neutral of a current transformer, or current transformers, connected in the power circuit of a normally grounded system.

65. Governor is the equipment which controls the gate or valve opening of a prime mover.

66. Notching or jogging device is a device which functions to allow only a specified number of operations of a given device or equipment or a specified number of successive operations within a given time of each other. It is also a device which functions to energize a circuit periodically, or which is used to permit intermittent acceleration or jogging of a machine at low speeds for mechanical positioning.

67. AC directional over-current relay is a relay which functions on a desired value of AC over-current flowing in a predetermined direction.

68. Blocking relay is a relay which initiates a pilot signal for blocking or tripping on external faults in a transmission line or in other apparatus under predetermined conditions, or cooperates with other devices to block tripping or to block re-closing on an out-of-step condition or on power swings.

69. Permissive control device is generally a two-position, manually operated switch which in one position permits the closing of a circuit breaker or the placing of equipment into operation, and in the other position prevents the circuit breaker or the equipment from being operated.

70. Electrically operated rheostat is a rheostat which is used to vary the resistance of a circuit in response to some means of electrical control.

71. Reserved for future application.

72. DC Circuit Breaker is a circuit breaker which is used to close and interrupt a DC power circuit under normal conditions or to interrupt this circuit under fault or emergency conditions.

73. Load-resistor contactor is a contactor that is used to shunt or insert a step of load-limiting, shifting, or indicating resistance in a power circuit, or to switch a space heater in circuit, or to switch a light or regenerative load resistor of a power rectifier or other machine in and out of circuit.

74. Alarm relay is a relay other than an annunciator, as covered under device No: 30, which is used to operate, or to operate in connection with, a visual or audible alarm.

75. Position changing mechanism is the mechanism which is used for moving a removable circuit breaker unit to and from the connected, disconnected, and test positions.

76. DC over-current relay is a relay which functions when the current in a DC circuit exceeds a given value.

77. Pulse transmitter is used to generate and transmit pulses over a tele-metering or pilot-wire circuit to the remote indicating or receiving device.

78. Phase-angle measuring or out-of step protective relay is a relay which functions at a predetermined phase angle between two voltages or between two currents or between voltage and current.

79. AC reclosing relay is a relay which controls the automatic reclosing and locking out of an AC circuit interrupter.

80. Reserved for future application.

81. Frequency relay is a relay which functions on a predetermined value of frequency—either under- or over- or on normal system frequency-or rate of change of frequency.

82. DC reclosing relay is a relay which controls the automatic closing and the reclosing of a DC circuit interrupter, generally in response to load circuit conditions.

83. Automatic selective control or transfer relay is a relay which operates to select automatically between certain sources or conditions in equipment, or performs a transfer operation automatically.

84. Operating mechanism is the complete electrical mechanism or servomechanism including the operating motor, solenoids, position switches, etc. for a tap changer, induction regulator, or any piece of apparatus which has no device function number.

85. Carrier or pilot-wire receiver relay is a relay which is operated or restrained by a signal used in connection with carrier-current or DC pilot wire fault directional relaying.

86. Locking out relay is an electrically operated hand, or electrically, reset relay or device which functions to shut down and hold an equipment out of service on the occurrence of abnormal conditions.

87. Differential protective relay is a protective relay which functions on a percentage or phase angle or other quantitative difference of two currents or of some other electrical quantities.

88. Auxiliary motor or motor generator is one used for operating auxiliary equipment such as pumps, blowers, exciters, rotating magnetic amplifiers, etc.

89. Line switch is a switch used as a disconnecting or isolating switch in an AC or DC power circuit, when this device is electrically operated or has electrical accessories, such as an auxiliary switch, magnetic lock, etc.

90. Regulating device is a device which function to regulate a quantity, or quantities, such as voltage, current, power, speed, frequency, temperature, and load, at a certain value or between certain limits, for machines, tie lines, or other apparatus.

91. Voltage directional relay is a relay which operates when the voltage across an open circuit breaker or contactor exceeds a given value in a given direction.

92. Voltage and power directional relay is a relay which permits or causes the connection of two circuits when the voltage difference between them exceeds a given value in a predetermined direction and causes these two circuits to be disconnected from each other when the power flowing between them exceeds a given value in the opposite direction.

93. Field changing contactor is a contactor which functions to increase or decrease in one step the value of field excitation on a machine.

94. Tripping or trip-free relay is a relay which functions to trip a circuit breaker, contactor, or equipment, or to permit immediate tripping by other devices; or to prevent immediate reclosure of a circuit interrupter, in case it should open automatically, even though its closing circuit is maintained closed.

95. Trip circuit supervision relay is used for the supervision of breaker trip coil supervision purpose.

96. Tripping relay for the special purpose in the protection system.

97. Not assigned and kept for future purpose

98. Not assigned and kept for future purpose

99. Over-flux and over-excitation relay for generator or transformer purpose.

SYMBOLS USED

Symbols	Meanings
	Over-current
V	Over-voltage
ů / V	Under-voltage
Z	Distance relay
X	Balanced or differential current
	Over-frequency
	Under-frequency
T	Over-temperature
	Balanced phase
X	Pilot wire
PW	Directional over-current
Z	Directional distance
CC	Carrier pilot
I	Over-current ground with instant element
B X	Bus current differential
B X	Bus ground differential

SYMBOLS AND DESIGNATIONS BASED ON THE INTERNATIONAL ELECTROTECHNICAL COMMISSION STANDARD(IEC 617-SERIES AND IEEE C37.2–1991)

General Blocks

Symbols	Meanings
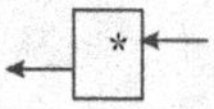	Protection relay
	Protection relay with enable input
	Protection relay with blocking input
	Coil with flag indication (OFF)
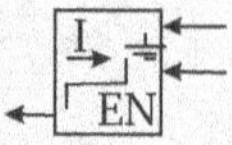	Direction E/F current relay with one input (current) and after (voltage), and one for block
	Coil with flag indication (ON)
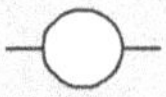	Coil with no flag
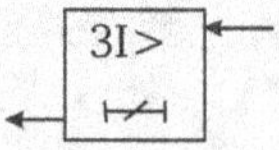	3 phase O/C relay with settable delay
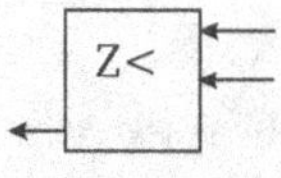	Under-impedance relay
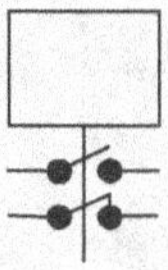	Relay with mechanical contacts (auto reset)

Parameters, Symbols and Functions

Symbols	Meanings
$I \leftarrow$	Reverse current
I_d	Differential current
I_{nf}	Current of nth harmonic
I_1, I_p	Positive sequence current
I_2, I_n	–ve sequence current
I_0, I_p	Zero sequence current
I_{rsd}	Residual current
⊢—⊣ I	Drop out delay
	Inverse time-lag characteristics
	Step
LO	Lock-out
TCS	Trip circuit supervision
O → I	Transition from off to on position (e.g., auto reclose)
I → O	Tripping
I_d	Current to frame
I_{N-N}	Current between neutral and two poly phase system
I_N	Current on neutral
I_{ub}	Unbalance current
	Thermal effect by current
U	Voltage
P	Active power

(Contd.)

Symbols	Meanings
P	Power at phase angle
Q	Reactive power
	Temperature
⚡	Fault, flashover
⊢—⊣	Delay
(coil / circle symbol)	PT, VT
(isolator symbol)	Isolator
(breaker symbol)	Breaker
(CVT symbol)	CVT
X/Y	Translation of signal
>	Operation above a set value
<	Operation below a set value
$\begin{smallmatrix} > \\ < \end{smallmatrix}$	Operation outside set limits (e.g., voltage regulation)
(make contact symbol)	Make contact (self-reset) (normally open)
(break contact symbol)	Break contact (self-reset) (normally closed)
(change over contact symbol)	Change over contact (break before make)

(Contd.)

Symbols	Meanings
	Time delay (drop-off)
	Time delay (pick-up)
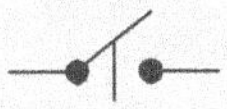	Normally open contact (hand reset)
	Normally close contact (hand reset)
	Positional contact (bold marked position is considered)
	Push button (normally open)
	Push button (normally closed)
	Fuse
	CT
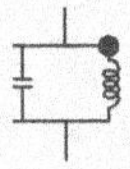	Wave trap

Note For study of drawing, the index, legends, etc. are to be referred always.

CODE OF PRACTICE BY M/s ABB LTD.
Terminal Blocks

External circuit		Internal circuit	
TB	**Purpose**	**TB**	**Purpose**
X_{01}	CT circuit	X_{11}	CT circuit
X_{02}	PT circuit	X_{12}	PT circuit
X_{03}	DC supply	X_{13}	DC supply
X_{04}	Control circuit	X_{14}	Control circuit
X_{05}	AC supply	X_{15}	AC supply
X_{07}	Annunciation, alarm circuit	X_{17}	Annunciation, alarm circuit

Equipment/Device Terminal Identification For M/s ABB Ltd.

Modular identification   It is designated by coordinate system of U and C/TE

where,

U = Height of module and

C/TE = Horizontal distance between mounting holes (width of module)

Note   Modular's upper left corner U and C/TE coordinate is taken.

Modular terminal identification

Example   101: 26   101, 137, 325—Modular number

137: 321   26, 321, 222—Terminal number

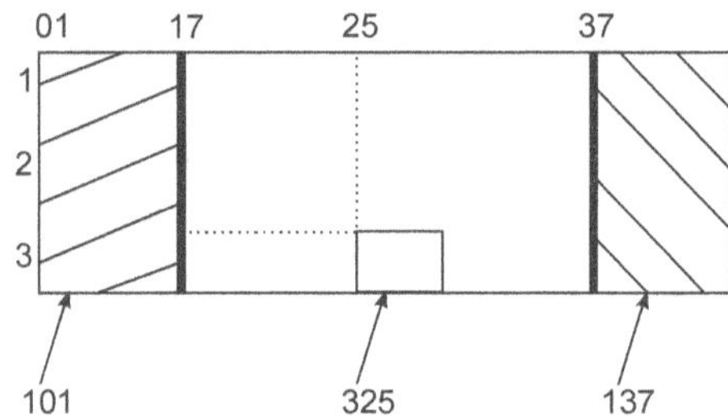

Note   For details refer combiflex identification in ABB manual.

DATA ON PROTECTION SCHEME

The protection scheme can be classified into:

- Apparatus protection (generator, motor, transformer, breaker, etc.)
- Bus protection
- Line protection

GENERATOR PROTECTION

Generator protections are broadly classified into three types.

CLASS A   This type of protection covers all the electrical actuated protections for the faults within the generating unit. The initiative relays send tripping signal to the master relay 86G, which in action trips the generator field breaker, generator breaker and turbine.

CLASS B   This type of protection covers all the mechanical actuated protections that are generated basically in the turbine. The initiative relays send tripping signal to the master relay 86T, which in turn trips the turbine breaker and in consequence generator will trip on reverse power/low forward power protections.

Class A Protection

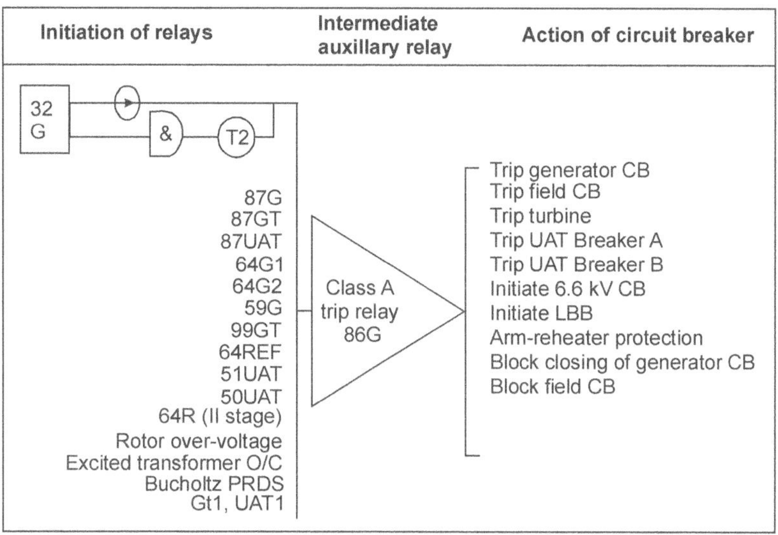

Class B Protection

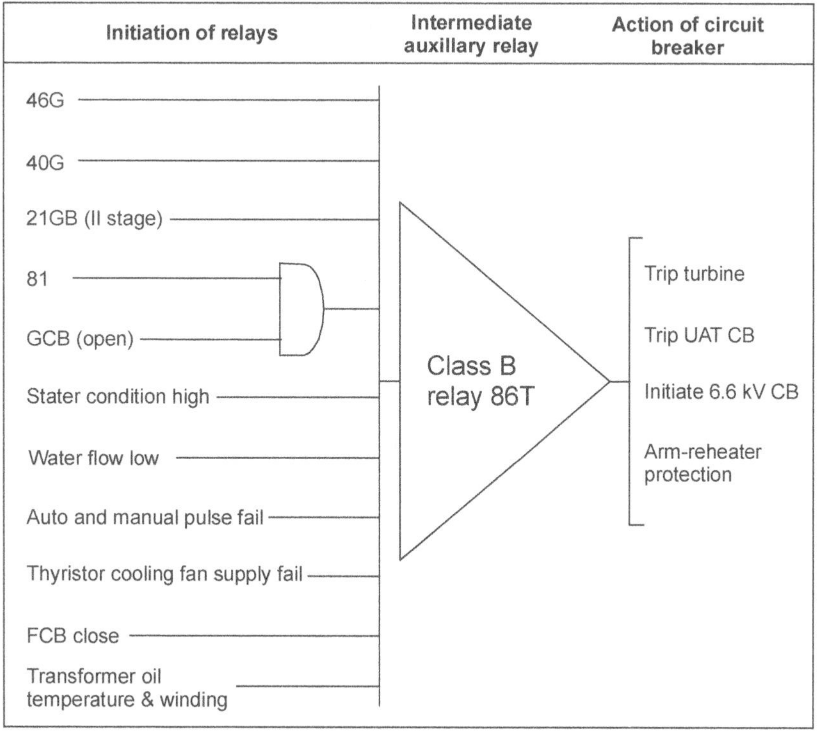

CLASS C   This type of protection covers the electrical actuated protections in the system for which generator breaker trips and generator gets un-loaded. The unit comes to house load and UAT is brought to service again. The initiative relays send tripping signal to the master relay 8GB, which in turn trips the generator and causes isolation from the system.

Class C Protection

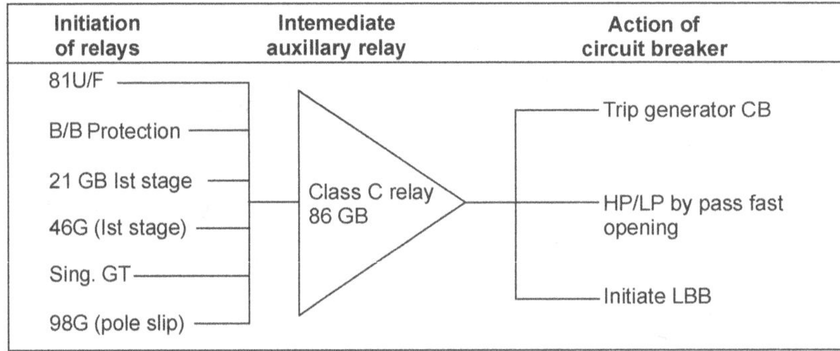

Proposed Protection Equipments for Different types of Generators

Generator size protection	I 0–4 (MVA)	II 4–15 (MVA)	III 15–50 (MVA)	IV 50–200 (MVA)	V Large turbo alternator
Rotor overload					X
Rotor E/F	X	X	X	X	X
Inter-turn fault		X 4	X 4	X 4	X
Differential generator		X	X	X	X
Differential block transformer		X	X	X	X
Under-frequency		X 3	X 3	X 3	4
Over-voltage	X	X	X	X	X
Stator E/F	X	X	X	X	X
Loss of excitation		X	X	X	X
Pole-slip (out of step)				X	X
Reverse power	X 1	X 5	X 5	X 5	X
Under-impedence		X 2		X	X
Unbalanced I^2 current)		X 7	X 7	X	X
Over-current (definite time)	X 6	X 6			
Stator overload					X

Generator size protection	I 0–4 (MVA)	II 4–15 (MVA)	III 15–50 (MVA)	IV 50–200 (MVA)	V Large turbo alternator
Over-current/Under-voltage	X	6	X	6	
Dead machine				X	X	X
Breaker failure				X	X	X

1—Only necessary for steam and diesel drives, 2—Only necessary for thyristor excitation from generator terminals, 3—Only necessary for pump operation, 4—Only necessary for when several bars of the same phase in the same slot, 5—Not necessary for Pelton turbines, 6—O/C should not be used with self-supported static excitation system, 7—When unbalance load is expected, X—Required

Protective Schemes for Various Generators

Functions	Steam and Gas Turbines			Hydro Turbines		
	Small (<10 MVA)	Medium (10–100 MVA)	Large (>100 MVA)	Small (<10 MVA)	Medium (10–100 MVA)	Large (>100 MVA)
Differential	Y	Y	Y	Y	Y	Y
95% Stator E/F	Y	Y	Y	Y	Y	Y
100% Stator E/F	N	Y/N	Y	N	Y/N	Y
Interturn faults	Y	Y	Y	Y	Y	Y
Backup impedance	N	Y	Y	N	Y	Y
Voltage controlled O/C	Y	N	N	Y	N	N
Negative sequence	Y	Y	Y	Y	Y	Y
Field failure	Y	Y	Y	Y	Y	Y
Reverse power	Y	Y	Y	Y	Y	Y
Pole slipping	N	N	Y	N	N	Y
Overload	N	N	N	Y	Y	Y
Over-voltage	Y	Y	Y	Y	Y	Y
Under-frequency	Y	Y	Y	Y	Y	Y
Dead machine	N	N	Y	N	N	Y
Rotor earth fault	Y	Y	Y	Y	Y	Y
Overfluxing	N	Y	Y	N	Y	Y

Main Protection (Distance Protection)

Standard 3 zone protection   The distance coverage of tripping zones for distance protection relay depends upon the selection practice and sensitivity of the protection scheme. There is no particular choice for the selection of the same by any utility. However maximum of the organization adopt the following methods as the normal standard for the case of distance protection relay to choose the distance coverage of protection.

Thermal Plant Unit Protections

Boiler protection	Turbine protection	Generator protection
Loss of unit critical DC power	Protection system power failure	Generator differential protection
Less than fire ball.	Low condenser vacuum (2 channels)	Overall differential protection
Loss of AC at any elevation in service.		
Very low drum water level for more than 5 sec.	Thrust bearing oil pressure high (2 channels)	UAT's differential protection
Very high drum water level for more than 10 sec.	High shaft vibrations	Over fluxing protection
Failure of both FD fans	Electrical over-speed	Loss of excitation protection
Failure of both ID fans	Main steam temperature very low	Negative phase sequence protection
Unit air flow as low as 30%	High exhaust steam temperature	Backup impedence protection
Very low furnace pressure	Over-frequency protection	Over-voltage protection
High furnace pressure	Digital electrohydraulic control power failure	Under-frequency protection
Loss of fuel (3 out of 4 nozzle valves open)	Failure of CWPS	Overload protection
Unit flame failure trip	All valves mainly main stop valve, re-heat stop valves, interrupt control valves closure	Pole stop protection
Loss of re-heater protection	Under-frequency protection	Generator under-voltage associated with loss of excitation protection
Loss of 220 V DC	Reverse power protection.	95% back charging stator E/F protection
	Mechanical over-speed protection.	95% stator E/F protection
	Low auto stop oil pressure.	100% stator E/F protection
	Reheat protection.	Excitation field protection
	ATRS emergency turbine trip	Dead machine relaying under independent GCB closure
		Protection of VT fuse
		Protection for GT restricted E/F protection

MOTOR PROTECTION

1. O/L protection
2. SC protections
3. E/F protections
4. Start protections
5. Prolonged starts
6. Stalling protections ──┬── During starting
                          └── During running
7. Loss of load
8. Under-voltage protection

Circuit for Differential Protection (Transformer)

Transformer winding connection	Secondary CT differential connection		Transformer winding connection	Secondary CT differential connection		Remark
Dy1	(Y1d	Yd11)	Y d1	D1y	Dy11	1. D1 connection corresponds to (I_R-I_B)
Dy11	(Y11d	Yd1)	Y d11	D11y	Dy1	2. D11 connection corresponds to (I_R-I_y)

Primary side P_2 terminal of CT conventionally towards transformer

Buchholtz Relay
Data

1. Works on principal of buoyancy of liquid
2. Position on pipe bent tank and conservator
3. Angle of inclination of pipe with horizontal—5° to 10°
4. Length of straight run pipe section
   i.   Transformer to relay = 5D (min.)
   ii. Relay to conservator = 3D (min.) (D = Diameter of pipe)
5. Vertical position of relay from tank = 8 cm (minimum)
6. Gas volume for alarm operation

Trans. size (MVA)	Pipe diameter (cm)	Setting range (cc)	Normal setting (cc)
Up to 1	2.5	100–120	110
1 to 10	5	185–250	220
10 MVA and above	7.5	220–280	250

TRANSFORMER PROTECTION

Requirement of relays

Capacity of transformer	Backup			Gas operated relay			Temperature protection		Diff. relay	REF. relay	Neutral displacement relay	Neutral O/C relay	Over-flux
	3 O/C	2 O/C	E/F	BUCH	OSR	PRV	OTI	WTI					
Up to 5 MVA	–	R	R	R	O	–	R	R	O	O	–	O	–
5 to 12.5 (MVA)	R	–	R	R	R	O	R	R	R	O	O	R	O
Above 12.5 (MVA)	R Dir.	–	R Dir.	R	R	R	R	R	R	R	O	O	R
Auto trans.	R Dir. Inst.	–	R Dir. Inst.	R	R	R	R	R	R	R	–	O	R

R—Required, O—Optional (—) Not required, Dir.—Directional, Inst.—Instant

7. Operating time 0.2 sec.
8. Diagnosis of troubles from colour of gas collected

Colour	Identification	Colour	Identification
Colourless	Air	Grey	Gas of over-heated oil due to melt of iron
White/milky	Gas of decomposed paper and cloth insulation	Black	Gas of decomposed oil due to electrical arc.
Yellow	Gas of decomposed wood insulation		

PRV (Pressure Release Valve)

Data

1. Required/universal for nitrogen-cushioned transformer and optional for conservator-type transformer.
2. Alarm—0.32 kg/cm^2/sec., Trip—0.6 kg/cm^2/sec.
3. Works on the principle of activation bellow due to pressure.

OTI (Oil Temperature Indicator)

Data

1. Works on the principle of volumetric change in bellow and corresponding conversion on scale due to temperature.
2. Temperature rise above ambient condition.

Type of winding	Top oil temperature rise on F.L. condition	Winding temperature rise on F.L. condition
OA, OW	50°C	55°C
OA/FA	50°C	55°C
OA/FA/FOA FOA FOW	45°C	50°C

WTI (Winding Temperature Indicator)

Data

1. Bellow (Bourdon tube) and shunt network, being connected to WCT (winding CT), simulates the temperature gradient, (proportionate with load current) and provides reading on the scale.

Temperature vs Current Signal Characteristics for RTD (Remote Thermal Device)
BS 1904 & DIN IEC 751

Temperature in °C	Resistance in (ohm)	Output signal 0–5 (mA)			Output signal 0–20 (mA)			Output signal 4–20 (mA)		
		L_o	Nom.	H_i	L_o	Nom.	H_i	L_o	Nom.	H_i
0	100.00	–	0	0.050	–	0	0.200	3.800	4.000	4.200
10	103.90	0.283	0.333	0.383	1.333	1.333	1.533	4.867	5.067	5.267
20	107.79	0.617	0.667	0.717	2.467	2.667	2.867	5.933	6.133	6.333
30	111.67	0.950	1.000	1.050	3.800	4.000	4.200	7.000	7.200	7.400
40	115.54	1.283	1.333	1.383	5.333	5.333	5.533	8.067	8.267	8.467
50	119.40	1.617	1.667	1.717	6.467	6.667	6.867	9.133	9.333	9.533
60	123.24	1.950	2.000	2.050	7.800	8.000	8.200	10.200	10.400	10.600
70	127.07	2.283	2.333	2.383	9.333	9.333	9.533	11.267	11.467	11.667
80	130.89	2.617	2.667	2.717	10.467	10.667	10.867	12.333	12.533	12.333
90	134.70	2.950	3.000	3.050	11.800	12.000	12.200	13.400	13.600	13.800
100	138.50	3.283	3.333	3.383	13.333	13.333	13.533	14.467	14.667	14.867
110	142.29	3.617	3.667	3.717	14.467	14.667	14.867	15.573	15.733	15.933
120	146.06	3.950	4.000	4.050	15.800	16.000	16.200	16.600	16.800	17.000
130	149.82	4.283	4.333	4.383	17.333	17.333	17.533	17.667	17.867	18.067
140	153.58	4.617	4.667	4.717	18.467	18.667	18.867	18.733	18.933	19.133
150	157.31	4.950	5.000	5.050	19.800	20.000	20.200	19.800	20.000	20.200

2. Maximum heater coil current $= 2$ A
3. Standard bellow heater $= (2.5 \pm 0.1)$ at 30°C

BREAKER PROTECTION

1. Pole discrepancy relay (PDR)
2. Local breaker backup relay (LBB)

Time Chart of PDR

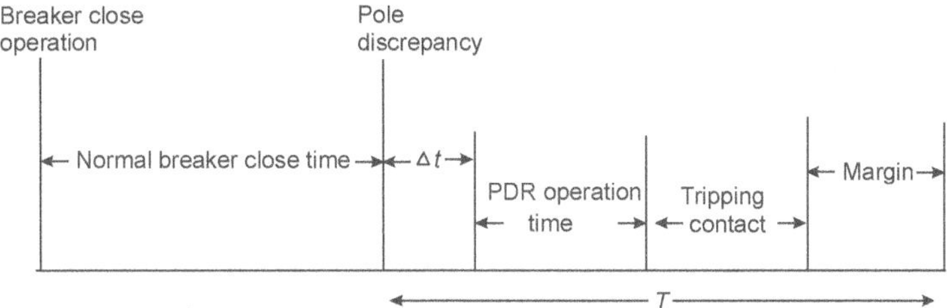

where,

$t =$ Mismatch allowable closing time between poles.

$T =$ Breaker trip time on PD (pole discrepancy) and it should be less than the timing of current unbalance relay and zone 2 time of DP relay.

Time Chart of LBB Relay

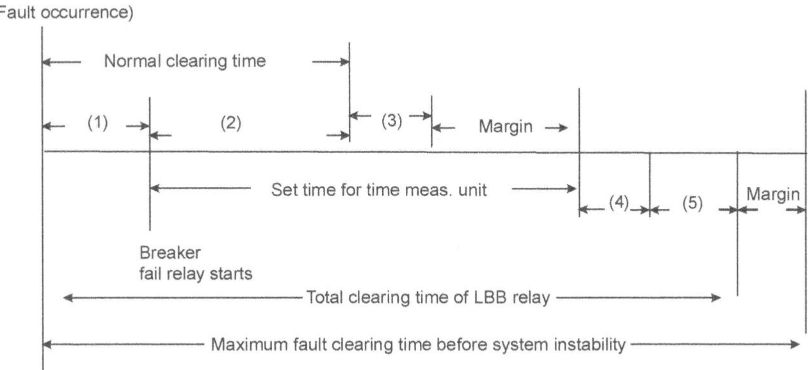

1—Protection relay operation time (e.g., DP relay)
2—Breaker interruption time
3—Resetting time current measuring unit
4—Trip relay time
5—Backup breaker interruption time

BUS PROTECTION

1. High impedance circulating current protection
2. Biased differential circulating current protection

LINE PROTECTION

1. Main protection (distance protection)
2. B/U protection (O/C, E/F protection)

Main Protection (Distance Protection)

Standard 3 zone protection

Method 1

Z_1 = Zone 1 (One) protection coverage = 80% of the protected line distance.

Z_2 = Zone 2 (Two) protection coverage

= (100% of protected line + 50% of shortest adjacent line)

Z_3 = Zone 3 (Three) protection coverage

= (100% of protected line + 100% of longest adjacent line)

Note

1. Maximum zone 2 reach should be within the minimum effective zone 1 reach of adjacent line.
2. Maximum effective zone 3 reach should cover the second section of line.

Method 2

When the current data at different stations are made available, the following method for the selection of distance coverage in distance protection relay can be chosen.

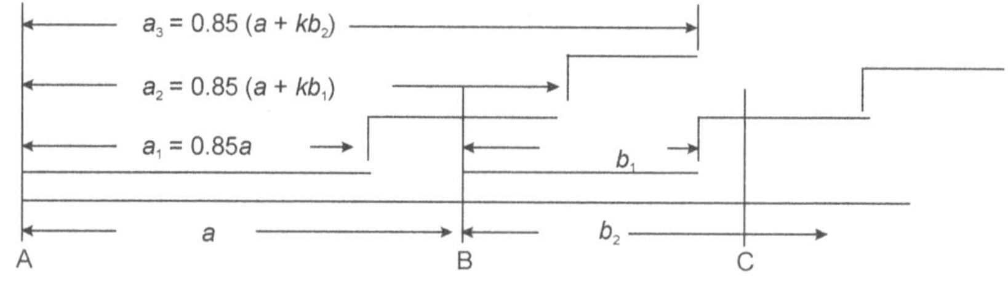

Zone 1 distance of station A = a_1 = 0.85 a

Zone 2 distance of station A = a_2 = 0.85 $(a + k\,b_1)$

Zone 3 distance = a_3 = 0.85 $(a + k\,b_2)$

a = distance of protected line.

b_1 = Zone 1 distance of station B

b_2 = Zone 2 distance of station B

k = Current ratio factor = $(I_A + I_B) / I_A$,

I_A = Service current on station A

I_B = Service current on station B

Time standard

Accepted fault clearance timing of EHV line (sec)		Selection of DP relay time	
System	**Time**	**Zone**	**Time**
400 kV	80 m	Zone 1	Instant relay operation time (30 to 40 ms)
220 kV	100 m	Zone 2	0.2 to 0.5 s
132 kV	150 m	Zone 3	0.4 to 1 s
		Reverse zone	1 to 2 s

Minimum de-energization time (for three-phase disconnection on transmission line)

Minimum de-energization time in sec.	Line (kV)	Minimum de-nergization time in sec.
0.1	220	0.28
0.15	275	0.3
0.17	400	0.5

Fault, voltage and current involvement in transmission line

Phase–earth fault			Phase–earth fault		
Voltage	**Current**	**Fault**	**Voltage**	**Current**	**Fault**
V_a	$I_a + 3KI_o$	$a - g$	V_{ab}	$I_a - I_b$	$a - b$
V_b	$I_b + 3KI_o$	$b - g$	V_{bc}	$I_b - I_c$	$b - c$
V_c	$I_c + 3KI_o$	$c - g$	V_{ca}	$I_c - I_a$	$c - a$

where,

K = Zero sequence compensation factor

= $(Z_0 - Z_1) / 3Z_1$

Z_0 = Zero sequence impedance

Z_1 = +ve sequence impedance

I_0 = Zero sequence current

Arc-resistance formula (A.R. van C. Warring Ton)

Method 1

$R_a = 28707 \, (L + ut)/I^{1.4}$

where,

L = Length of arc (length of insulator string) in m.

u = Wind velocity in km/hr.

I = Actual fault current in A.

Method 2

$$R_a = \frac{(440 \quad \text{arc length in feet})}{\text{fault current}}$$

Data on arc on the transmission line

Arc resistance of line			Secondary voltage due to arc for 220 kV line, arc distance 1.12 m and PTR 2000	
Line voltage (kV)	Length (m)	R_a (arc resistance) ( )	Fault current (kA)	Secondary voltage (mV)
33	1.22	2.5	40	221
132	3.66	8.6	20	292
220	6.7	12.8	10	386
400	8.3	18	5	509
			2	735
			1	970

Types of DP relay

   i. Impedance

  ii. Reactance

 iii. Mho or admittance

 iv. Ohm

  v. Offset mho

 vi. Modified impedance

vii. Elliptical

viii. Quadrilateral

 ix. Parallelogram

DP relay setting calculation

Data—assumed line parameter data

Line voltage	R ohm/ph/km	X ohm/ph/km	Yc μ mho/ph/km
132 kV panther	0.151	0.401	1.43
220 kV zebra	0.078	0.402	1.44
400 kV twin moose	0.031	0.327	1.73

Formulae for parameter calculation

1. R_1 = $1/a$ ohm/ph/km

   R_1 = +ve sequence resistance (usually given in conductor data)

2. x_1 = $2\,f \times 2 \times 10^{-4} \ln\,(D_m/D_s)$ ohm/ph/km

   x_1 = +ve sequence reactance

   D_m = Mutual geometric mean distance

   = $(D_{mab} \times D_{mbc} \times D_{mca})^{1/3}$

   D_{mab} = Mutual GMD between, 'A' and, 'B' conductor of the line

   D_{mbc} = Conductor between 'B' and 'C'

   D_{mca} = Conductors between 'C' and 'A'

   D_s = Self-GMD

   Note   For double circuit, loop reactance = $2X_1$

3. R_0 = Zero sequence resistance = $(R_1 + 0.00477f\,/1.609)$ ohm/ph/km

4. X_o = Zero sequence reactance ohm/ph/km

   = $(0.\,01397\,f/1.609) \times \log_{10} D_e/(GMR_{conductor} \times GMD^2_{sep})^{1/3}$ ohm/ph/km

   D_e = Equivalent depth return in m   658 $\sqrt{(\ /f)}$

   = Earth resistivity in ohm m

   $GMR_{conductor}$ = Self radius = $0.7788\,r$

   GMD_{sep} = $(D_{ab} \times D_{bc} \times D_{ca})^{1/3}$

   D_{ab} = Distance between A and B conductor,

   D_{bc} = Between B and C, D_{ca} = Between C and A

5. R_{om} = Zero sequence mutual resistance

$$R_{om}\;\frac{0.00477\,f}{1.609}\,ohm\,/\,ph\,/\,km$$

6. X_{om} = Zero sequence mutual reactance

$$= (0.01397\, f/1.609) \times \log_{10} (D_e/\text{GMD}_{eq}) \text{ ohm /ph/km}$$

$(\text{GMD})_{eq}$ = Ninth root of products of all nine possible distance between two circuits.

Minimum Load Impedance (Z_L)

$$Z_L = (U^2/MVA)$$

where,

U = Line to line voltage and

MVA = Maximum permitted load

$$(U_{min} / \sqrt{3}\ I_{max})$$

Conversion to secondary value

$$Z\, secondary \quad Z\, primary \quad \frac{CTR}{PTR}$$

$$Base\, impedance\, value \quad PU \quad \frac{KV^2}{100}$$

Zero sequence compensation factor (K_N)

i. $K_N = (Z_0 - Z_1)/3Z_1$, (when mutual zero sequence impedance is not considered.)
$= (X_0 - X_1)/3X_1$,

ii. K_0 angle $= \tan^{-1} (X_0 - X_1)/(R_0 - R_1) - \tan^{-1} (X_1)/R_1$

iii. $K_{NP} = (X_0 + X_{om} - X_1)/3X_1$, (when parallel system in normal operation)

iv. $K_{NG} = (X_0^2 - X_{om}^2 - X_oX_1)/3X_oX_1$ (for parallel system is out of service and grounded at both ends).

Possible length of line and optimum power to be transmitted

Possible optimum power transmission		Possible length (km)	
Line voltage (kV)	Line loading (kW-km)	Length minimum	Line Maximum
11	24×10^3	–	–
33	200×10^3	–	–
66	600×10^3	40	120
110	11×10^6	50	140
132	20×10^6	50	160
166	35×10^6	80	180
230	90×10^6	100	300

Suitability of relay performance

Method 1

Minimum voltage at relay

$$\text{SIR (System impedance ratio)} \quad \frac{\text{Source impedance}}{\text{Relay setting impedance}}$$

CIR (characteristics impedance ratio) = Maximum value of system impedance ratio.

$$\frac{E \quad V}{V}$$

$$E \quad = \text{Secondary voltage of PT and}$$

$$V \quad = \text{Minimum voltage at relay}$$

$$Z_s \quad = (KV^2)/(MVA)_{\text{Fault/source}} = \text{Source impedance}$$

$$I_f \quad = \frac{KV}{\sqrt{3}(Z_s \quad Z_L)} \text{ for 3 fault current}$$

$$Z_L \quad = \text{Line impedance}$$

$$V_{\text{relay}} = E/(1 + Z_s/Z_L)$$

Method 2

i. Ph–Ph fault

$$V_{\text{relay}} \quad (\sqrt{3}\, Z_L \quad I_F)\,/\,\text{PTR or VT ratio}$$

ii. Ph–earth fault

$$V_{\text{relay}} = (I_F \times Z_{\text{re}})/\text{VT ratio}$$

$$Z_{\text{re}} = \text{Earth loop impedance} = Z_{L1}\,(1 + (K-1)/3)$$

where,

$$K = Z_{Lo}/Z_{L1},$$

$$Z_{Lo} = \text{Zero sequence impedance}$$

$$Z_{L1} = +\text{ve sequence impedance}$$

Selection of power swing block

1. (Block Z_1 = off), (Block $Z_2\, Z_3$ = on)
2. During PS ($Z/\ t$)   Slow

O/C and E/F Protection

1. Non-directional B/U protection
2. Directional B/U protection

Type

(a) Definite time relay

(b) Inverse definite minimum time lag relay (IDMTL)

    i. Normal inverse (NI)

    ii. Very inverse (VI)

    iii. Extremely inverse (EI)

    iv. Long time inverse (LTI)

Operating characteristics (IDMTL relay) (IEC 255-4 BS 142, 3.2)

$$t = \frac{K \; T_m}{(I / I_s)^C - 1}$$

where,

t = Operating time in seconds

I = Fault current in A

I_s = Start current = 1.1 I_b in A

T_m = Time multiplier

Value of K and C

Type	*K*	*C*
NI	0.14	0.02
VI	13.5	1.0
EI	80	2.0
LTI	120	1.0

Errors of IDMTL relay (as per BS 142 limits)

Operating current (multiple of plug setting)	Any PS and 1.0 time setting		100% plug setting and 1.0 time setting	
	(%)	(sec.)	(%)	(sec.)
2	± 24.15	± 2.42	± 16.65	± 1.67
5	± 15.98	± 0.69	± 8.48	± 0.36
10	± 15.08	± 0.45	± 7.58	± 0.23
20	± 15.00	± 0.33	± 7.5	± 0.17

O/C + E/F relay high set setting

Maximum 3 short-circuit current

i. (At beginning of line) I_{max} $U / \sqrt{3}\, Z_s$,

ii. (At end) I_{max} $U / \sqrt{3}\,(Z_s\ Z_L)$

iii. (Source impedance) $= Z_s = KV^2/MVA$

Minimum setting of relay

$I_{min} = 1.3\, I_{max}$

Safety factor $= 1.2$

$I_{relay} = 1.2\, I_{min}$

O/C Highest

Relay setting $= (I_{relay})/\text{CTR}$

E/F High set

$$\frac{O/C \text{ high set}}{3}$$

Directional B/U relay (O/C)

Relationship between MTA (Maximum Torque Angle) and relay connection of O/C relay

Relay connection angle	MTA	Connection					
		Aph		Bph		Cph	
		Current coil	Voltage coil	Current coil	Voltage coil	Current coil	Voltage coil
30°	0°	I_a	V_{ac}	I_b	V_{ba}	I_c	V_{cb}
60° Type 1	0°	I_{ab}	V_{ac}	I_{bc}	V_{ba}	I_{ca}	V_{cb}
60° Type 2	0°	I_a	$-V_c$	I_b	$-V_a$	I_c	$-V_b$
90°	30° Lead	I_a	V_{bc}	I_b	V_{ca}	I_c	V_{ab}
90°	45° Lead	I_a	V_{bc}	I_b	V_{ca}	I_c	V_{ab}

Directional B/U relay (E/F)

Relay characteristic angle	Connection	
	Current coil	Voltage coil
12.5° (lag)	Residual current	Residual voltage
14° (lag)	$I_0 = I_A + I_B + I_C$	= Open delta secondary voltage
45° (lag)		
60° (lag)		

Contacts, Terminals for other Relays of GEC ALSTOM Make

Standard	Contacts		Current terminal	Voltage terminal	Auxillary DC
	Alarm	Trip			
Two unit/case	23-25 9-11	15-17 1-3		13-14 27-28	220–250 V
Three unit/case	2-4	1-3	23-24, 25-26, 27-28 #		# 24-26-28 (shorted)
Single unit/case	2-4	1-3	27-28		
Single unit/case	6-7 8-10	1-2 15-16	27-28		13-14
Three unit/case	36-37 33-35	29-30 41-42	21-22 23-24, 25-26 #		13-14 # 22-24-26 (shorted)
Single unit/case	9-11	1-3 2-4	23-25-27		
				23-25-27	
			25-23-21	Short 37 & 42, 41 & 46, 38 & 45	
			27-28	25-23-21	
	7-9 18-20				13-14
	7-9	6-8			13-14
	4-16			27-28	
	4-6	1-3	27-28		13-14

Contacts, Terminals for B/U Relays of Different Manufacturers

Make	Type	Standard	Alarm	Trip		Current	Voltage	Auxillary
---	---	---	---	Normal	Instant	terminal		DC
ER	IDMT Non-dir	Single unit/case	1, 2	3,4	11,12	9,10	–	5,6
	TJM10, 2TJM10, TJM20,	Multi unit/case 3O/C	1, 2	3,,4	11,12	(5,6–R), (7, 8–Y), (9, 10–B)	–	
	TJM11, 2TJM10, TJM21	Multi unit/case 2O/C, E/F	1, 2	3, 4	11, 12	(5, 6–R), (7, 8–E/F), (9, 10–B)	–	
ER	IDMT dir, TJM12, 2TJM12,	Single unit/case O/C	1, 2	3, 4	11, 12	9, 10	5, 6	–
	TJM22	Single unit/case E/F	1, 2	3, 4	11, 12	9, 10	5, 6	–
ER	IDMT non-dir E/F type TJM60	Single unit/case E/F	1, 2	3, 4	–	9, 10	–	–
ER	IDMTL programme DCD,MIT	Multi unit/case	1, 2	3, 4	3, 4	(5, 6-R), (7, 8-Y), (9, 10-B), (11, 12-E/F)	–	
EE ALSTO M	IDMT non-dir CDG11, 12, 13, 14	Single unit/case	1, 2	3, 4	3, 6	9, 10	–	3, 8

Make	Type	Standard	Alarm	Trip Normal	Instant	Current terminal	Voltage	Auxillary DC
EE ALSTOM	IDMT non-dir CDG31, 32,33,34 CDG61, 62	Multi unit/case 3O/C	1,2	3,4		(5,6-R), (7,8-Y), (9,10-B)	–	3,12 OR20
		Multi unit/case 2O/C, E/F	1,2	3,4		(5,6-R), (7,8-E/F), (9,10-B)	–	3,12 OR20
EE ALSTOM	IDMT dir CDD 21, 23, 24,26 CDD31,33, 34, 36	Single unit/case	1,2	3,4		9,10	7,8	3,6 OR 20
EE ALSTOM	DEF. time delay CTU32	Multi element/ case	1,2	3,4	3,17	(5,6-R), (7,8-E/F), (9,10-B)	–	19,20 30V DC
EE ALSTOM	Instant CAG11, 12,13,14, 17, 19	Single unit/case	1,2	–	3,4	9,10	–	–
JVS	IDMT non-dir JRC053	Multi element/ case	11,12	13,14	–	(1,2-R), (3,4 -Y), (5,6-B), (7,8-E/F)	–	20,19
JYOTI	IDMT non-dir	Single unit/case	9,10	11,12	–	7,8	–	1,2
ABB	IDMT non-dir ICM21P	Single unit/case	5,6	7,8	–	1,2	–	3,4
UE	IDMT non-dir R-1156	Single unit/case	9,10	1,2	–	7,8	–	1,4

Contacts, Terminals for Differential Relays

Make	Type	Standard	Alarm	Trip	HT side (common)			LT side (Bias)			Operating point			Auxillary volt DC
					R	Y	B	r	y	b	R_0	Y_0	B_0	
ER	4C21	Single unit/case	1, 2	3, 4	10	10	10	9	9	9	7	7	7	-
	DUO-BIAS M	Multi unit/case	Module 2 (RL_1, RL_2) General Command		3	6	13	3	6	13	1+1 During test both HT & LT terminals to be shorted (e.g., 1 + 1)	4+4	11+11	220 V DC
ABB	RADSB	"	15, 16	17,18	6	7	8	3	4	5	12	13	14	1,2
EE ALSTOM	DTH31	"	1,2	3,4	14	18	10	11	15	7	12	16	8	5–110 V 6–220 V 19–30V 20 –ve
	DTH32	"	1,2	3,4	7	17	27	10	20	30	8 or 9 Short (8 +9)	18 or 19 Short (18 +19)	28 or 29 Short (28 +29)	11–30 V 13–110 V 15–220 V 12 –ve
EE ALSTOM	DDT	Single unit/case	1,2	3,4	10	10	10	5	5	5	7	7	7	3 +ve 8 –ve
	DDT	Multi unit/case	1,2	3,4	5	18	13	10	20	11	6	16	14	3 +ve 8 –ve
	MBCH 12 (2 winding)	Single unit/case	9, 11	1,3	25	25	25	23	23	23	27 Short (24 + 26 + 28)	27	27	13 +ve 14 –ve 220 V
	MBCH 13 (3 winding)	"	9,11	1,3	25	25	25	23 or 21	23 or 21	23 or 21	27	27	27	13 +ve 14 –ve 220 V

DATA ON ELECTRICAL SYSTEM EQUIPMENTS

Breakers

	Technical particulars			
Particulars	**33 kV**	**132 kV**	**220 kV**	**400 kV**
System voltage (kV)	36	145	245	420
System frequency (Hz)	50	50	50	50
Quenching medium	Vacuum, oil	Oil, air blast SF_6	Oil, air blast SF6 (puffer)	SF_6 (puffer) air blast
Operating medium	Spring	Spring hydraulic air pressure	Spring hydraulic air pressure	Air pressure
Insulation standard				
i. Lightning impulse voltage (kVp) $(1.2/50\ \mu)$	170	650	1050	1425
ii. Power frequency withstand voltage (kV) 1min./50 Hz)	70	275	460	520
iii. Minimum disruptive voltage (kV)	28	105	176	320
Normal current (A)	1250	1250/1600	2000	2000/3150
Short-time current withstand capacity (kA) (3 sec.)	25	40	40	40
Fault rating				
i. Making capacity (kA)	70	100	100	100
ii. Breaking capacity (kA)	25	40	40	40
iii. Breaking current out of ph (kA)	6.5	10	10	10
iv. Rated time charging current (A)	50	50	125	400
v. Over-voltage factor for switching	3.0	3.0	3.0	3.0
Operating sequence				
i. Normal		O-10s-CO-3min-CO		
ii. Auto reclose		O-0.3s-CO-3min-CO		

(Contd.)

Technical particulars				
Particulars	**33 kV**	**132 kV**	**220 kV**	**400 kV**
TRV (transient recovery voltage) first phase to clear factor	1.5	1.5	1.3	1.3
Breaker operating time				
i. Maximum break time (open) ms	60	50	50	40
ii. Maximum close time (ms)	100	150	150	120
iii. Maximum close-open time (ms)	–	80	80	60
iv. Maximum time open interval between 1st and last phase (ms)	5	3.3	3.3	3.3
v. Maximum time close interval between poles	5	2	1	1

Circuit Breaker

132 kV SF$_6$ Gas Circuit Breaker

Particulars	Rating/value	Particulars	Rating/value
Make		Rated lightning impulse withstand voltage	650 kVp
Type		Rated short-circuit breaking current	31.5 kA
Rated voltage	145 kV	Rated operating pressure	15.5 kg/cm^2
Rated frequency	50 Hz	First pole to clear factor	1.5
Rated normal current	3150 A	Rated duration of short-circuit current	31.5 kA 3 sec.
Rated closing voltage	220 V DC	Rated line charging breaking current rated SF6 gas pressure	50 A
Rated opening voltage	220 V DC	Rated voltage and frequency for auxillary circuit	415 V AC 50 Hz
Gas pressure SF6	6.0 bar at 20°C	Rated operating sequence	0–0.3S-CO-3 min-CO
Total weight with gas	2000 kg	Gas weight	9 kg
Sl. No.		STD. IEC	56
Month/year of manufacturing			

220 kV SF$_6$ Gas Circuit Breaker

Particulars	Rating/value	Particulars	Rating/value
Make		Rated lightning impulse withstand voltage	1050 kV$_p$
Type		Rated short-circuit breaking current	40 kA
Rated voltage	245 kV	Rated air pressure	21.5 kg/cm^2
Rated frequency	50 Hz	First pole to clear factor	1.3
Rated normal current	2000 A	Rated duration of short-circuit current	40 kA 3 sec.
Rated closing voltage	220 V DC	Rated line charging breaking current rated SF$_6$ pressure	125 A
Rated opening voltage	220 V DC	Rated voltage and frequency for auxillary circuit	1-ph (230 V & 3 ph 415 VAC 50 Hz
Gas pressure SF$_6$	7.0 bar at 20° C	Rated operating sequence	0–0.3S-CO-3 Min-CO
Total weight with gas	3800 kg	Sl. No.	
Month/year of manufacturing			

33 kV Vacuum Circuit Breaker

Particulars	Rating/value	Particulars	Rating/value
Make		Impulse withstand voltage	170 kVp
Type		Short-time current	25 kA
Rated voltage	36 kV	Rated air pressure	
Rated frequency	50 Hz	Making capacity	62.5 kAp
Rated normal current	1250 A	System breaking capacity with Dur.	25 kA for 3 sec.
Shunt trip coil	220 V DC	p.f withstand voltage	70 kV
Spring release coil	220 V DC	Specification	IS 2156/ IEC 56
Rated operating sequence	0.3 min. - CO	Month/year of manufacturing	
Sl. No.		Total weight	1000 kg

CURRENT TRANSFORMER

Specifications

Ratio of the CT				Number of cores		Rated burden and factors			Class of accuracy	
Items	≤ 33 kV Indoor	≥ 33 kV Outdoor	CT rating	No. of cores	Core	Burden (VA)	Factors		Core	Accuracy class
Ratio	Single ratio	Multi ratio	≤ 33 kV	3 cores (metering, protection, differential)	Metering	2.5, 5, 7.5,10, 15, 30	ISF (5, 10, 20) ALF (5,10,15,20, 30) Voltage across CT = (burden × ALF)/rated current		Metering	0.1, 0.2, 0.5, 1, 3, 5
Primary current	Suitably (10, 15, 20, 30, 50, 75) multiple or fraction			5 Cores (metering, protection, differential, bus protection, distance protection)	Protection	2.5, 5, 7.5, 10, 15, 30			Protection	(5P, 10P, 15P) *
			≥ 66 kV							
Secondary current	1 A or 5 A					Rated short-time current I_{st} = 150 I_p for 1 sec.			Selective protection	PS **

* Accuracy class is usually followed by ALF (5P10, 5P15, etc.)

** For accuracy minimum knee-point voltage (V_K) and permissible magnetizing current (I_{mag}) to be considered

Note

1. $V_K = K_{Is} (R_{CT} + R_b)$,
    where,

       K = Parameter depends upon system fault level and characteristics of the relay

       I_s = Secondary reflected current,

       R_{CT} = CT secondary resistance at 75°C

       R_b = Resistance of secondary circuit with lead

2. $I_{mag} = P_{mA}$ at V_K/F_M, where, I_{mag} = Maximum allowable magnetic current (mA)

       P_{mA} = Permissible magnetizing current (mA)

       F_m = Factor to be chosen 2 or 4, depending upon the application.

Voltage Class and Insulation Level

Nominal system voltage kV (RMS)	Highest system voltage kV (RMS)	Power frequency withstand voltage kV (RMS)	Lightning impulse withstand voltage kV Peak	
			List 1	List 2
Up to 0.6	0.66	3	–	–
3.3	3.6	10	20	40
6.6	7.2	20	40	60
11	12	28	60	75
33	36	70	145	170
66	72.5	140	325	325
110	123	185	450	
		230	550	
132	145	230	550	
		275	650	
220	245	360	850	
		395	950	
		460	1050	
400	420	950*	1175	
		1050*	1300	
		1050*	1425	
525	524	1050*	1425	
		1175*	1550	

* Switching impulse withstand voltage in kV (peak)

Errors in CT

Metering core

Accuracy class	± % Current ratio error at % of rated current					± Phase angle displacement error in minutes at % of rated current				
	1%	5%	20%	100%	120%	1%	5%	20%	100%	120%
0.1	–	0.4	0.2	0.1	0.1	–	15	8	5	5
0.2	–	0.75	0.35	0.2	0.2	–	30	15	10	10
0.5	–	1.5	0.75	0.5	0.5	–	90	45	30	30
1.0	–	3.0		1.0	1.0					
0.2s	0.75	0.35	0.2	0.2	0.2	30	15	10	10	10
0.5s	1.5	0.75	0.5	0.5	0.5	90	45	30	30	30

Note 1 Test to be done at 25% to 100% of rated burden.

Note 2 PF of 0.8 lag, for all cores above 5 VA and unity PF for 1 VA to 5 VA to be taken during test.

Protection core

Accuracy class	Current error at rated primary current (%)	Phase displacement at rated primary current (minutes)	Composite error at rated accuracy primary current (%)
5P	± 1	± 60	5
10P	± 3	–	10
15P	± 5	–	15

Relay Details for Selection of Instrument Transformers

Transformers differentials

Alstom make

1. Relay type: DTH 31/32
   $V_k > 40 \times I \, (R_{CT} + 2R_L)$;
   Example: $V_k > 40(1)(3+4) > 280 \text{ V}$
2. Relay type: MBCH 12/13
   $V_k > 24 \, I_n \, (R_{CT} + 2R_L)$
   where
   V_k = Knee-point voltage

I_n = Relay rated current

R_L = Total lead resistance

I_e = <3% I_n at $V_k/4$ for both above types of relays, i.e., 0.03 I

i.e., 30 mA at $V_k/4$

3. Relay type: KBCH, MiCOM P630 (numerical)

Application	Knee-point voltage V_k	Through-fault stability X/R	I_f
Transformers, generators, generator transformers, motors, shunt reactors, series reactors also.	24 I_n [R_{CT} +2R_I]	40	15 I_n
Overall generator-transformer units	48 I_n [R_{CT} + 2R_I]	120	15 I_n
Transformers connected to a mesh corner, having two sets of CTs each supplying separate relay inputs.	48 I_n [R_{CT} + 2R_I]	40 120	40 I_n 15 I_n

ABB make

1. Relay type: RADSB (Static)(Medium impedance)

   V_k >30(R_{CT} + 2R_L +R_{re}) I_n, >30(4 + 4 + 3)1, >330 V

   Note: Over-current factor of 30 recommended

   Excitation current: Not applicable*

2. Relay type: SPAD346C (Stabilized differential relay)

   V_k > 4 × I_{max} × (R_{in} + R_L)/n

   where,

      n = Transformation ratio of CT >(R_{CT} + R_L + 0.5/sq. of I_{sn})

      R_{in} = Secondary resistance of CT

      2R_L = Control cable ('To & fro') resistance

      I_{max} = I_d/I_n>>set on relay (range available 5 to 30, default set is 10)

3. Relay type: RET316 (Stabilized differential relay)

   $n' = n$ ($Pr + Pe$)/($Pb + Pe$),

   where,

   n = ALF

   n' = Effective over-current factor, is a function of fault current

   I_k = Frequency and time constant of network, and read from graph in RET manual

Pb = Connected burden at rated current and Pe = CT losses of secondary windings

Pr = Rated CT burden, DC time constant assumed is 300 msec.

* Relay provided with "magnetizing inrush restraint" based on second harmonic content of the inrush current and hence 'I_{mag}' calculation is not applicable.

Easun rerolle

1. Relay type: 4C21 (Static)(Low impedance)

   CT Class : PS, $V_k > 2I_f = (R_{CT} + R_L + R_{ict}$ (P)) + ($I_{CT} V_k \times I_{CT}$ ratio)

   Example: $V_k > 2 \times 10.9375 \ (2 + 3 + 1) + (14.43/0.875) \times 0.577 > 140.75$ volts

   where,

   R_{CT} = Main CT resistance

   R_{ict} (P) = ICT primary winding resistance

   R_L = Lead resistance

   I_f = Maximum through-fault current

2. Relay type: Duobias M(Numeric), (Differential and restricted earth fault)

   $V_k > 4 \times I(A + C)$,

   where,

   I = Either max 3-phase through-fault current referred to secondary (as limited by transformer impedance)or high-set setting, whichever is greater

   A = Secondary winding resistance of each star-connected CT

   C = CT secondary loop resistance for internal faults

       CT Class recommended-PS, X to BS 3938, TPS to IEC-44

Generator differential protection

ALSTOM

1. Relay type: CAG34 (High impedance scheme)

   $V_k > 2I_f (R_{CT} + 2R_L)$

   Example $V_k > 2 \times 10(3 + 4) > 140$ V

   where,

   I_f = secondary equivalent of fault current

   $I_e = I_s - I_r = (0.15 - 0.10)/2 = 25$ mA at $V_K/2$

2. Relay type: LGPG, MiCOM 340 (Numerical)

   For voltage dependent, over-current, field failure and negative phase, sequence protection

   i. $V_k > 20I_n (R_{CT} + 2R_L)$

   ii.   For stator earth fault protection

       $V_k > I_s (R_{CT} + 2R_L + R_R)$

   iii. For generator differential protection

Low impedance differential $V_k > 50 I_n (R_{CT} + 2R_L)$

High impedance differential $V_k > 2 V_s$

where,

$V_s = 1.5 I_f (R_{CT} + 2R_L), R_s = V_s / I_s$

3. Relay type: YTGM15, YCG15AA, ZTO11(Generator backup)

   $V_k > 2 I_f (R_{CT} + 2R_L + M + CM),$

where,

CM = Connected burden.

ABB

1. Relay type: RADHA /RADHD (High impedance)

   $V_k > 2 I_k (R_{CT} + R_L), > 2 \times 25(4 + 3), > 350$ V,

   R_L in case of generator is longer, i.e., $2R_L = 6$

   I_k will be higher considering "X_d"(0.2 pu) and CT sec. of 5A

   Excitation current: Not applicable*

Excitation current is kept low for increasing the primary sensitivity.

* Relay provided with "Magnetizing inrush restraint" based on second harmonic content of the inrush current and hence 'I_{mag}' calculation is not applicable.

Easun reyrolle

1. Relay type: 4B3 (EM)/DAD 3 (Static)/Argus-1

   (Numeric)(High impedance scheme)

   CT class :PS

   $V_k > 2 I_f (R_{CT} + 2R_L)$, Example: $V_k > 2 \times 10(3 + 4) > 140$ volts

where,

I_f = Maximum through-fault current

R_{CT} = Main CT resistance and

R_L = Lead resistance between CT to relay.

2. Relay type: GAMMA (Numeric) (High impedance)

   For two off 3 phase inputs (line end and neutral end) and for neutral-earthed CTs.

   In case of low impedance bias differential functions

   i. $V_k > 50 \times I_n (R_{CT} + 2R_L + R_R)$

       where, maximum through-fault current $= 10 \times I_n$ with maximum $X/R = 120$.

   ii. $V_k > 30 \times I_n (R_{CT} + 2R_L + R_R)$

   where,

   maximum through-fault current $= 10 \times I_n$ with maximum $X/R = 60$

   I_n = Rated current secondary $X/R = X/R$ ratio for maximum through-fault condition.

   R_{CT} = Secondary resistance of CT,

R_L = Lead resistance between CT and relay

R_R = Resistance of any other protection functions sharing the CT

Bus differential protection

ABB

1. Relay type: RADHA/RADHD (High impedance scheme)
   $V_k > 2I_k (R_{CT} + R_L)$, $> 2 \times 40 (4 + 4)$, > 640 V
2. Relay type: RADSS (Medium impedance scheme)
   Depending on different ratios, for 1A CT, V_k shall be 500 V.
   Excitation current: Not applicable*

   * Relay provided with "Magnetizing inrush restraint" based on second harmonic content of the inrush current and hence 'I_{mag}' calculation is not applicable.

Alstom

1. Relay type: CAG34 (High impedance scheme)
   $V_k > 2I_f (R_{CT} + 2R_L)$, Example: $V_K > 2 \times 10(3 + 4)$, > 140 V
2. Relay type: DIFB –DIFBCL
   $V_k > K \times I_n (RTCP + R_F + R_d/n_2)$,
   where,

K	=	$(1.2/40) \times (I_{CC}/I_N)$
I_N	=	Main CT primary rated current,
I_{CC}	=	Maximum short-circuit current delivered to bus bar via the input where MCT is installed.
$RTCP$	=	Restistance of secondary of MCT,
R_f	=	Restistance of link loop between MCT and auxiliary CT
n	=	Ratio of auxiliary CT and R_d/n_2 = Value of differential resistance transposed to ACT primary.

3. Relay type: MCTI 34 (numerical)
   $V_k > 1.6 V_S$, $V_S = 1.25 \times I_f (R_{CT} + 2R_L)$
   where,

   R_{CT} = CT resistance,

   R_L = Maximum lead resistance from CT to common point and

   I_f = Maximum internal secondary fault current.

Easun rerolle

1. Relay type: B3 (EM)/DAD3 (static)
   CT Class: PS, $V_k > 2I_f (R_{CT} + R_L)$
   Example $V_k > 2 \times 10(3 + 4) > 140$ V

where,

I_f = Maximum through-fault current

R_{CT}= Main CT resistance

R_L = Lead resistance between CT and relay

Distance protection

Easun rerolle

1. Relay type: THR (Static)

   CT Class : PS, $V_k > I \times (R_L + R_2 + X/Rx(R_3 + R_2))$

   Example $V_k > 10(3.8 + 7 + 4(1.2 + 7)) > 436$ V

   where,

   R_L = Burden of relay (3.8   max.)

   R_2 = Resistance of leads plus resistance of CT secondary

   X/R = Ratio of reactance to resistance of the system for fault at the end of zone 1

   R_3 = Constant depending on impedance setting of zone 1 (1.2   max.) and

   I = Secondary fault current for fault at end of zone 1

   Note $X/R = 4$ for 132 kV system in above.

   $X/R = 7$ for 220 KV

   $X/R = 11$ for 400 KV

2. Relay type: Ohmega (Numeric)

   V_k should be equal or greater than the higher of following two expressions.

   a) $V_k > K \times (I_p/N(1+X_p/R_p)) \times (0.03 + R_{CT} + R_L)$ for phase–phase faults

   b) $V_k > K \times (I_e/N(1+X_e/R_e)) \times (0.06 + R_{CT} + R_L)$ for phase–earth faults

where

    I_p = Phase fault current calculated for X_p/R_p ratio at the end of zone 1

    I_e = Earth fault current calculated for X_e/R_e ratio at the end of zone 1

    N   = CT ratio

    X_p/R_p= Power system reactance to resistance ratio for the total plant including the feeder line parameters calculated for phase fault at the end of zone 1

    X_e/R_e = Similar ratio to above but calculated for an earth fault at the end of zone 1

    R_{CT} = CT resistance

    R_L = Lead burden CT to relay and

    K   = Factor chosen to ensure adequate operating speed which is >1.0.

Alstom

1. Relay type: Micromho, Quadrmaho

   $V_k > I_f (X/R)(M + R_{CT} + nR_L)$

Example $V_k > 10(4)(10.2+3+4) > 40(17.2) > 688$ V

$I_e < 3\%\ I_n$ at $V_k/2 < 30$ mA at $V_k/2$

where, M = Relay resistance (phase fault).

2. Relay type: MiCOM 430/441/442, EPAC, LFZR, LFZP, PD521, PD932 (numerical)

$V_k > I_f \times (1+X/R) \times (R_{CT} + R_L + R_B)$

where,

X/R = The primary system ratio

R_B = Relay burden and

R_L = Rest of cable connecting CT to relay (lead and return for ground faults, lead only for phase faults).

ABB

1. Relay type: RAZOA/REL511 (Static)(Numerical)

Secondary limiting voltage $> (I_k \times I_{sn}/I_{pn}) \times a \times (R_{CT} + R_L + 0.5/(I_{sn}/I_{pn})2)$

where,

a = Factor for the DC time constant (approximately 10 for about 100 m sec.)

Excitation current $< 0.2\ I_{sn} < 0.2$ A < 200 mA

Feeder differential protection

Easun reyrolle

1. Relay type: Solkor-M and Microphase-FM (Numerical) (Current differential)

$V_k > K \times X/R \times I_f/N \times (R_{CT} + 2R_L + R_b)$

where,

K = Stability factor = 0.8 for micro phase-FM

X/R = X/R ratio for the maximum through-fault conditions.

(The value of this transient factor depends upon the sum of the source and transmission circuit impedances.)

R_b = Burden of relay. The AC burden of the relay per phase is 0.05 VA at 1A,

tap = 0.05   and 0.30 VA at 5A, tap = 0.012

The values of magnetizing currents of CTs at two ends should not differ by more than $I_n/20$ for output voltages up to $50/I_n$ volts.

Alstom

1. Relay type: MBCI

Translay 'S' Differential (for feeder and transformer)

   i. For plain feeders

$V_k > 0.5 \times N \times K_1 \times I_n(R_{CT} + XR_L)$

where,

V_k = KPV of CTs for through-fault stability

R_L = Restistance of CT secondary circuit

X = 1 for core wire connections between main CT and the relay and

X = 2 for six wires connection

N = Relative neutral turns on summation transformer winding

K_1 = The selected time-dependent constant.

ii. For all application at or above 220 kV where X/R ratio are large

$V_k > N \times K_1 \times I_n(R_{CT} + XR_L)$

Magnetizing current$< 0.05 \times I_n$ at $10/I_n$ V

iii. For transformer feeder differential

   i. $V_k > 50 \times I_n(2.2/I_n 2 + R_{CT} + R_L)$ for star-connected CTs

   ii. $V_k > 50 \times I_n/\sqrt{3}(9.7/I_n 2 + R_{CT} + R_L)$ for delta-connected CTs

2. Relay type: MiCOM P540 (Numerical)

$V_k > K \times I_n (R_{CT} + 2R_L)$

where,

K is a constant depending on I_F = The maximum value of through-fault current for stability and is determined as follows:

i. For relays set at I_{s_1} 20%, I_{s_2} $2I_n$, $k_1 = 30\%$, $k_2 = 150\%$:

$K = 40 + (0.07 \times (I_f \times X/R))$ and

$K = 65$. This is valid for $(I_f \times X/R) < 1000 I_n$

For higher $(I_f \times X/R)$ up to $1600 I_n$:

$K = 107$

ii. For relays set at I_{s_1} 20%, I_{s_2} $2I_n$, $k_1 = 30\%$, $k_2 = 100\%$:

$K = 40 + (0.35 \times (I_F \times X/R))$ and

$K = 65$. This is valid for $(I_F \times X/R) < 600 I_n$

For higher $(I_f \times X/R)$ up to $1600 I_n$:

$K = 256$

Over-current and earth fault relay

Alstom

1. Relay type: CDG11 (IDMT)

   This relay has 3.5 VA burden. So total VA burden requirement is 10 or 15 VA.

   ALF factor of 10 is sufficient.

   If backup protection scheme is envisaged, ALF of 15 is required.

   The time current setting characteristic of IDMT relay becomes straight line after 15 times setting; therefore, time discrimination is ineffective after 15 times current setting.

2. Relay type: MiCOM P120, P140 (numerical)
    Class: 5P10, Burden: 5 VA

ABB

1. Relay type: SPAJ 140 (Numerical)
    This relay requires generally CT with 5P10/5P20 CT with very low burden, e.g., 0.1 VA

Easun reyrolle

1. Relay type: ARGUS/MIT (Numerical)
    Class: 5P10, Burden: 5 VA
2. Relay type: Solkor-R/RF (Pilot wire differential protection)
    CT class: PS $V_k = 50/I_n + (I_f/N)(R_{CT} + R_L)$
    I_n = Rated current
    I_f = Maximum primary steady state through-fault current
    N = CT ratio, R_{CT} = CT resistance, R_L = Lead resistance.

Name Plate Details

132 kV current transformer

Particulars	Rating/value
Make	
Ref. standard	IS 2705-1992
Rated primary current	600-300-150 A
Insulation level (kV)	275 RMS/ 650 Peak
Frequency	50 Hz
Minimum creep age	3625 mm
Sl. No	
Type	
Normal system voltage	132 kV
Highest system voltage	145 kV
S.Ty. current kA/sec	18.2/3
Wt. of oil/CT kg	120/550
Drg. No	
Suitable for hotline washing	

Caution

1. Secondary terminals must be shorted before burden is disconnected
2. Pf testing terminal to be earthed during operation

Core	Ratio/ ampere	Secondary connection	Rated VA	Accuracy class	V_k (V) (min.)	$I_{exc}@ V_k$ (mA) (max.)	R_{CT} at 75°C (max.)
1	600/1	1s1-1s3	–	PS	1200	10	5.0
	300/1	1s1-1s2	–	PS	600	20	2.5
	150/1	1s1-1s2	–	PS	600	20	2.5
2	600/1	2s1-2s3 (S1-S3)	15	0.5 Fs<5	–	–	–
	300/1	2s1-2s2 (S1-S2)	15	0.5 Fs<5	–	–	–
	150/1	2s1-2s2 (S1-S2)	15	0.5 Fs<5	–	–	–
3	600/1	3s1-3s3	–	PS	1200	10	5.0
	300/1	3s1-3s2	–	PS	600	20	2.5
	150/1	3s1-3s2	–	PS	600	20	2.5

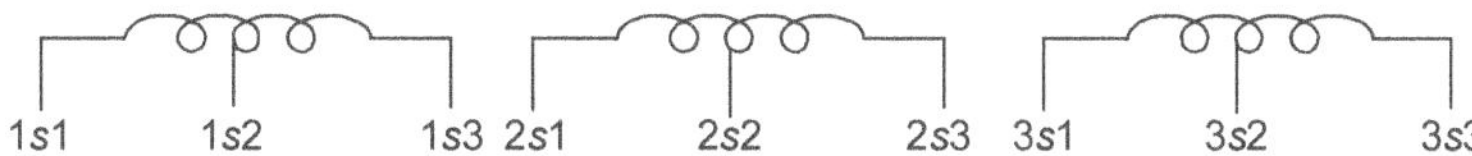

220 kV current transformer

Particulars	Rating/value
Make	
Ref. standard	IS 2705-1992
BIL	1050/460 kV
Frequency	50 Hz
Making capacity	100 kAp
Sl. No/year	
Type	
HSV/NSV	245/ 220 kV
S.T.current	40/1 kA/sec.
Wt. of Oil	350 kg
Total weight	1200 kg

Ratio	400-200-100/1-1-1-1-1				
Primary/secondary current (A)	400/1	200/1	100/1		
Primary connection	P_1C_1–P_2C_2	C1–C2	C1–C2		
Secondary connection					
Core 1	1S1–1S3	$1S_1$–$1S_3$	$1S_1$–$1S_2$		
Core 2	2S1–2S3	$2S_1$–$2S_3$	$2S_1$–$2S_2$		
Core 3	3S1–3S3, S′	3S1–S3, S′	$3S_1$–$3S_2$, S′		
Core 4	$4S_1$–$4S_3$	$4S_1$–$4S_3$	$4S_1$–$4S_2$		
Core 5	$5S_1$–$5S_3$	$5S_1$–$5S_3$	$5S_1$–$5S_2$		
Core	1	2	3	4	5
Output*	–	–	40	–	–
Accuracy class*	PS	PS	0.5	PS	PS
ISF/ALF*	–	–	5	–	–
V_k (V) min*	1200	1200	–	1200	1200
$I_{exc}@V_k$ *(mA) max.*	25	25	–	25	25
R_{CT} at 75° C (max)*	5	5	–	5	5

* At 400/1 and 200/1 ratio only

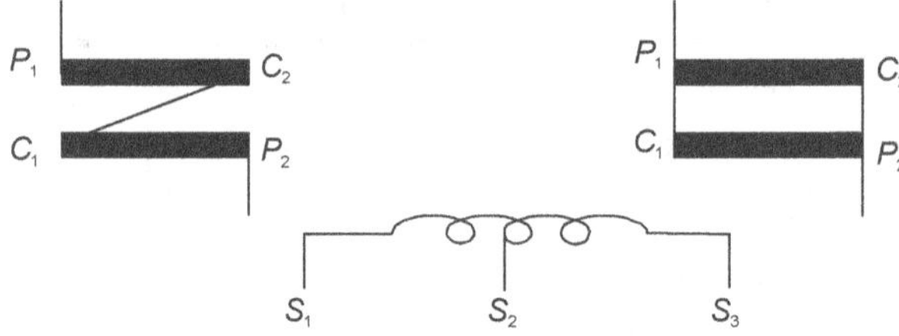

33 kV current transformer

Particulars	Rating/value	Particulars	Rating/value
Make		Type	
Reference standard	IS 2705-1992	Frequency	50 Hz
Ratio	400-200-100/1-1-1A	Highest System voltage	36 kV
Insulation level	70/170 kV	S.T. current kA/sec	25 kA/1
Sl. No			

Core	Rated VA	Accuracy class	SF	V_k (V) (min.)	$I_{exc}@V_k$ (mA) max	ISF	R_{CT} at 75° C (max.)
I	30	5P	10		–	–	–
II	30	0.5	–	–	–	< 5	–
III	–	PS		1000–1200	25/15		<3 /<6

Connection diagram

Ratio	Primary connection	Secondary connection
100/1	$C_1 + C_2$	$S_1 - S_2$
200/1	$P_1 + C_1$ and $P_2 + C_2$	$S_1 - S_2$
400/1	$P_1 + C_1$ and $P_2 + C_2$	$S_1 - S_3$

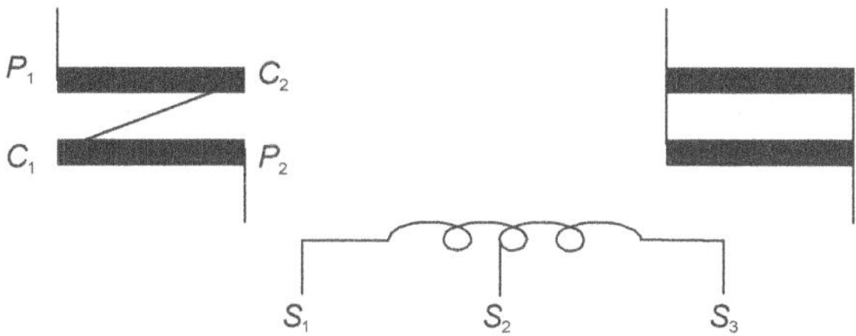

Fault Finding Study for Star-Connected CT Protection Circuitry

Current in the CT secondary	Expected Faults
R = Y = B = x A, N = 0 A	No fault in the circuits
R = Y = B = x A and N = 2x A	Any one of the phase 'CT' polarity reversed
If (R+Y) = (Y + B) $\sqrt{3x}$ and (B + R) = x	Y phase reversed
If (R+Y) = (B + R) $\sqrt{3x}$ and (Y + B) = x	R phase reversed
If (Y+B) = (B + R) $\sqrt{3x}$ and (R +Y) = x	B phase reversed
i. If R = 0 A & Y = B= N = x A	R phase primary side open
Then check for all other R phase CT secondary cores, if values obtained as If R = 0 A and Y = B = N = x A	R phase primary side open
ii. Similarly for Y phase and B phase also.	Corresponding phase primary side open
iii. If R = 0 A & Y = B = N = x A For only in one core, then	R phase secondary is shorted or R phase is mixed with other cores or for use of auxillary CT, any one of the side might be shorted.
iv. Similarly for Y phase and B phase also.	Corresponding phase
R = Y = B = x A N = 3x A	All phases have been connected to one CT only instead of different cores as 1st, 2nd, 3rd cores, etc. as R phase cores, Y phase cores and B phase cores or primary side has been connected from a single source
i. R = Y = x/2 A, B = x A, N= 0 A	R & Y phases of CT secondary similar polarities have been shorted.
ii. Y = B = x/2 A, R = x A, N = 0 A	Y & B phases of CT secondary similar polarities have been shorted.
iii. B = R = x/2 A, Y = x A, N = 0 A	B & R phases of CT secondary similar polarities have been shorted.
i. R = x A, R = B = 0 A, N = x A	Y & B phases of CT secondary have been shorted.
ii. Y= x A Y = B = 0 A N = x A	B & R phases of CT secondary have been shorted.
iii. B = x A R = Y = 0 A N = x A	R & Y phases of CT secondary have been shorted.
R = Y =B =N = 0 A	All the 3 CTs are shorted.
If the values are resulted other than the above readings as described.	1. CTR may be different 2. Wrong primary link connection 3. Phase angle problem 4. CT saturation problems

SPECIFICATIONS OF VOLTAGE TRANSFORMER

Ratio of PT			Rated burden	Class of accuracy	
Item	≤ 22 kV	≥ 33 kV	1. The rated burden at a p.f = 0.8 (lag) shall be chosen as (10, 25, 50, 100, 200, 400, 500) VA/phase for 3 phase transformer	Winding	Accuracy class
Unit	Multi phase in single unit	3 single phase PT in star connection separate		Metering	0.1, 0.2, 0.5, 1.0, 3
Winding connection	Y/Y, V/V	Y/Y, Y/Y – open delta	2. Two independent secondary windings are to be provided for metering and protection core		
Primary voltage	Rated voltage $/ \sqrt{3}$			Protection	3P, 6P
Secondary voltage	110 V$/\sqrt{3}$				

Errors in PT

Metering core		
Accuracy class	**% Voltage ratio error**	**Phase angle displacement error in minutes**
0.1	0.1	5
0.2	0.2	10
0.5	0.5	20
1.0	1.0	40
3.0	3.0	–

Note Errors at any voltage between 80 to 120% of rated voltage, with burdens between 25 to 100% of rated burden at p.f 0.8 (lag)

Protection core		
Accuracy class	**% Voltage ratio error**	**Phase angle displacement error in minutes**
3P	3	120
6P	6	240

Note 1 Errors at 5% rated voltage and voltage multiplied by voltage factor (1.2, 1.5 or with burdens between 25 to 100% of rated burden at p.f 0. 8 (lag).

Note 2 Errors at 2% rated voltage shall be twice as high as given in the table with similar burdens to Note 1.

Name Plate Details

132 kV Capacitor voltage transformer (1 ph)

Particulars	Rating/value	Particulars	Rating/value
Make		Sl. No.	
Type		Rated voltage	$132\,kV / \sqrt{3}\,kV$
Highest system voltage	145 kV	Rated insulation level	275/650 kV
Total weight	420 ± 10% kg	Rated frequency	50 Hz
Cap oil	25 ± 10% kg	Standard	IS 3156
EMU oil	85 ± 10% kg	HV (primary) capacitance	6511+ 10% –5% pF

Particulars	Rating/value	Particulars	Rating/value
Intermediate V (sec.) Capacitance	35418 + 10% –5% pF	Equivalent Capacitance (Cn) for PLCC	5575 + 10% –5% pF
Nominal intermediate voltage	13 kV	Total secondary burden/class	150 VA/0.5
Total thermal burden	300 VA	Month/year of manufacturing	
Voltage factor	1.2 Continuous/ 1.5 for 30 sec.	1 phase solidly earth connection	

Rated secondary voltage	Terminal marking	Rated burden VA	Accuracy class
$110/\sqrt{3}$ V	1a–1n	100	0.5
$110/\sqrt{3}$ V	2a–2n	100	3P

33 kV potential transformer (1 ph)

Particulars	Rating/value	Particulars	Rating/value
Make		Type	
Ref. Standard	IS 3156 (PT I, II, III)	Sl.no.	
Ratio	$33/\sqrt{3}$ kV $/110/\sqrt{3}$ V	Insulation level	70/170 kV
Voltage factor	1.2 continuous/ 1.5 for 30 sec.	Highest system voltage	36 kV

Core	Rated VA	Accuracy class	Terminal marking
W1	100	0.5	1a–1n
W2	100	0.5	2a–2n

220 kV capacitor voltage transformer (1 ph)

Particulars	Rating/value	Particulars	Rating/value
Make		Sl. no.	
Type		Rated voltage	$220/\sqrt{3}\,$kV
Highest system voltage	245 kV	Rated insulation level	245/460 /1050 kV
Total creepage	6125 min. nom.mm	Rated frequency	50 Hz
Weight of oil	140 kg	Standard	IEC : 60186/ IS: 3156
Total weight	750 kg	HV (pri.) capacitance	4840 pF
Intermediate V (sec.) capacitance	48400 pF	Equipment capacity (C_n) for PLCC	4400 + 10% –5% pF
Nominal intermediate voltage	$20/\sqrt{3}\,$kV	Temperature category	–5 to 55° C
Total thermal burden	750 VA	Class of insulation	A
1 ph solidly earth connection		Suitable for hot line washing	
Month/year of manufacturing		Voltage divider ratio	$220/\sqrt{3}\,$kV $/20/\sqrt{3}\,$kV
Voltage factor	1.2 continuous/ 1.5 for 30 sec.	G.A Drg. No.	

Rated secondary voltage	Terminal marking	Rated burden VA	Accuracy class
$110/\sqrt{3}\,$V	1a–1n	150	0.5
$110/\sqrt{3}\,$V	2a–2n	150	3P
$110/\sqrt{3}\,$V	3a–3n	50	3P

LIGHTNING ARRESTER

Maximum Switching Surge Level in pu *(pu* $\sqrt{2}\,V_{line(max)}/\sqrt{3}$*)*

Highest system voltage kV, (RMS)	Typical switching surge pu	Highest system voltage kV, (RMS)	Typical switching surge pu
12–36	< 4	525	2.25
123–145	<3	765	2.0
245	3	1500	1.5 (projected)
420	2.5		

Technical Particulars of Station Class Arresters From 11 To 33 kV (As Per IEEMA 20-2000)

Particulars	System voltages in kV							
	11	11	11	11	22	22	33	33
Rating kV (RMS)	9	9	9	9	18	18	30	30
MCOV (RMS)	7.2	7.2	9.6	9.6	15	15	25	25
Discharge current	10 kA							
Line discharge class	1	2	1	2	1	2	1	2
Rated frequency	50 Hz							
i. IR at MCOV	Less than 400 mA							
ii. IG at MCOV	About 1200 mA							
iii. Reference current, (mA)	1 to 5 mA							
iv. Reference volt at reference current	Greater than rated voltage							
Maximum RDA (kV_p) at								
i. 5 kA	27	26	36	33	51	52	90	86
ii. 10 kA	30	28	38	36	60	56	95	90
iii. 20 kA	34	30	42	40	68	60	105	100
Max. switch IMP R V (kV_p)								
500 A	24	22.4	30.4	28.8	48	44.8	76	72
1000 A	–	–	–	–	–	–	–	–

Particulars	System voltages in kV							
	11	11	11	11	22	22	33	33
Max. steep current impulse RDV (kVp)	36	34	42	40	60	56	105	100
High current impulse withstand				100 kAp				
TOV (kV_p)								
i. 0.1 sec.	15	16	21	21	32	32	53	53
ii. 1.0 sec.	15	15	20	20	30	30	51	51
iii. 10.0 sec.	14	14	19	19	29	29	49	49
iv. 100.0 sec.	13	13	18	18	28	28	47	47
Insulation withstand								
i. Lightning IMP	75	75	75	75	125	125	170	170
ii. Power frequency	28	28	28	28	50	50	70	70
iii. Switching IMP	–	–	–	–	–	–	–	–
Partial discharge				Less than 50 pC				
PR relief class				Class A				
PR relief class kA (RMS)				40 kA				
Total creepage distance in mm	270	270	300	300	600	600	900	900
Maximum cantilever strength in kgm	325	325	325	325	325	325	325	325

Technical Particulars of Station Class Arresters from 132 To 400 KV

Particulars	System voltages in kV							
	66	66	110	132	220	220	400	400
Rating kV (RMS)	60	60	96	120	198	216	360	390
MCOV (RMS)	51	51	81	102	168	175	292	303
Discharge current				10 kA				
Line discharge class	2	3	3	3	3	3	3	3/4
Frequency				50 Hz				
a) IR at MCOV	Less than 400 μA				Less than 500 μA			
b) IG at MCOV	About 1200 μA				About 1500 μ A			
a) Reference current, mA				1 to 5 mA				
b) Reference volt at reference current				Greater than rated voltage				
Max. RDA (kV_p) at								
i) 5 kA	170	160	251	320	518	567	820	860
ii) 10 kA	180	170	272	340	550	600	880	950
iii) 20 kA	200	119	307	380	610	668	925	1000
Maximum switch IMPRV (kV_p)								
500 A	–	–	–	–	–	–	–	–
1000 A	144	136	217	272	455	496	830	850

Particulars	System voltages in kV							
	66	66	110	132	220	220	400	400
Maximum steep current impulse RDV (kVp)	200	190	298	372	600	654	1000	1050
High current impulse withstand				100 KAp				
TOV (kV_p)								
i. 0.1 sec.	106	106	170	212	350	382	636	689
ii. 1.0 sec.	102	102	163	204	336	366	610	661
iii. 10.0 sec.	98	98	156	195	322	351	585	634
iv. 100.0 sec.	94	94	149	187	308	336	560	607
Insulation withstand								
a) Lightning IMP	325	325	550	650	1050	1050	1425	1425
b) Power frequency	140	140	230	275	460	460	630	630
c) Switching IMP	–	–	–	–	–	700	1050	1050
Partial discharge				Less than 50 pC				
PR relief class				Class A				
PR relief class kA (RMS)				40 kA				
Total creepage distance in mm	1800	1800	3075	3625	6125	6125	10500	10500
Max. cantilever strength in kgm	500	500	500	1000	1000	1000	1000	1000

CLASS III LIGHTNING ARRESTER TECHNICAL PARTICULARS Ref. Standard IEC 99-4, 1999

Particulars	Unit	36 kV	72.5 kV	145 kV	245 kV
System BIL	kVp	170	325	650	1050
Rated voltage (RMS), $\dfrac{(\text{System voltage} \times \sqrt{2})}{\sqrt{3}}$	kV	30	60	120	198
Max. continuous operating voltage (MCOV)	kV	25	52	102	168
Nominal discharge current (NDC)	kAp	10	10	10	10
High current withstand	kAp	100	100	100	100
Protection levels					
1) Impulse residual voltage	(kVp)				
i. Steep current @ NDC		94	185	380	600
ii. Lightning current					
@ 0.5 NDC		80	156	330	510
@ 1.0 NDC		84	165	350	550
@ 2.0 NDC		94	185	390	610
iii. Switching current	@ 1 kAp	70	136	290	450

Particulars	Unit	36 kV	72.5 kV	145 kV	245 kV
2) Temperature over-voltage	(kVp)				
i. For 0.1 sec.		60	120	254	404
ii. For 1 sec.		55	111	234	372
iii. For 10 sec.		52	105	222	353
iv. For 100 sec.		51	102	215	31
Insulation data					
i. Wet power frequency withstand	kV	70	140	275	460
ii. Dry lightning impulse	kVp	170	325	650	1050
iii. Creepage	mm	900	1800	3625	6125
Overall dimension					
i. Outer Diameter	mm	280	280	280	280
ii. Height	m	0.68	0.940	1.540	2.785
iii. Weight	kgs	45	60	150	250
Mounting arrangement					
i. PCD	mm	368	368	368	368
ii. Number of holes		4	4	4	4
iii. Diameter of holes	mm	15	15	15	19
Terminal connector					
Suitable conductor	ACSR	Single rabbit	Single Panther (66 & 132 kV)		Single zebra

Classification of Over-Voltage and Surge Impedance $\sqrt{L/C}$

Classification of over-voltage				Surge impedance ($\sqrt{L/C}$)	
Particulars	**Temporary over-voltage**	**Switching over-voltage**	**Lightning over-voltage**	**Objects**	**Surge impedance** ($\sqrt{L/C}$)
Magnitude	1 to 2 pu	1.5 to 5 pu	Hundreds of kV to several tens of MV	Tower	$Z_T = 100$ to $150\ \Omega$
Duration	mS to tens of sec.	Tens of μS to tens of mS	Few tens to hundred of μS.	OH ground wire	$Z_G = 400\ \Omega$
Effect	P.D. causes retardation of life of insulation	Partial discharge on insulation	Influences transformer insulation and break down of weaker section	OH phase conductor	$Z_T = 325$ to $400\ \Omega$
Testing evaluation	–	$250\ \pm\ 100\ \mu S$ $2500\ \pm\ 100\ \mu S$	$1.2\ \mu S\ \pm\ 30\%$ $50\ \mu S\ \pm\ 20\%$	Source surge impedance	$Z_T = 1500$ to $3000\ \Omega$

Lightning Arrestor Classification

Range of voltage	Impulse current kA 8 × 20 sec.	High current	Long duration current	
			Magnitude (A)	Duration (sec.)
Low voltage or secondary arrestor (175 to 660 V)	2.5	25	50	1000
Distribution class (3 kV to 18 kV)	5.0	50	75	1000
Intermediate class (3 kV to 110 kV)	5.0	50	75	1000
Station class (light duty) (11 kV to 198 kV)	10.0	65	150	2000
Station class (heavy duty) (198 kV and above)	10 15 20	65	300	3000

WAVE TRAP

Technical Particulars of 0.5 mH/1250 A Wave Trap for 220 kV Line

Particulars	Value
Type	Outdoor, air cored, air cooled
Continuous current rating at 50° C ambient	1250 A
Continuous current rating at 65° C ambient	1125 A
Max. symmetrical short-circuit current for 1 sec.	31.5 kA
Asymmetrical peak value of first half of rated short- time current	80.5 kA
Rated inductance	0.5 H
Blocking range	150–500 KHz
Min. resistive component in blocking frequency range	570
Radio interference voltage	< 500 μV
Mounting	Suspension

(Contd.)

Particulars	Value
Basic insulation level	32.37 kV$_p$
Standard nominal discharge current for 8/20 sec. wave impulse	10 kA
Rated voltage of arrestor	6 kV
Max. 1.2/50 sec. impulse spark over-voltage	21.6 KkVP
Min. value of power frequency spark over-voltage	9 kV (RMS)
Virtual steepness and max. front of wave impulse spark over-voltage	49.8 kV/ s 24.9 kV$_p$
Tuning	Broad band
Visual corona extinction voltage	156 kV
Max. residual discharge voltage for 8/20 sec. impulse discharge current	
1. 5000 A	21.6 kV$_p$
2. 10000 A	21.6 kV$_p$
No. of turns in main coil	28 (2 in parallel)

ISOLATORS

Technical Particulars for 245 kV Isolator

Particulars	Value
Type	Air break, off load
Standards used	1SS 9921/85
Highest system voltage	245 kV
Nominal system voltage	220 kV
Max. continuous current rating	2000 A
Rated short-time current for 3 sec.	40 kA
Max. magnetizing current	0.7 A at 0.15 pF
Rated peak short-time current	100 kAp
1.2/50 μsec. impulse withstand voltage	1200 kV
Radio interference voltage	1000 μV
Earthing switch rated current capacity	50% of main switch

(Contd.)

Particulars	Value
Minimum clearance in	
1. Between live parts and ground	2400 mm
2. Fixed contact and blade in open condition	1600 mm
Operating time	
1. Opening	10–12 sec.
2. Closing	10–12 sec.
Continuous rating of auxillary contact	10 A
Temperature rise	55 above ambient
Insulation level	530 kV

TRANSFORMER

Current Rating of Transformer (3~)

Thumb rule current = $(600 \times MVA)/kV$

Actual rule current = $(575 \times MVA)/kV$

where,

MVA = Rating of transformer and

kV   = Rated voltage of transformer.

Supply voltage, rated voltage (kV)	Actual rule current/MVA	Thumb rule current/MVA	Thumb rule current/MW (for cos Φ = 0.9)
11	52.3	54.55	60
33	17.43	18.2	20
66	8.72	9.1	10
132	4.36	4.55	5
220	2.615	2.73	3
400	1.44	1.5	1.66

Permissible Flux Density

Note Maximum flux density = 2 weber/m^2

Working flux density = 1.2 to 1.48 weber/m^2 at the knee of core saturation curve.

$B_{\mathrm{m}} = E_1/4.44\, fN_1 A$

where,

B_m = Maximum flux density (weber/m^2)
E_1 = EMF induced in volt
f = Frequency in Hz
N_1 = Number of turns
A = Cross sectional area (m^2)

Permissible Over-excitation Capacity

% Over-excitation (% U/f)	(U/f) / (U_N/f_N)	Time in seconds
110	100	
120	109	82
130	118	19
140	127	9.9
150	136	6.0

Permissible Exciting Current:
(No Load Current, Magnetizing Current)

3 kVA	Voltage class (Full insulation)						
	2.5 kV	15 kV	25 kV	69 kV	138 kV	161 kV	230 kV
500	3.7%	3.7%	3.8%	4.9%	–	–	–
1000	3.3%	3.3%	3.6%	4.3%	–	–	–
2500	–	3.1	3.2%	3.8%	–	–	–
5000	–	–	2.8%	3.1%	2.5%	4.1%	–
10000	–	–	3.0%	3.1%	2.4%	3.6%	4.0%*
25000	–	–	2.2%	2.4%	3.1%	3.9%	3.5%*
50000	–	–	–	–	3.1%	3.9%	2.2%*

* Reduced insulation

Notes for testing engineer

1. No load current should be maximum 3 to 4% of rated current.
2. The exciting current varies directly with voltage rating and inversely to KVA rating.

Permissible Inrush Current (Transient Current)

Inrush currents of transformer at no-load with cylindrical windings, at energization zero-point of supply voltage.

Rated power in kVA	Grain-oriented laminations		Non-grain oriented laminations		Decay to 1/2 the value in cycle
	Switching on		Switching on		
	Outer winding (A)	Inner winding (A)	Outer winding (A)	Inner winding (A)	
500	11.0	16	6.0	9.4	8 to 10
1000	8.4	14	4.8	7.0	8 to 10
5000	6.0	10	3.9	5.7	10 to 60
10000	5.0	10	3.2	3.2	10 to 60
50000	4.5	9	2.5	2.5	60–3600

1. In actual practice the value may vary 8 to 10 times the maximum value. It depends upon the cycle of energization of supply wave.
2. (Maximum value of inrush current)/(Peak value of rated current) = Ratio.

Short-Circuit Current

Permissible short-circuit currents and duration of 3 phase power transformer with two windings.

Rated KVA	U_b 36 kV			U_b> 36 kV		
	I_{SS}/I_N	t	UC%	I_{SS}/IN	t	UC%
Up to 630	25	2	4	–	–	–
630–3150	16.7	4	6	–	–	–
3150–10000	12.5	5	8	10	6	10
10000–40000	10.0	6	10	9.1	7	11
Above 40000	–	–	–	8	8	12.5

U_b = * System highest voltage (RMS) on HT side

I_{SS} = Max. permissible short-circuit current (RMS)

I_N = Rated current

T = Maximum permissible duration of short-circuit

UC% = Corresponding % S.C. voltage

*System highest voltage = 110% of rated voltage

Theoretical SC calculation

SC current $I_{sh} = (I_{FL} \times V_A \times 100)/(\%Z \times V_N)$

I_{FL} = Rated full load (FL) current, whose side S.C. current is to be calculated

= (Rated MVA $\times 10^6$)/{ $\sqrt{3}$ $\times$ rated voltage (volt)}

V_A = Applied voltage during the test in volt

$\%Z$ = % Impedance

V_N = Rated voltage of that side to which testing supply is given in volt.

Thermal Resistivity of Materials used in Transformer

Material	Thermal resistivity (°C/W/inch³)	Material	Thermal resistivity (°C/W/inch³)
Water	70	Aluminum	0.30
Air	1710	Mica	110
Transformer oil	245	Pressboard (untreated)	400–500
Copper (pure)	0.0896	Pressboard (oil-treated)	250–300
Copper (commercial)	0.1125	Varnished cambric (sheet form)	200–250
Wrought iron	0.50	Varnished cambric (tap ½ lap wraps)	250–300
Caste iron	0.984	Porcelain and cement	40
Steel (laminated with grain)	2.4	Steel (laminated across grain)	25.0

Temperature Rise Above Ambient

Type of cooling	Top oil final temperature rise on F.L. condition	Winding temperature (final) rise on F.L. condition
OA, OW	50°C	55°C
OA / FA	50°C	55°C
OA / FA / FOA	45°C	50°C
FOA	45°C	50°C
FOW		

where,

OA = Oil natual air natural
OW = Oil natual water cooled
FA = Forced air

FOA = Forced oil air natual

FOW = Forced oil water cooled

Standard Range of Impedance for Two-winding Power Transformer, Rated at 55°C Rise

High voltage winding insulation class	Low voltage winding insulation class	%	
		Min.	**Max.**
25	15	5.5	8
34.5	15	6	8
	25	6.5	9
46	25	6.5	9
	34.5	7	10
69	34.5	7	10
	46	8	11
92	34.5	7.5	10.5
	69	8.5	12.5
	46	10	15
196	92	11.5	17
	161	12.5	19
	46	11	16
230	92	12.5	18
	138	14	20
	161	14	20

Cooling Order Symbols

1st letter kind of medium	2nd letter kind of circulation	3rd letter kind of medium	4th letter kind of circulation
Indicating the cooling medium that is in contact with the winding		Indicating the cooling medium that is in contact with the external cooling system	

E.g. ONAN: Oil natural air natural (Oil insulation is provided internally and air medium externally).

Reduced Temperature Rises for Transformer Designed to Work at High Altitude

Type		Reduced by %	
(i)	Oil-immersed, natural air cooled	2%	Above 1000 m sea level and reduction is for each 500 m (1650 ft.) above 1000 m sea level
(ii)	Dry type natural air cooled	2.5%	
(iii)	Oil-immersed, forced air cooled	3%	
(iv)	Dry type, forced air cooled	5%	

Temperature Rise Limits for Oil–Immersed Type Transformer

Part	Cooling method	Oil circulation	Temperature rise
Top oil (measured by thermometer)	(All cases)	(All cases)	60°C when transformer is sealed or equipped with conservator. 55°C when transformer is neither so sealed nor equipped.
Core and other parts	″	″	The temperature in no case to reach a value that will injure the core or its adjacent material
Winding	Natural air, forced air, water, internal cooler	Natural	65°C
	Forced air Water, external cooler	Forced	65°C

Note

1. Average air temperature/day $\leq$ 30°C
   Average air temperature/year $\leq$ 20°C
   Maximum air temperature 40°C
   Lowest air temperature –25°C
2. Height of working area $\leq$ 1000 m (3300 ft.) above the sea level.

Losses (No Load Loss And Full Load Loss) For Two Winding Transformer

Rated power (kVA)	No load loss (kW)	Load loss at 75°C (kW)
500	1.66	6.92
1000	2.8	11.88
2000	3.2	21
3150	4.6	28
5000	8	41
6300	9.3	48
10000	12.5	65
12500	15.0	81
20000	20.0	112
31500	27.0	155
40000	32.0	185
Above 40,000	Rise of 0.55/MVA approx.	Rise of 3.45/MVA approx.

Efficiency and Short-circuit Voltage

1. Efficiency $= x\,P\cos\phi\,/(xP\cos\phi + Wi + x^2 W_{cu})$
   where,

   P = Rated FL kVA,

   x = Loading ratio = Loading kVA/FL kVA,

   $\cos\phi$ = Power factor of load

   Wi = Core loss in kW and W_{cu} = F.L. copper loss in kW.

Table showing efficiency and short-circuit voltage table

Rated power (kVA)	* Short-circuit voltage in % of rated voltage	FL % efficiency #
500	6.0	98.31
1000	6.0	98.55
2000	6.0	98.80
3150	6.0	98.97
5000	8.0	99.02
6300	8.0	99.09
10000	8.0	99.23
12500	10.0	99.24

(Contd.)

Rated power (KVA)	* Short-circuit voltage in % of rated voltage	F.L. % efficiency #
20000	11.0	99.34
31500	11.0	99.42
40000	11.0	99.46
Above 40,000 to 100,000	12.0	(99.5 to 99.6)

Efficiency based on unity power factor.

*   Voltage required to draw full load current on short-circuit of the transformer is called short-circuit voltage.

Calculation for Regulation of Transformer

% regulation = (% R cos ϕ $\pm$ % X sin ϕ),

Note   (+) for lagging p.f. and (–) for leading p.f.

Insulation Resistance

Conversion factor for IR value (Ref = 60°C)

Temp in °C	MF	Temp in °C	MF	Temp in °C	MF
10	0.05	24	0.12	30	0.165
15	0.067	25	0.125	31	0.17
20	0.088	26	0.13	32	0.18
21	0.091	27	0.135	33	0.195
22	0.097	28	0.14	34	0.21
23	0.11	29	0.15	35	0.22

Temp in °C	MF	Temp in °C	MF	Temp in °C	MF
36	0.235	51	0.58	66	1.4
37	0.25	52	0.64	67	1.5
38	0.265	53	0.67	68	1.7
39	0.29	54	0.69	69	1.75
40	0.30	55	0.74	70	1.85
41	0.31	56	0.78	71	1.95
42	0.34	57	0.84	72	2.1
43	0.37	58	0.90	73	2.2

Temp in °C	MF	Temp in °C	MF	Temp in °C	MF
44	0.39	59	0.93	74	2.35
45	0.40	**60**	**1.00**	75	2.45
46	0.45	61	1.05	76	2.6
47	0.48	62	1.11	77	2.75
48	0.49	63	1.12	78	2.95
49	0.52	64	1.25	79	3.12
50	0.55	65	1.35	80	3.3

Note 1   Thumb rule for every 10°C change (reduction). IR value changes by ratio 2/1.

Note 2   60°C temperature has been considered as reference mark for comparison.

Insulation condition of transformer (as per IEEE)

P.I (IR at 600/IR at 60 s	Insulation condition	P.I (IR at 600s/IR at 60 s	Insulation condition
Less than 1	Dangerous	Between 1.25 and 2.0	Fair
Between 1 and 1.1	Poor	Above 2	Good
Between 1.1 and 1.25	Questionable		

PI = Polarisation index

Insulation resistance and PI value

i. Minimum insulation resistance is obtained by following formula
$$R \quad CE / \sqrt{kVA}$$
where,

R = IR of winding with ground when other windings are grounded in M .

C = Constant = 0.8 for ONAN at 20°C.

= 16 for dry, compound filled or untanked oil filled,

kVA = Rated capacity of winding under test.

E = 1 voltage (Y) and line voltage for (D) in volt.

Minimum IR value

Rated voltage of winding	Minimum safe IR in M			
	30°C	**40°C**	**50°C**	**60°C**
66 kV and above	600	300	150	75
22 kV and 33 kV	500	250	125	65
6.6 kV and 11 kV	400	200	100	50
Below 6.6 kV	200	100	50	25

The IR value should be taken with all windings earthed except the tested winding.

Minimum PI value

$PI_1 = R_{15}/R_0$	$PI_2 = R_{60}/R_{15}$	$PI_3 = R_{600}/R_{60}$
2.5 to 3	1.5 to 2	1.2 to 1.5

Transformer vector symbol

1. First symbol   HV winding connection,
2. Second symbol   LV winding connection
3. Third symbol   Phase displacement expressed as the clock hour number

    Ex: (Dy1)—D   HV winding is delta, y   LV winding is star

    1   Phase displacement is –30° (represents clock hour number 1)

Note   HV winding is always taken as reference.

Explanation

1. HV winding is connected in delta fashion.
2. LV winding is connected in star fashion.
3. LV winding has been displaced by 30° and lagging to HV winding. So it has been represented by clock hour number 1.

    (From 12 O' clock to 1 O' clock displacement is 30° clock wise.)

Standard vector group

Group	Lag	Type	Symbol	Winding wiring
	Wye Reference Winding		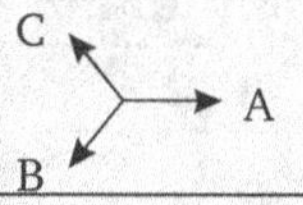	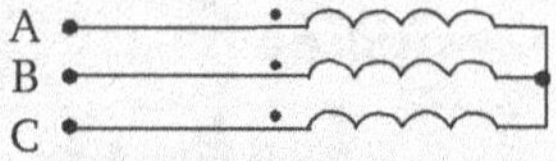
0	0°	Yy0	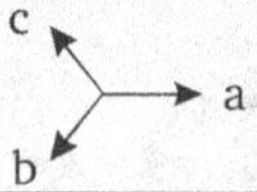	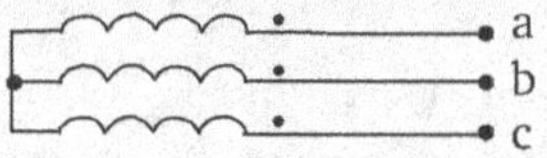
1	30°	Yd1	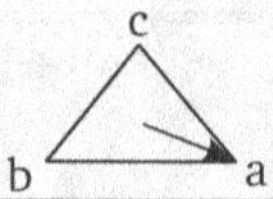	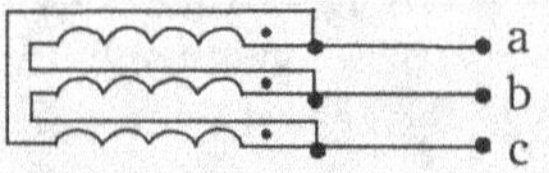
2	60°	Yy2	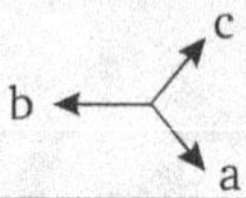	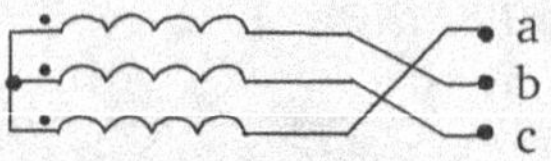
3	90°	Yd3	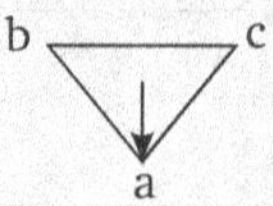	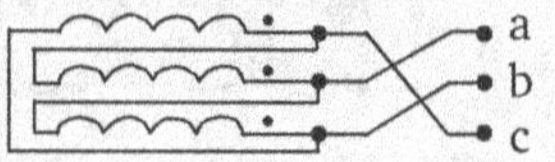
4	120°	Yy4	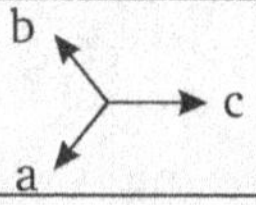	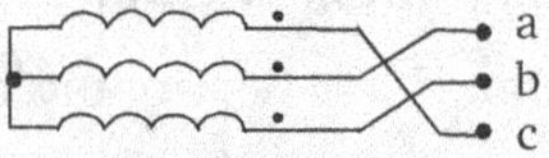
5	150°	Yd5	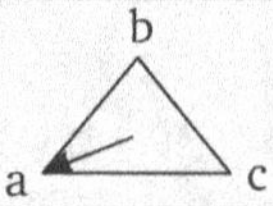	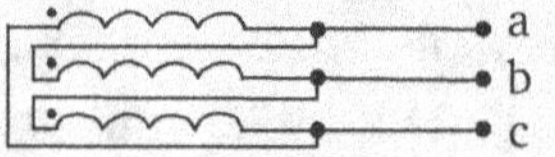
6	180°	Yy6	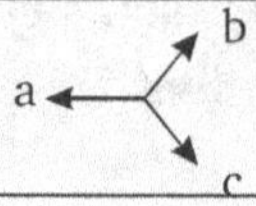	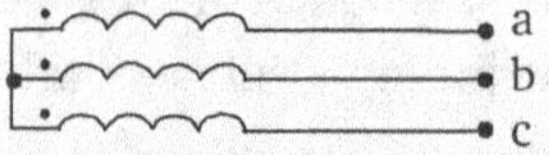
7	210°	Yd7	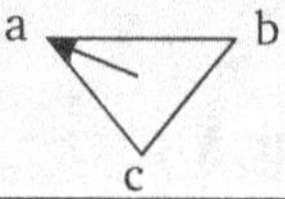	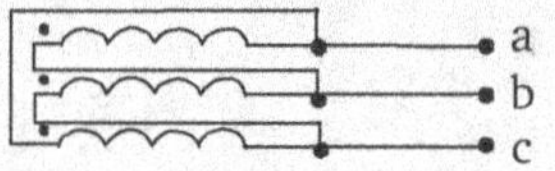
8	240°	Yy8	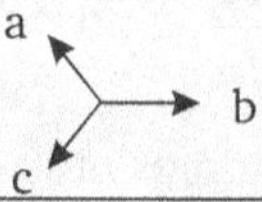	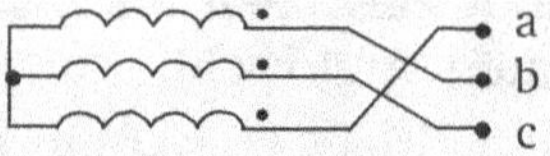
9	270°	Yd9	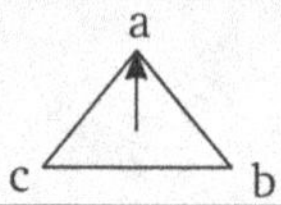	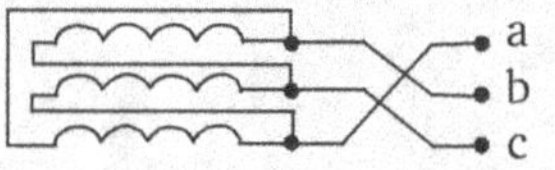
10	300°	Yy10	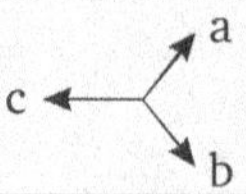	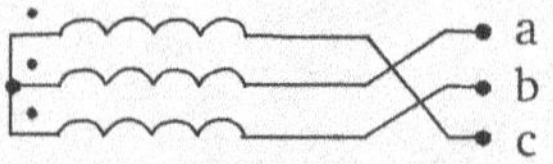
11	330°	Yd11	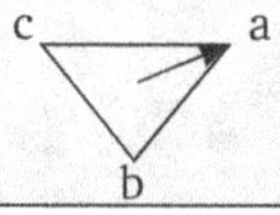	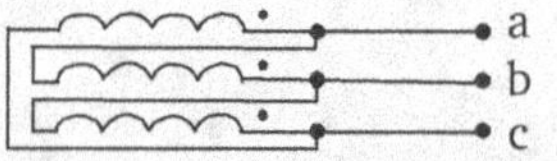

Group	Lag	Type	Symbol	Winding wiring
Delta Reference Winding				
0	0°	Dd0		
1	30°	Dy1		
2	60°	Dd2		
3	90°	Dy3		
4	120°	Dd4		
5	150°	Dy5		
6	180°	Dd6		
7	210°	Dy7		
8	240°	Dd8		
9	270°	Dy9		
10	300°	Dd10		
11	330°	Dy11		

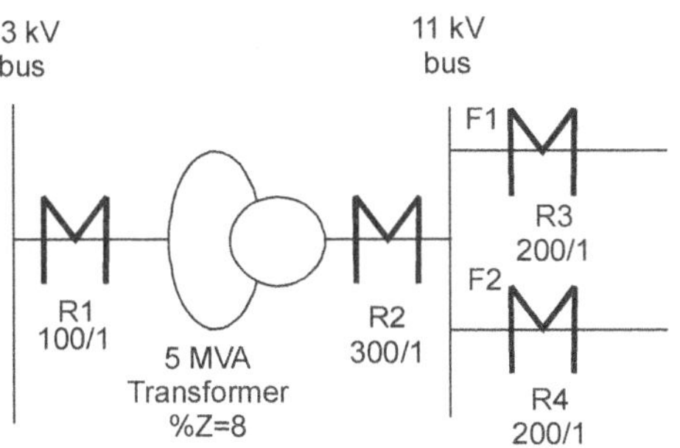

Parallel Operation of Transformer

Terminal marking (viewed from HT side)

1. 1 transformer subscripts are marked in descending order from left to right.
2. 3 transformers neutral is on extreme left and then phases are in sequence (R, Y, and B).
3. Autotransformer neutral is on extreme left and then phases are in sequence (R, Y, and B).

Conditions for parallel operation

  i. Same inherent phase angle difference between primary and secondary terminals
 ii. Same voltage ratio
iii. Same frequency
iv. Same polarity
 v. Same phase rotation

Load Sharing by Transformers in Parallel

1. For two transformers

$$P_A = P(Q_A \cdot Z_B)/(Z_A \cdot Q_B + Z_B \cdot Q_A)$$
$$P_B = P(Q_B \cdot Z_A)/(Z_A \cdot Q_B + Z_B \cdot Q_A)$$

where,

P_A = Load shared by transformer (A)
P_B = Load shared by transformer (B)
Q_A = Rating of transformer (A)
Q_B = Rating of transformer (B)
Z_A = % Impedance of transformer (A)
Z_B = % Impedance of transformer (B) and
P = Total load

Note Currents are also shared by same proportionate as loads.

2. For three transformers

$$P_A = P(Q_A \cdot Z_B Z_C)/ \quad , P_B = P(Q_B \cdot Z_C Z_A)/ \quad , P_c = P (Q_C Z_A \cdot Z_B) /$$
where,
$$= (Q_A \cdot Z_B Z_C + Q_B \cdot Z_C Z_A + Q_C \cdot Z_A \cdot Z_B)$$

Note Currents are also shared by same proportionate as loads.

Circulating Currents

Two Transformers in Parallel

 i. For impedance having same ratio (R/X)

$$I_{Cr} = (V_A - V_B)/(Z_A + Z_B)$$

ii. For impedance having different ratio (R/X)

$$I_{Cr} = (V_A - V_B)/Z,$$

where,

$$Z = \sqrt{(R_A \quad R_B)^2 \quad (X_A \quad X_B)^2}$$

V_A = Secondary terminal voltage of transformer (A) (lower ratio)
V_B = Secondary terminal voltage of transformer (B) (higher ratio),
$Z_A = (V_{ZA} \cdot V_A)/(100 \times I_A)$, I_A = FL current
V_{ZA} = % Impendance voltage drop at F.L. rating

Three transformers in parallel

i. For impedance having same ratio (R/X)

$$I_{CrA} = (V_A - M)/Z_A$$
$$I_{CrB} = (V_B - M)/Z_B$$
$$I_{CrC} = (V_C - M)/Z_C$$

where, $\quad M = (V_A \cdot Z_B \cdot Z_C + V_B \cdot Z_C \cdot Z_A + V_C \cdot Z_A \cdot Z_B)/(Z_A \cdot Z_B + Z_B \cdot Z_C + Z_C \cdot Z_A)$

Transformer Tappings

Note

1. For 2-winding transformer, tap changer is generally provided on the HT side, i.e., LT side voltage remains constant.

2. For autotransformer, tap changer is generally provided on the LT side, i.e., HT side voltage remains constant.

Transformer		% Winding undergone tap change	Standard no. of tap from normal tap
1.	Distribution transformer	10	+5% up, –5% below
	(11/0.4 or 11/6.6 kV)	15	+5% up, –10% below
2.	33/11 kV	20	+10% up, –10% below
		20	+5% up, –15% below
3.	132/33 kV	20	+10% up, –10% below
	132/11 kV	20	+5% up, –15% below
4.	220/132 kV	20	+10% up, –10% below
	220/33 kV	20	+5% up, –15% below

Transformer Oil Data

Characteristic requirements of IS, IEC, & BS specification for uninhibited transformer oil data

Characteristics	IS:335 1993	IEC-296 Class-1	IEC-296 Class-II	BS-148 Class-I	BS-148 Class-II
Appearance	Oil should be clear, transparent, free from suspended matter and sediments				
Density gm/cc max@ 29.5° C	0.89	0.895	0.895	0.895	0.89
Kinematic visco. CSt max	27 @27°C	16.5 @ 40°C 800 @ −15°C	11.0 @ 40°C 1800 @ −15°C	16.5 @ 40°C 800 @ −15°C	11.0@ 40°C 1800 @ −15°C
Interfacial tension (N/mmin.)	0.04	NR	NR	NR	NR
Flash point °C min.	140	140	130	140	130
Pour point °C min.	−6	−30	−45	−30	−45
BDV kV/cm min.					
i. New unfiltered	30	30	30	30	30
ii. After filter	60	50	50	As delivered	As delivered
Tan delta @ 90°C max.	0.002	0.005	0.005	0.005	0.005
Resistivity Ohm-cm					
@ 90°C min.	35×10^{12}	NR	NR	NR	NR
@ 27°C min.	1500×10^{12}				
Oxidation stability 164 hrs.					
1. Neutralization value (mg KOH/g max)	0.4	0.4	0.4	1.2	1.2
2. Sludge cont. (%/wt. max)	0.1	0.1	0.1	0.8	0.8

Characteristics	IS:335 1993	IEC-296 Class-I	IEC-296 Class-II	BS-148 Class-I	BS-148 Class-II
Neutralization value 1. Total acidity (mg KOH/gm)	0.03 max.	0.03 max.	0.03 max.	0.03 max.	0.03 max.
2. Inorganic acidity	Nil	–	–	–	–
Corrosive sulphur	Non-corrosive	Non-corrosive	Non-corrosive	Non- corrosive	Non-corrosive
Oxidation inhibitor	0.05% max.	Not detectable	Not detectable	Not detectable	Not detectable
Water content (ppm) max. bulk /drum delivery	50	40	40	20/30	20/30
S.K. value	Under consideration	–	–	–	–
Accl. ageing test (open-beaker method with copper catalyst) 1. Tan delta @ 90° C max.	0.2				
2. Resistivity Ohm-cm. @ 90° C	0.2×10^{12} min.	NR	NR	NR	NR
3. Resistivity Ohm-cm. @ 27° C	2.5×10^{12} min.				
4. Total acidity	0.05				
5. Sludge cont. (%/wt max.)	0.05				
Gassing tendency at 50 Hz after 120 min. mm^3/min. max	–	–	–	+5	+5
Total PCB content mg/kg	–	–	–	Not detectable	Not detectable
Total furan mg/kg max.	–	–	–	1.0	1.0
Polycyclic aromatic wt%, max.	–	–	–	3.0	3.0

NR—Not required

Characteristic requirements of IS, IEC, & BS specification for inhibited transformer oil data

Characteristics	IS: 12463/1988	IEC-296 Class-1A	IEC-296 Class-IIA	BS-148 Class-IA	BS-148 Class-IIA
Appearance	Oil should be clear, transparent, free from suspended matter and sediments				
Density g/cc max@ 29.5° C	0.89	0.895	0.895	0.895	0.89
Kinematic viscosity CSt max	27 @27° C	16.5 @ 40° C 800 @ –15° C	11.0@ 40° C 1800 @ – 15° C	16.5 @ 40° C 800 @ - 15° C	11.0@ 40° C 1800 @ –15° C
Interfacial tension (N/m) min.	0.04	0.04 @ 25° C	0.04 @ 25° C	–	–
Flash point °C min.	140	140	130	140	130
Pour point °C min.	–6	–30	–45	–30	–45
BDV kV/cm min.					
1. New unfiltered	30	30	30	30	30
2. After filter	60	50	50	As delivered	As delivered
Tan delta @ 90°C max.	0.002	0.005	0.005	0.005	0.005
Resistivity Ohm-cm .					
@ 90° C min	35×10^{12}	–	–	–	–
@ 27° C min.	1500×10^{12}				
Neutralization value					
1. Total acidity (mg KOH/gm max.)	0.03	0.03	0.03	0.08	0.08
2. Inorganic acidity	Nil	–	–	–	–
Corrosive sulphur	Non-corrosive	Non-corrosive	Non-corrosive	Non-corrosive	Non-corrosive

Characteristics	IS: 12463/1988	IEC-296 CLASS-IA	IEC-296 CLASS-IIA	BS-148 CLASS-IA	BS-148 CLASS-IIA
Oxidation stability 164 hrs					
1. Neutralization value (mg KOH/gm max).	0.4	0.4	0.4	0.25	0.25
2. Sludge cont. (%/wt. max.)	0.1	0.1	0.1	0.01	0.01
3. Volatile acidity max	–	0.28	0.28	–	–
Oxidation stability 500 hrs.					
1. Neutralization value (mg KOH/g max.)	–	–	–	1.5	1.5
2. Sludge cont. (%/wt. max.)	–	–	–	1.0	1.0
Oxidation inhibitor	0.3	0.15–0.4	0.15–0.4	0.3	0.3
Water content (ppm) max. bulk/drum delivery	50	30/40	30/40	20/30	20/30
Accl. ageing test (open beaker method with copper catalyst)		NR	NR	NR	NR
1. Tan delta @ 90°C max.	0.2				
2. Resistivity Ohm-cm. @ 90°C	0.2×10^{12} min.				
3. Resistivity Ohm-cm. @ 27°C	2.5×10^{12} min.				
4. Total acidity	0.05				
5. Sludge cont. (%/wt. max.)	0.05				
Gassing tendency at 50 Hz after 120 min., mm^3/min. max.	–	–	–	+8	+8
Total PCB content mg/kg	–	–	–	Not detectable	Not detectable
Total furan mg/kg max.	–	–	–	1.0	1.0
Polycyclic aromatic wt%, max.	–	–	–	3.0	3.0

NR—Not required)

High quality transformer oils for power transformers

Characteristics	ASTM D 3487 Type-I	HVDC SPEC.	DIN 57370 VDE 0370	Australian AS 1767.1 Class I/II
Physical properties				
Appearance		Clear and bright		
Density gm/cc max.	0.91 @ 15°C	0.885 @ 20° C	0.895 @ 20° C	0.895 @ 20°C
Kinematic Viscosity CS t (SUS) max.				
@ 100° C	3 (36)	–	–	–
@ 40° C	12 (66)	11	25 @ 20°C	16.5/11
@ 0° C	76 (350)	–	–	–
@ –15° C	–	–	–	800/-/1800
@ –30° C	–	1800	1800	
IFT @ 25° C N/m min.	0.04	0.04	-	0.04
Flash point °C min. (PMCC)	145	140	130	140/130
Pour point °C max.	–40	–30		–30/–45
Electrical properties				
BDV kV min. before/after filtration	30/50	30/50	30/50	30/50
BDV impulse, @ 25° C, kV min. needle neg. to sphere grounded (25.4 mm gap)	145	150	–	–
Gassing tendency at 50 Hz after 120 min., mm³/min. max.	+15/+30			
Tan delta @60 Hz	0.05 @ 25° C	0.005 @ 90° C	0.005 @ 90° C	0.005 @ 90°C

Characteristics	ASTM D 3487 Type-I	HVDC SPEC.	DIN 57370 VDE 0370	Australian AS 1767.I ClassI/II
Chemical properties				
Oxidation stability max.				
1. 72 hrs. sludge % mass	0.15	–	–	–
TAN mg KOH/gm	0.5	–	–	–
2. 164 hrs. sludge % mass	0.3	0.03	0.6	0.1
TAN mg KOH/gm	0.6	0.15	0.3	0.4
Ageing resistance as per Badder (140 hrs/110°C)				
Saponification no. mg KOH/gm	–	–	0.6	–
Sludge cont. wt%, max.	–	–	0.05	–
Tan delta @ 90° C max.	–	–	0.18	–
Oxidation stability (rotating bomb test) minutes minimum	195	–	–	–
Oxidation inhibitor % mass/max.	0.08–0.3	–	–	0.15–0.4
Corrosive sulphur	Non-corrosive	Non-corrosive	Non-corrosive	Non-corrosive
Water content (ppm) max.	35	10	–	30(bulk)/40 (drum)
Total acidity (mg KOH/gm max.)	0.03	0.03	0.03	0.03
PCB content ppm	Nil	Nil	Nil	Nil
Total aromatic content % max.	–	12	–	–
Total sulphur content % max.	–	0.15	–	–

Test on transformer oil in service

Characteristics	Voltages	Test Method	Periodicity	Permissble limit
BDV	145 kV and Above 72.5 to < 145 kV < 72.5 kV	IS 6792/1992 (Ave. of 6 values with 2.5 mm gap)	After filling or refilling prior to energizing then after 3 months and after one year	50 kV (min.) 40 kV (min.) 30 kV (min.)
Water content	145 kV and above Below 145 kV	IS 335/1993	After filling or refilling prior to energizing then after 3 months and after one year	25 ppm 35 ppm
Specific resistance @ 90° C in Ohm-cm	All voltages	IS 6103 /1971	After filling or refilling prior to energizing then after 3 months and after 2 years	0.1×10^{12} min.
Tan delta @ 90° C max.	145 kV and above Below 145 kV	IS 6262/1971	After filling or refilling prior to energizing then after 2 years	0.2 1.0
Neutralization value Total acidity max.	All voltages	IS 1448 Part-2 1967	"	0.5 mg KOH/gm
Sediment or precipitate sludge	All voltages	Appendix-A IS 1866/1983	"	No sludge
Flash point	-do-	IS 1448 Part-21, 1992	"	Decrease of 15° C from initial value, min. 125° C
IFT @ 27° C min.	-do-	IS 6104 /1971	"	0.018 N/m
DGA	145 kV & above	1S 434/1992	After filling or refilling prior to energizing then after 3 months and after one year	Refer IS 10593/1993/DGA Chart

Dissolved Gas Analysis (Interpretation of Results)

Doernenburge Ratio Method

Suggested fault diagnosis	CH_4/H_2	C_2H_2/C_2H_4	C_2H_2/CH_4	C_2H_6/C_2H_2
Thermal decomposition	> 1.0	< 0.75	< 0.3	> 0.4
Corona (low intensity PD)	< 0.1	–	< 0.3	> 0.4
Arching (high intensity PD)	> 0.1 < 1.0	> 0.75	> 0.3	< 0.4

Roger's Ratio Method

Method 1

Suggested fault diagnosis	C_2H_2/C_2H_4	CH_4/H_2	C_2H_4/C_2H_6
Normal	0.1 to 1.0	< 0.1	<1.0
Low energy density arching	<0.1	<0.1	<1.0
Arching (high intensity P.D)	0.1 to 3.0	0.1 to 1.0	>3.0
Low temp. thermal	<0.1	0.1 to 1.0	1.0 to 3.0
Thermal >700° C	<0.1	>1.0	1.0 to 3.0
Thermal <700° C	<0.1	>1.0	>3.0

Method 2

Suggested fault diagnosis	CH_4/H_2	C_2H_6/CH_4	C_2H_4/C_2H_6	C_2H_2/C_2H_4
If CH_4/H_2 <0.1, then P.D otherwise normal deterioration	0	0	0	0
Slight over-heating below 150° C	1	0	0	0

Suggested fault diagnosis	CH_4/H_2	C_2H_6/CH_4	C_2H_4/C_2H_6	C_2H_2/C_2H_4
Slight over-heating below 150°C to 200°C	1	1	0	0
Slight over-heating below 200° C to 300° C	0	1	0	0
General conductor over-heating	0	0	1	0
Circulating currents/over-heated	1	0	1	0
Flashover without power flow current	0	0	0	1
Tap changer selector breaking current	0	1	0	1
Arc with power flow	0	0	1	1

Remarks

Ratio <1.0 is taken as 0, Ratio >1.0 is taken as 1.

A given ratio can be taken for diagnosis if the concentration of one gas is at least equal to the limit values as below.

Value in ppm.

H_2 = 200, CH_4 = 50, C_2H_6 = 15, C_2H_4 = 60, C_2H_2 = 15

Key gas method

Suggested fault diagnosis	Major key gas	Minor key gas
Over-heating of oil Thermal degradation/ decomposition of oil	C_2H_4 > 150 ppm (60–70 %)	C_2H_6 (10–20%) CH_4 (10–20%)
Power discharge Arching in oil Electric discharge	H_2 (60–70%), >100 ppm C_2H_4 (30–40%), > 30 ppm	CH_4 (5–10%), C_2H_4 (3–5 %) C_2H_6 (1.5–3%)
Internal corona Partial discharge	H_2 (80–90%) > 100 ppm	CH_4 (10–15%), C_2H_4 (0.1–0.5%), C_2H_6 (0.5–1.0 %)
Hot spot in oil	CH_4 (50–60%) > 50 ppm	H_2 (40–60%), C_2H_4 (0.1–0.5%), C_2H_6 (0.5–1.0%)

Suggested fault diagnosis	Major key gas	Minor key gas
Over-heating of solid insulation Thermal ageing of oil	CO (90–95 %) > 350 ppm	H_2 (5–10 %), C_2H_6 (1–2 %), CH_4 (2–5%)
Arching in cellulose Decomposition of insulation	CO (50–60 %), > 350 ppm CO_2 ((50–60 %), > 350 ppm	H_2, CH_4, C_2H_2, C_2H_4, C_2H_6 Rest %

Total dissolved combustible gas limits (TDCG)

TDCG limits in ppm	Interpretations
0–720	Satisfactory operation—unless individual gas acceptance values are exceeded.
721–1920	Normal ageing/slight decomposition unless individual gas acceptance values are exceeded.
1921– 4630	Significant decomposition fault is to be monitored.
>4630	Very substantial decomposition of oil. Immediate action to be taken.

Permissible limit (in ppm) of gases for service transformers

Gases	Transformer within 4 years	Transformer 4–10 years	Transformer after 10 years
CO_2	2000–3000	3000–4000	4000–6000
CO	150–300	300–450	400–550
C_2H_4	100–150	150–200	200–300
C_2H_2	20–30	30–50	30–60
H_2	100–150	150–250	200–300
CH_4	50–70	70–150	70–150

Other Informations

Disintegration of transformer oil at ordinary temp.		Relation of evolved gas with temperature		National specifications	
Gas	%	Temp. in °C	Gas	Country	Specification
CO_2	1.17	>120	Methane (CH_4)	USA	ASTM D 1040–73
Heavy hydrocarbon	4.86	>120	Ethane (C_2H_6)	France	N.F Cir. C 1.03
O_2	1.36	>150	Ethylene (C_2H_4)	Germany	VDE–0370
CO	19.21	>700	Acetylene (C_2H_2)	India	IS 335
H_2	59.10			Sweden	SEN-14
N_2	10.10			Italy	A.E.1.7
CH_4	4.2			Switzerland	S.E.V.124
Total	100			UK	BS 148
				USSR	GOST-981
				IEC	IEC: 296

Typical Value of Capacitance and Tan δ of Transformers

Voltage rating	Configuration	Capacitance (nF)	Tan δ
400/220 kV	HV–LV	4–5	0.002–0.005
	HV-tank	13–15	0.007–0.009
	LV-tank	23–24	0.004–0.008
220/132 kV	HV–LV	5–7	0.003–0.006
	HV-tank	10–12	0.005–0.010
	LV-tank	19–22	0.004–0.010
132/33 kV	HV–LV	5–7	0.003–0.006
	HV-tank	10–12	0.005–0.010
	LV-tank	19–22	0.004–0.010
132/11 kV	HV–LV	5–7	0.003–0.006
	HV-tank	10–12	0.005–0.010
	LV-tank	19–22	0.004–0.010
66/11 kV	HV–LV	5–6	0.003–0.006
	HV-tank	2.9–3.5	0.006–0.008
	LV-tank	6–10	0.005–0.008
33/11 kV	HV–LV	8–10	0.015–0.018
	HV-tank	11–13	0.015–0.020
	LV-tank	15–17	0.015–0.020

Duration of Overloading of Oil-immersed Transformer

Previous continuous loading as % rated load	Oil temperature at the beginning of the overload in 0°C for cooling method		Duration of overload in minutes				
	ONAN ONAF	OFAN OFAF OFW	10%	20%	30%	40%	50%
50	55	49	160	80	60	30	15
75	68	60	120	60	30	15	8
90	78	68	60	30	15	8	4

Note Type of insulation—A class for above table

Permissible Overloading Capacity Transformer

Cooling air temp. (°C)	All day heavy load (continuous load)	Long period		Medium period		Short period	
		16 hrs. heavy load	Remaining 8 hrs. light load	8 hrs. heavy load	Remaining 16 hrs. light load	3 hrs. heavy load	Remaining 21 hrs. light load
0	120	125	105	130	105	150	105
5	115	120	100	125	100	145	100
10	110	115	94	120	94	140	94
15	105	110	88	115	88	135	88
20	100	105	82	110	82	130	82
25	94	100	76	105	76	125	76
30	88	94	70	100	70	120	70
35	82	88	64	94	64	115	64
40	76	82	57	88	57	110	57
45	70	76	49	82	49	105	49
50	64	70	40	76	40	100	40

Note Type of cooling—ON (oil natural)
Type of insulation—A class.
Maximum oil temperature—80°C
Maximum winding temperature—95°C.

Daily Peak Loads per unit of Name plate rating to give Normal Life Expectancy (Cooling—Self-Cooled or Water-Cooled *(OA or OW)*$

Peak load time (hrs)	Cooling—self-cooled or water-cooled (OA or OW)$																	
	Continuous equivalent load in % of rated kVA preceding the peak load																	
	50 %						70 %						90 %					
	Ambient in °C						Ambient in °C						Ambient in °C					
	0	10	20	30	40	50	0	10	20	30	40	50	0	10	20	30	40	50
1/2	2.00	2.00	2.00	1.89	1.70	1.52	2.00	2.00	1.95	1.78	1.60	1.41	2.00	1.99	1.81	1.64	1.46	1.24
1	2.00	1.88	1.73	1.58	1.41	1.23	1.95	1.80	1.65	1.49	1.32	1.14	1.86	1.70	1.55	1.39	1.20	1.99
2	1.76	1.64	1.51	1.37	1.22	1.00	1.72	1.59	1.46	1.32	1.16	0.99	1.66	1.53	1.39	1.24	1.08	0.90
4	1.54	1.43	1.33	1.19	1.06	0.94	1.52	1.41	1.29	1.17	1.04	0.89	1.50	1.39	1.26	1.13	1.00	0.84
8	1.41	1.30	1.19	1.08	0.96	0.84	1.40	1.30	1.19	1.07	0.95	0.83	1.39	1.29	1.18	1.06	0.94	0.82
24	1.33	1.22	1.11	1.00	0.89	0.78	1.33	1.22	1.11	1.00	0.89	0.78	1.33	1.22	1.11	1.00	0.89	0.78

Note $—Subtract 5° C from each of the ambient column heading for water-cooled transformer. Minimum water temperature must be above 0°C

Daily peak loads per unit of max. name plate rating to give normal life expectancy (cooling–forced air-cooled rated 133% or less of self-cooled rating or water-cooled (OA/FA))

Peak load time (hrs)	Cooling—forced air-cooled rated 133% or less of self-cooled rating or water-cooled (OA/FA)																	
	Continuous equivalent load in % of rated KVA preceding the peak load																	
	50 %						70 %						90 %					
	Ambient temperature (°C)						Ambient temperature (°C)						Ambient temperature (°C)					
	0	10	20	30	40	50	0	10	20	30	40	50	0	10	20	30	40	50
1/2	2.00	2.00	1.97	1.82	1.66	1.49	2.00	2.00	1.89	1.74	1.58	1.40	2.00	1.92	1.77	1.61	1.43	1.25
1	1.90	1.77	1.64	1.50	1.35	1.19	1.84	1.71	1.57	1.43	1.28	1.11	1.77	1.63	1.49	1.35	1.19	1.00
2	1.64	1.53	1.42	1.29	1.16	1.02	1.61	1.50	1.38	1.26	1.12	0.97	1.58	1.46	1.34	1.21	1.08	0.91
4	1.46	1.36	1.26	1.15	1.03	0.90	1.45	1.35	1.24	1.13	1.01	0.88	1.44	1.34	1.23	1.11	1.00	0.85
8	1.37	1.27	1.17	1.07	0.96	0.84	1.37	1.27	1.17	1.07	0.96	0.83	1.36	1.27	1.17	1.06	0.95	0.83
24	1.31	1.21	1.11	1.00	0.89	0.78	1.31	1.21	1.11	1.00	0.89	0.78	1.31	1.21	1.11	1.00	0.89	0.78

Note The peak load in this table are calculated on the basis of all cooling in use during the period preceding the peak load. When operating without fans, use this table for OA transformer

Daily peak loads per unit of max. name plate rating to give normal life expectancy (cooling—forced air-cooled Rated 133% or less Of self-sooled rating or water-cooled (OA/FA))

Peak load time (hrs)	Cooling—forced air-cooled rated 133 % or less of Self-cooled rating or water-cooled (OA/FA)*																	
	Continuous equivalent load in % of rated KVA preceding the peak load																	
	50 %						70 %						90 %					
	Ambient in °C						Ambient in °C						Ambient in °C					
	0	10	20	30	40	50	0	10	20	30	40	50	0	10	20	30	40	50
1/2	2.00	1.91	1.78	1.65	1.52	1.37	1.96	1.84	1.71	1.58	1.43	1.28	1.89	1.77	1.64	1.50	1.35	1.19
1	1.73	1.62	1.51	1.38	1.25	1.12	1.68	1.58	1.46	1.33	1.20	1.06	1.64	1.53	1.41	1.28	1.15	1.01
12	1.53	1.43	1.33	1.22	1.11	0.98	1.51	1.41	1.30	1.19	1.07	0.95	1.49	1.39	1.28	1.17	1.06	0.93
4	1.40	1.31	1.21	1.11	1.00	0.89	1.40	1.31	1.21	1.10	1.00	0.88	1.39	1.30	1.20	1.09	0.99	0.87
8	1.34	1.35	1.16	1.06	0.96	0.84	1.34	1.25	1.16	1.06	0.96	0.84	1.34	1.25	1.15	1.05	0.95	0.84
24	1.30	1.20	1.10	1.00	0.90	0.79	1.30	1.20	1.10	1.00	0.90	0.79	1.30	1.20	1.10	1.00	0.90	0.79

* Subtract 5°C from each of the ambient column heading for water-cooled transformer. Minimum water temperature must be above 0°C

Example   Assume a load cycle which resolves to a constant value of 50% followed by a 100% peak load for 2 hrs. Using the above table if the ambient temperature is 30° C, a self-cooled (OA), a water-cooled (OW) transformer will carry 1.32 times name plate rating for 2 hrs, following an equivalent continuous load up to 70 % of name plate rating, if the equivalent 2 hrs. peak load from the load cycle is 10 MVA. The constant equivalent load before the peak is 5 MVA which is 66.6 % of the name plate rating of the transformer. Therefore a 7.5 MVA transformer is suitable for this daily load cycle.

Per cent Change in kVA Load for Each Degree Centigrade Change in Average Ambient Temperature

Type of cooling	Air above 30°C average or water above 25°C	Air below 30°C average or water below 25°C
Self-cooled	–1.5% per degree	1% per degree
Water-cooled	–1.5% per degree	1% per degree
Forced air-cooled	–1% per degree	0.75% per degree
Forced oil-cooled	–1% per degree	0.75% per degree

Comparison of Tolerances of BSS 171:1970 and ISS 2026:1926

Particulars	BSS 171	ISS 2026
Total loss	+1/10 of total loss	+10% of the guaranteed value
Component losses	+1/7 of each component losses, provided that the tolerance for the total losses is not exceeded.	+10% of the guaranteed value
Voltage ratio at no load on the principal tapping (rated voltage ratio)	1/200 of declared ratio or % of the declared ratio equal to 1/10 of the actual % voltage at rated current	Same as BSS
Impedances voltage (a) Principal tapping		
i. Two winding transformer	1/10 of the declared impedance voltage for that tapping	Same as BSS
ii. Multi-winding transformer	1/10 of the declared impedance voltage for specified pair of winding. 1/7 of the declared impedance voltage for the second specified pair of winding.	±15%
b) For tapping other than the principal tapping	1/7 of the stated value for each tapping within 5% of principal tapping declared impedance voltage for the second specified pair of winding.	-
No load current	+3/10 of the declared no load current	No tolerances

Oil Handling Procedure in Transformer

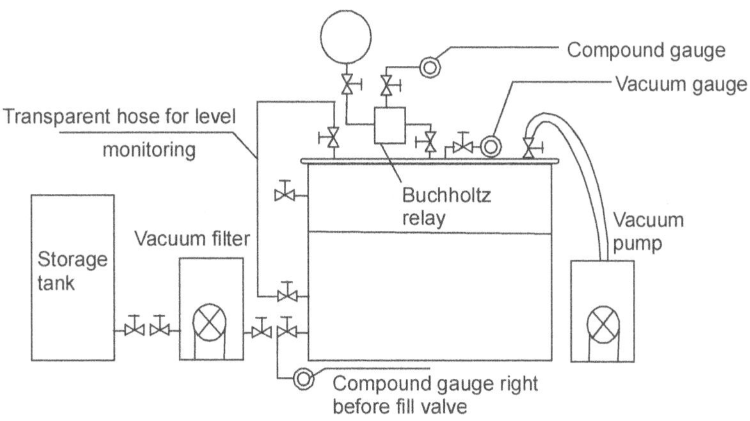

Figure 4.1   Vacuum oil filling on site

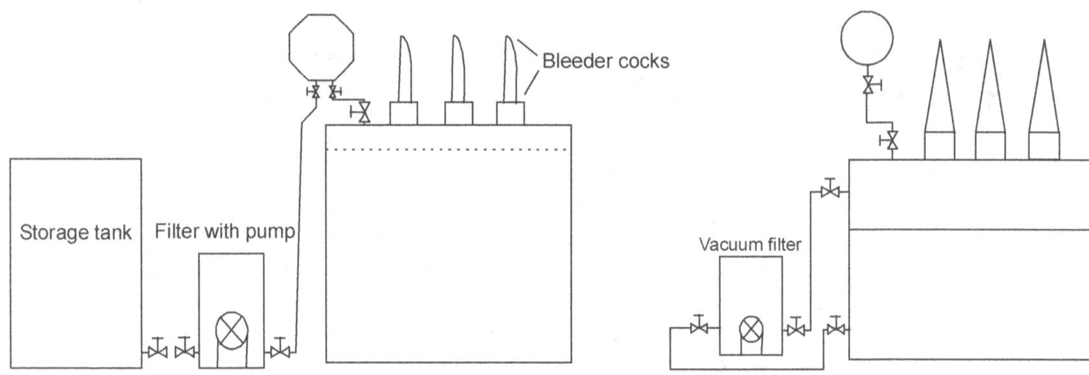

Figure 4.2 Oil topping on site   **Figure 4.3** Oil filtration on site

Typical Value of Capacitance and tan of Bushings

Voltage	Capacitance (nF)	tan
400 kV	420–480	0.002–0.005
220 kV	280–400	0.002–0.005
132 kV	180–300	0.002–0.005
66 kV	180–300	0.002–0.004
36 kV	260–280	0.002–0.004

DGA for Normal Bushing Oil (as per LEC–36 AWG3)

H_2 = 100 ppm, CH_4 = 30 ppm, C_2H_2 = 2 ppm,
C_2H_4 = 300 ppm, C_2H_6 = 50 ppm, CO = 1000 ppm, CO_2 = 3000 ppm

Explanation   Bushings of transformers are generally filled with oil and condition of oil normally deteriorates very slowly on comparison to transformer body oil. The risk of damage to bushing increases if the oil gets contaminated and the rate of deterioration quickly rises. So, it is necessary to monitor the status of oil. The values indicated above are the guide lines for the protection engineer to know the limitation of gas percentage, that normal oil should contain.

SQUIRREL CAGE INDUCTION MOTOR

Data

% Voltage$_{dip}$ = (Starting KVA $\times$ 100)/fault KVA at point of supply

Size of motor (400V, 50 Hz)	Approx. current (A) at 0.8 p.f.	Size of PVC aluminum cable (mm^2)	Starting KVA (approx.) for condition of starting			
			D.O.L 6 F.L current	Star/delta 2 F.L current	Auto transformer	
					1.5 F.L	1.35 F.L
5 H.P	8	2.5	33.25	–	–	–
10 H.P	15	4	62.35	20.78	–	–
15 H.P	22	10	91.44	30.48	–	–
20 H.P	29	16	120.54	40.18	30.14	25.11
30 H.P	42	25	174.58	58.19	43.65	36.37

Capacitor Rating (KVAR) for Motors

Motor rating (kW)	< 4	4	5.5	7.5	11	15	18.5	22	≤ 30
Capacitor rating (KVAR)	1	2	2.5	3	5	6	8	10	Approx. 35% of motor rating

Capacitor Required for Desired Pf

Existing pf	Capacitor in Kvar with 1 kW active power, for below pf					
	0.75	**0.8**	**0.85**	**0.9**	**0.95**	**1.0**
0.20	4.02	4.15	4.28	4.42	4.57	4.90
0.40	1.41	1.54	1.67	1.81	1.96	2.29
0.50	0.85	0.98	1.11	1.25	1.4	1.73
0.6	0.45	0.58	0.71	0.85	1.00	1.33
0.7	0.14	0.27	0.40	0.54	0.69	1.02
0.8	–	–	0.13	0.27	0.42	0.75
0.85	–	–	–	0.11	0.26	0.59
0.9	–	–	–	–	0.16	0.48
0.95	–	–	–	–	–	0.35

Capacitor required for induction motors in kVAR (Orissa Gazette notification No. 23191-Com-VI–3186 (Vol- II) Dt. 19.01.1988

MOTOR HP	750 RPM	1000 RPM	1500 RPM	3000 RPM
3	1	1	1	1
5	2	2	2	2
7	3	3	3	3
10	4	4	4	4
15	6	5	5	4
20	8	7	6	5
25	9	8	7	6
30	10	9	8	7
40	14	12	11	10
50	16	16	13	11
60	20	20	16	14
75	24	23	19	16
100	30	30	24	20
125	39	38	31	26
150	45	45	36	30

BATTERY
Data

Data/particulars	Values
Recommended end voltage	1.85/cell
Specific gravity (fully charged)	1200 at 27° C
Specific gravity (fully discharged)	1190 at 27° C
Allowable difference in maximum and minimum specific gravity	30 points/cell
Normal operating voltage	2.15/cell
Fully charged voltage	2.4–2.5/cell
Equivalent specific gravity at 27° C	Hydrometer reading at $t°C + 0.0007\,(t-27)$
Rate of charging	(% AH rating of battery) in A
1. Finishing rate at 10 hrs	4
2. Normal rate	10
3. Equalizing charge at 10 hrs	2
Float charge	
1. Float charge current	$(AH \times 2)/2400$ + sub-station load
2. Tickle charge current	50 to 100 mA/100 AH battery capacity
3. Set voltage	2.15 to 2.17 V/cell

Notes for Maintenance of Battery

1. Battery should be normally kept on float charge
2. *Equalizing charge*
   1. For float charge (2.16 to 2.2)/cell, equalizing charge to be given once in 3 months
   2. For float charge (2.06 to 2.16)/cell, equalizing charge to be given once in a month for 220 V battery
3. *Boost charge*
   1. Allow BOOST charge after test discharge.
   2. Allow BOOST charge for low sp. gravity of major cells (below 1190) and low voltage (below 1.9 V).

4. *Condition charge*

Discharge the battery to end voltage 1.85 V and again charging is called condition charging. It should be done once in a year.

Maintenance Schedule of Battery

Periodicity	Items to be checked	Periodicity	Items to be checked
Daily	1. Measure and record the pilot cell voltage, sp. gravity and electrolyte temperature 2. Battery voltage by switching off charger 3. Hourly reading of DC voltage, charger output current and trickle charge current	Monthly	1. Sp. gravity, voltage of each cell and electrolyte temperature 2. Give equalizing charge 3. Switch-off charger and test tripping/ closing of any one feeder from battery source 4. Check all connection of battery and charger
Weekly	1. Cleaning of terminals, topping up distilled water if required 2. Check pilot cell reading and adjust the trickle charge current if required	Yearly	1. Allow condition charging

Correction Factor for Actual Power/Energy (as per IEEMA Publication)

True power/energy $= K \times$ meter reading $\times ARCT \times ARPT$

$$K \, \frac{\cos}{\cos(\qquad)} \, ARCT \, \frac{\text{Nominal ratio}}{1 \ (e/100)} \, ARPT \, \frac{\text{Nominal ratio}}{1 \ (p/100)}$$

where,

$ARCT =$ Actual ratio of CT

$ARPT =$ Actual ratio of PT

       $=$ Load factor angle

       $=$ Phase angle error of CT

       $=$ Phase angle error of PT

e $=$ CT ratio error

p $=$ PT ratio error.

ENERGY METERS

Percentage Error Limits (Single Phase Meters and Polyphase Meters)

Table for accuracy class (0.2, 0.5, 1.0, and 1.5)

	IEC 687: 1992			**CBIP 88 : 1996**					**IS 14697: 1999**		
% Load	**p.f**	**Class**		**p.f.**	**Class**				**p.f.**	**Class**	
(I_b)	**(Cos ϕ)**	**0.2**	**0.5**	**(Cos ϕ/sin ϕ)**	**0.2**	**0.5**	**1.0**	**1.5**	**(Cos ϕ)**	**0.2S**	**0.5S**
					Balanced load						
1–<5	1.0	±0.4	±1.0	1.0	±0.4	±1.0	–	–	1.0	±0.4	±1.0
2–<5	–	–	–	1.0	–	–	±2.0	±3.0	–	–	–
5–<I_{max}	1.0	±0.2	±0.5	1.0	±0.2	±0.5	±1.0	±1.5	1.0	±0.2	±0.5
2–<10	0.5 Lg	±0.5	±1.0	0.5 Lg	±0.5	±1.0	±2.0	±3.0	0.5 Lg	±0.5	±1.0
	0.8 Ld	±0.5	±1.0	0.8 Ld	±0.5	±1.0	±2.0	±3.0	0.8 Ld	±0.5	±1.0
10–<I_{max}	0.5 Lg	±0.3	±0.6	0.5 Lg	±0.3	±0.6	±1.2	±1.8	0.5 Lg	±0.3	±0.6
	0.8 Ld	±0.3	±0.6	0.8 Ld	±0.3	±0.6	±1.2	±1.8	0.8 Ld	±0.3	±0.6
*Special request	0.25Lg	0.5	1.0	0.25Lg	±0.5	±1.0	±2.5	±3.5	0.25 Lg	±0.5	±1.0
	0.5 Ld	0.5	1.0	0.5 Ld	±0.5	±1.0	±2.5	±3.5	0.5 Ld	±0.5	±1.0
10–<I_{max} (20– < I_{max}) for CBIP 88: 1996											
					Unbalanced loads						
5–< 120	1.0	±0.3	±0.6	1.0	±0.3	±0.6	±1.5	±2.5	–	–	–
5–<I_{max}	–	–	–	–	–	–	–	–	1.0	±0.3	±0.6
10–<I_{max}	0.5 Lg	±0.4	±1.0	0.5 Lg	+0.4	+1.0	+2.0	+3.0	0.5 Lg	±0.4	+1.0

Table for Accuracy Class (1.0, 2.0)

<table>
<tr><td colspan="9" align="center">Balanced load</td></tr>
<tr><td colspan="4" align="center">IS 13779: 1993</td><td colspan="5" align="center">IEC 1036:1996</td></tr>
<tr><td rowspan="2">% Load (I_b)</td><td rowspan="2">pf (Cos ϕ)</td><td colspan="2">Class</td><td colspan="2">% Load (I_b)</td><td rowspan="2">pf (Cos ϕ)</td><td colspan="2">Class</td></tr>
<tr><td>1.0</td><td>2.0</td><td>Direct connected</td><td>T/F operated</td><td>1</td><td>2</td></tr>
<tr><td>5–<10</td><td>1.0</td><td>±1.5</td><td>±2.5</td><td>5–<10</td><td>2–<5</td><td>1.0</td><td>±1.5</td><td>±2.5</td></tr>
<tr><td>10–<I_{max}</td><td>1.0</td><td>±1.0</td><td>±2.0</td><td>10–<I_{max}</td><td>5–<I_{max}</td><td>1.0</td><td>±1.0</td><td>±2.0</td></tr>
<tr><td>10–<20</td><td>0.5 Lg</td><td>±1.5</td><td>±2.5</td><td>10–<20</td><td>5–<10</td><td>0.5 Lg</td><td>±1.5</td><td>±2.5</td></tr>
<tr><td></td><td>0.8 Ld</td><td>±1.5</td><td>–</td><td></td><td></td><td>0.8 Ld</td><td>±1.5</td><td>–</td></tr>
<tr><td>20–<I_{max}</td><td>0.5 Lg</td><td>±1.0</td><td>±2.0</td><td>20–<I_{max}</td><td>10–<I_{max}</td><td>0.5 Lg</td><td>±1.0</td><td>±2.0</td></tr>
<tr><td></td><td>0.8 Ld</td><td>±1.2</td><td>–</td><td></td><td></td><td>0.8 Ld</td><td>±1.0</td><td>–</td></tr>
<tr><td>*Special Request 20–<100</td><td>0.25 Lg</td><td>±3.5</td><td>–</td><td>* Special request 20–<100</td><td>* Special request 10–<100</td><td>0.25 Lg</td><td>±3.5</td><td>–</td></tr>
<tr><td></td><td>0.5 Ld</td><td>±2.5</td><td>–</td><td></td><td></td><td>0.5 Ld</td><td>±2.5</td><td>–</td></tr>
<tr><td colspan="9" align="center">Unbalanced load</td></tr>
<tr><td>10–<I_{max}</td><td>1.0</td><td>±2.0</td><td>±3.0</td><td>10–<I_{max}</td><td>5–<I_{max}</td><td>1.0</td><td>±2.0</td><td>±3.0</td></tr>
<tr><td>(20–I_{max}</td><td>0.5 Lg</td><td>±2.0</td><td>±3.0</td><td>(20–<I_{max}</td><td>10–<I_{max}</td><td>0.5 Lg</td><td>±2.0</td><td>±3.0</td></tr>
</table>

* These tests are normally not conducted in practice. But on special request and on common agreement the test on specified load condition can be tested.

Reading of Electrical Meters (Multiplying Factor (MF))

Meters	MF for meter readings	Meters	MF for meter readings
Ammeters	(line CTR/meter CTR) × meter MF	Energy meters	(line CTR/meter CTR) × (line PTR/meter PTR) × meter MF
Voltmeters	(line PTR/meter PTR) × meter MF	Power meters	(line CTR/meter CTR) × (line PTR/meter PTR) × meter MF

Example for Calculation of MF Energy Meters

1. WH meter has data in name plate (Meter CTR = $-/1$, Meter PTR = 11 kV/110 V, Meter MF (not given))

   Suppose this meter is connected to a 33 kV system of line CTR = 400/1,

   Line PTR = 33 kV/110 V

   Then MF for meter reading = (Line CTR/Meter CTR) × (Line PTR/Meter PTR) × meter MF

   $$\frac{400/1}{/1} \quad \frac{300}{100} \quad 1$$

   $$400 \quad 3$$

   1200 for WH reading (because meter is WH meter)

Different Connection of Energy Meters

Type of meters	Terminals		Remarks
	CT	PT	
3 , 3 wire meters	R, B	R, Y, B	$W_1 = V_L I_L \cos(30° -)$ $W_2 = V_L I_L \cos(30° +)$
3 , 4 wire meters	R, Y, B	R, Y, B, N	$W_1 = V_{PH} I_{PH} \cos$ $W_2 = V_{PH} I_{PH} \cos$ $W_3 = V_{PH} I_{PH} \cos$
1 , 2 wire meters	Ph	Ph, N	$W = V_{PH} I_{PH} \cos$

DATA ON TRANSMISSION LINE

ELECTRICAL SPAN FOR TRANSMISSION LINE

Voltages (kV)	Span (MTR)
66	240, 250, 275
132	315, 325, 335
220	335, 350, 375
400	400, 425
800	400, 450

Wind span—It is the sum of two half spans adjacent to the support under consideration

PARAMETERS FOR DIFFERENT SOILS FOR TOWER FOUNDATION

Type of soil	Safe bearing Capacity (kg/m²)	Density (kg/m³)	Angle of repose degree ( )
Normal dry	11000	1440	30
Black cotton	5500	1000	0
Soft rock	44000	1700	20 to 30
Hard rock	87000	2000	45
Fully submerged	5500	1000	15

WIND ZONE

Wind zone	Basic wind speed (m/sec.)	Wind zone	Basic wind speed (m/sec.)
1	33	4	47
2	39	5	50
3	44	6	55

PERMISSIBLE WEIGHT SPAN (Horizontal Span between Lowest points of the conductors on the two adjacent spans)

Terrain/tower type		400 kV		220 kV		132 kV		66 kV	
		Normal wire Max./Min.	Broken wire Max./Min.	Normal wire Max./Min.	Broken wire Max./Min.	Normal wire Max./Min.	Broken wire Max./Min.	Normal wire Max./Min.	Broken wire Max./Min.
Plain	Suspension	600/200	360/100	525/200	315/100	488/195	195/104	375/163	150/75
	Small/medium angle	600/0	360/−200	525/0	315/−200	488/0	195/−200	375/0	150/−150
	Large angle	600/0	360/−300	525/0	315/−300	488/0	195/−300	NA	NA
Hilly	Suspension	600/200	360/100	525/200	315/100	488/208	192/104	375/163	150/75
	Small/medium /large angle	1000/−1000	−600/−600	1000/−1000	−600/−600	960/−960	576/−576	750/−750	450/−450

DISC INSULATOR FOR EHT LINES

Voltage	Suspension		Tension	
	No. of discs	UTS in KN	No. of discs	UTS in KN
66 kV	5	45	6	50
132 kV	9	70	10	120
220 kV	14	90	15	120
400 kV	23	115	2 × 23	165

UTS: Ultimate tensile strength

Insulator Strings for 132 kV, 220 kV and 400 kV

Particulars	Disc insulator (standard type)				Disc insulator (anti-fog type)			
	70 KN	90 KN	120 KN	160 KN	70 KN	90 KN	120 KN	160 KN
Size and designation of the ball pin shank (mm)	16	16	20	20	16	16	20	20
Diameter of the disc (mm)	255/280	255/280	280	280	280/ 305	255 /280	280	280
Tolerance on the diameter +/– (mm)	11/13	11/13	13	13	13/15	11/13	13	13
Ball to ball spacing between disc (mm)	145	145	145	170	145	145	145	170
Tolerance on ball to ball spacing +/– (mm)	4	4	4	5	4	4	4	5
Minimum creepage distance of a single disc@	292	292	315	330	430	292	315	330
Steepness of the impulse voltage which the disc unit can withstand in steep wave front test (kV per μ sec.)	2500	2500	2500	2500	2500	2500	2500	2500
Purity of zinc used for galvanizing (in %)	99.95	99.95	99.95	99.95	99.95	99.95	99.95	99.95
No. of dips in standard preece test								
(i) Cap socket	6	6	6	6	6	6	6	6
(ii) Ball pin	6	6	6	6	6	6	6	6

@ The minimum creepage distance of single composite insulator unit shall be such that it matches with the total creepage distance of the respective strings with disc insulator units.

AIR CLEARANCE AND SWING ANGLES
(Live Conductor to Earthed Metal Part)

System voltage/line voltage (kV)	Single suspension insulator		Jumper	
	Swing from vertical (degree)	Minimum clearance (mm)	Swing from vertical (degree)	Minimum clearance (mm)
72/66	Nil	915	Nil	915
	15	915	10	915
	30	760	20	610
	45	610	30	610
145/132	Nil	1530	Nil	1530
	15	1530	10	1530
	30	1370	20	1070
	45	1220	30	1070
	60	1070	–	–
245/220	Nil	2130	Nil	2130
	15	1980	10	2130
	30	1830	20	1675
	45	1675	–	–
420/400	Nil	3050	Nil	3050
	22	3050	20	3050
	44	1860	40	1860
800 Zone-I & II	Nil	5600/5100	Nil	5100
	22	4400	15	4400
	45	1300	30	1300
800 Zone-III & IV	Nil	5600/5100	Nil	5100
	27	4400	20	4400
	55	1300	40	1300
800 Zone-V & VI	Nil	5600/5100	Nil	5100
	30	4400	22	4400
	60	1300	45	1300
500 DC	Nil @ (V–string)	3750	40	1600

FORMULAE FOR VERTICAL AND HORIZONTAL CLEARANCE IN METRES

1. Vertical clearance $0.75\sqrt{d_{75}\ l_s}\ (V/150)$

2. Horizontal clearance $0.62\sqrt{d_{75}\ l_s}\ (V/150)$

   where,

   d_{75} = Sag at 75° C, (m)
   l_s = Length of insulator string (m)
   V = Voltage in kV.

TYPICAL SAG-TENSION CALCULATION OF CONDUCTOR (AAAC-ZEBRA)

Properties of Conductor

Sectional area—487.5 mm²
Overall Diameter—28.71 mm
Weight—1.3455 kg/m
Modulus of elasticity (E)—5608 × 10⁶ kg/m²
Coefficient linear expansion ( )—23 × 10⁻⁶/°C
Minimum ultimate tensile tension—131.63 KN = 134522 kg
Normal span—350 m
Wind pressure—52 kg/m²
Min. temperature—0° C
Max. temperature—90° C
Every day temperature—32° C
Factor of safety—4

Calculation

Working tension (T_1) = Minimum ultimate tensile tension/FOS= 134522 kg/4 = 3355.5 kg
Working stress (f_1) = 3355.5/487.5 × 10⁻⁶ = 6.88 × 10⁶ kg/m²

Sag and tension at 90° C (still air)—hot sag

$$F = [f_1 - (w^2 l^2 E/24\ T_1{}^2)] - E(\ _2 - \ _1)$$

$$= [6.88 \times 10^6 - (1.3455^2 \times 350^2 \times 5608 \times 10^6/24 \times 3355.5^2)] - 23 \times 10^{-6}$$
$$\times 5608 \times 10^6\ (90 - 32)$$

$$= -5.20041 \times 10^6$$

T_2 = Tension = $f_2 \times A$

But $f_2^2 (f_2 - F) = G = w^2 l^2 E/24\, A^2$

$f_2^2(f_2 + 5.20041 \times 10^6) = 1.3455^2 \times 350^2 \times 5608 \times 10^6/24 \times 487.5 \times 10^{-12}$

Now $f_2 = 4.7006 \times 10^6$ kg/m²

$T_2 = \text{Tension} = f_2 \times A = 4.7006 \times 10^6 \times 487.5 \times 10^{-6} = 2292$ kg

$\text{SAG} = D_2 = wl^2/8\, T_2 = 8.989$ m

Sag and tension at 0° C (still air)—cold sag

$F = [f_1 - (w^2 l^2 E/24\, T_1^2)] \quad E(\theta_2 - \theta_1)$

$\quad = [6.88 \times 10^6 - (1.3455^2 \times 350^2 \times 5608 \times 10^6/24 \times 3355.5^2)] -23 \times 10^{-6} \times 5608$
$\quad\quad \times 10^6 (0 - 32)$

$\quad = 6.40815 \times 10^6$

$T_3 = \text{Tension} = f_3 \times A$

But $f_3^2 (f_3 - F) = G = w^2 l^2 E/24\, A^2$

$f_3^2(f_3 - 6.40815 \times 10^6) = 1.3455^2 \times 350^2 \times 5608 \times 10^6/24 \times 487.5 \times 10^{-12}$

Now $f_3 = 9.0629 \times 10^6$ kg/m²

$T_3 = \text{Tension} = f_3 \times A = 4418$ kg

$\text{SAG} = D_3 = wl^2/8\, T_3 = 4.663$ m

SPAN VERSUS SAG TABLE (AAAC–ZEBRA)

Span	Hot sag (D_2) $w\,l^2/8\,T_2$	Cold sag (D_3) $w\,l^2/8\,T_3$	Span	Hot sag (D_2) $w\,l^2/8\,T_2$	Cold sag (D_3) $w\,l^2/8\,T_3$
10	0.00733	0.0038	300	6.60422	3.20157
50	0.18345	0.09517	350	8.98907	4.663415
100	0.73380	0.38068	400	11.7408	6.090991
150	1.65105	0.85654	450	14.8594	7.708910
200	2.93520	1.52274	500	18.3450	9.51717
250	4.58626	2.37929	550	22.1975	11.51578

LINE LOADING (SIL—SURGE IMPEDANCE LOADING)

Voltage (kV)	No./Size of conductor (mm²)	SIL (MW)
765	4/686	2250
765	4/686	614
400	2/520	515
400	4/420	614
400	3/420	560
400	2/520	155
220 ZEBRA	420	132
132	200	50

TRANSMISSION TOWERS

Type and Shape of Towers

	Types	Utilization
Suspension (with I or V insulator)	Tangent tower(0°),	Straight run
	Intermediate tower (0° to 2°), A type	Straight run and up to 2° deviation
	Light angle tower (0° to 5°)	Straight run and up to 5° deviation
Tension tower with tension string	(0° to 15°) B type	Line deviation (0° to 15°)
	(15° to 30°) C type	Line deviation (15° to 30°)
	(30° to 60°) D type	Line deviation (30° to 60°)
	Dead end	Dead end or anchor tower
	Large angle and dead end	Line deviation (30° to 60°) or dead end

Note 1. **Transposition tower** Two multiple tension insulator strings are connected back-to-back through a strain plate, a single suspension insulator string with almost the double insulator discs and air gap is suspended.

2. **Special Tower** Used for river crossing, valley crossing, creek crossing and power line crossing.

Tower Spotting Data (Reference) (E.g., 220 kV DC Tower)

Data	Tower type			
	A	B/sec.	C/sec.	D/DE
Deviation not to exceed	2°	15°/0°	30°/0°	60°/15°
Insulator string	Suspension	Tension	Tension	Tension
Vertical load limit weight span	**min. (max.)**	**min. (max.)**	**min. (max.)**	**min. (max.)**
Ground wire				
Effect of both span	525 (100)	525 (100)	600 (100)	600 (100)
Effect of one span	315 (100)	315 (100)	360 (100)	360 (100)
Conductor				
Effect of both span	525 (100)	525 (100)	600 (100)	600 (100)
Effect of one span	315 (100)	315 (100)	360 (100)	360 (100)
Weight of wires				
Ground wire				
Effect of both span	227(43)	227(43)	258 (43)	258 (43)
Effect of one span	136 (43)	136 (43)	155 (43)	155 (43)
Conductor				
Effect of both span	706 (135)	706 (135)	807 (135)	807 (135)
Effect of one span	424 (135)	424 (135)	484 (135)	484 (135)

Data	Tower type							
	A		B/sec.		C/sec.		D/DE	
Permissible sum of adjacent of one span								
For deviation angles	2°	700	15°	700	30°	700	60°	700
	1°	810	14°	809	29°	806	59°	795
	0°	910	13°	918	28°	913	58°	891
	–	–	12°	1027	27°	1020	57°	988
	–	–	11°	1137	26°	1127	56°	1085
	–	–	10° and below	1247	25° and below	1234	55° and below	1182
	–	–						
Design load tension in kg								
Ground wire								
32° C (full wind)	1640		1626/1640		1584/1640		1428/1640	
0°C (2/3 full wind)	1712		1697/1712		1653/1712		1483/1712	
Conductor								
32° C (full wind)	2122		4287/4843		4098/4243		3675/4243	
0° C (2/3 full wind)	2406		4771/4812		4648/4812		4167/4812	

Slopes of Tower Legs and Shield Angles

Voltage	Towers	Slope	Voltage level	Shield angle
Up to 220 kV	Suspension	4°–9°	66, 110, 132, 220 kV	30°
	Angle	7°–11°	400 kV SC (horizontal) outer phase	20°
	Dead end	8°–13°	400 kV SC (vertical)	20°
400 kV and above	Suspension	8°–12°	400 kV DC	20°
	Angle	10°–17°	800 kV SC (horizontal) outer phase	15°
	Dead end	11°–15°	Inner phase	45°

Standard Steel Section used for Towers

Type	Section (mm)	Unit weight kg/m	Remarks (use)
HT	150×150×20	44.1	Leg
HT	150×150×16	35.8	Leg
HT	150×150×15	33.8	Leg
HT	150×150×12	27.3	Cleat
HT	130×130×12	23.5	Cleat
HT	130×130×10	19.7	Cleat
HT	120×120×10	18.2	Leg
MS	110×110×8	13.4	Leg
HT	110×110×8	13.4	X A Rm
HT	110×110×10	16.6	Leg
MS	100×100×8	12.1	Leg
MS	100×100×7	10.7	Leg
MS	100×100×6	9.2	Leg
HT	90×90×7	9.6	Bracing
MS	90×90×6	8.2	Cleat
HT	90×90×6	8.2	Leg
MS	80×80×6	7.3	Cleat

(Contd.)

Type	Section (mm)	Unit weight (kg/m)	Remarks (use)
HT	80×80×6	7.3	Bracing
MS	75×75×6	6.8	Cleat
HT	75×75×6	6.8	Bracing
MS	70×70×5	5.3	Bracing
HT	70×70×5	5.3	Bracing
MS	65×65×5	4.9	Redundant
HT	65×65×5	4.9	Bracing
HT	65×65×4	4.0	Redundant
HT	65×65×4	4.0	Bracing
MS	60×60×5	4.5	Redundant
HT	60×60×4	3.7	Redundant
HT	55×55×4	3.3	Redundant
MS	55×55×4	3.3	Bracing
HT	50×50×4	3.0	Redundant
MS	50×50×4	3.0	Redundant
MS	45×45×4	2.7	Redundant
HT	45×45×4	2.7	Redundant

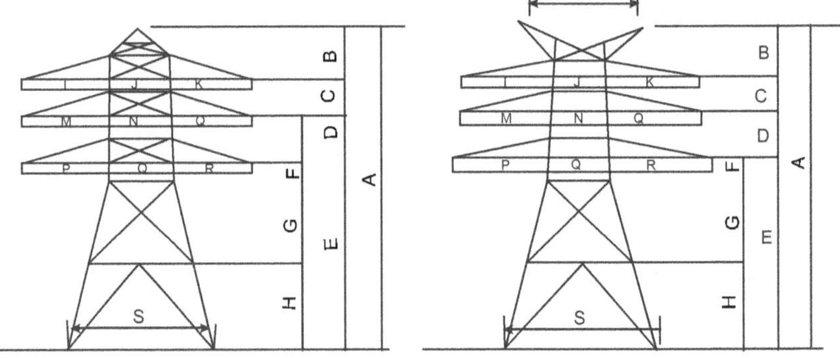

Tower dimension codal representation

Dimensions on Various Categories of Towers [Refer Schematic Diagram]

132 kV (Normal span-300 m, conductor-37/3, 15 mm AAA, earth wire-7/3.15 mm GS wire) all in mm

Tower dimension	A	B	C	D	E	F	G	H	I
Type of Tower									
DA (0 to 2°)	28908	4193	4170	4230	16315	2455	7606	6100+150	3150
DB (2 to 15°)	27550	5890	3900	3900	13860	0	7606	6100+150	2600
DC (15 to 30°)	27810	6150	3900	3900	13860	0	7606	6100+150	2660
DD (30 to 60 °)	28280	5990	4130	4300	13880	0	7607	6100+150	2475

	J	K	M	N	O	P	Q	R	S	L
DA (0 to 2°)	850	3150	3028	1024	3028	3045	1200	3045	5890	
DB (2 to 15°)	1600	2600	2525	1750	2525	2560	1900	2560	6700	
DC (15 to 30°)	1800	2360	2770	1900	2130	2905	2000	2100	7000	
DD (30 to 60°)	1860	2475	2291	2217	2291	2665	2600	2665	8600	

220 KV(Normal span-350 m, conductor-37/4 mm AAA, earth wire-7/3.15 mm GS wire)

	A	B	C	D	E	F	G	H	I
DA (0 to 2°)	35506	5540	5175	5275	19515	2768	9582	7015+150	3600
DB (2 to 15°)	34490	7940	4900	4900	16750	0	9582	7015+150	3706
DC (15 to 30°)	35390	7940	5350	5350	16750	0	9582	7015+150	3466
DD (30 to 60°)	35255	8406	5050	5060	16750	0	9582	7015+150	3650

(Contd.)

220 kV (Normal span-350 m, conductor-37/4 mm AAA, earth wire-7/3.15 mm GS wire)

Tower dimension

Type of tower	J	K	M	N	O	P	Q	R	S	L
DA (0 to 2°)	1400	3600	3591	1598	3591	3900	1800	3900	7502	
DB (2 to 15°)	1750	3025	3612	1975	3012	3475	2200	3096	7830	
DC (15 to 30°)	2068	3166	3757	2285	2667	3950	2500	3180	8500	
DD (30 to 60°)	2400	3650	3450	2800	3450	3250	3200	3250	9900	

400 kV (Normal span-400 m, conductor-61/3.45 mm AAA, earth wire-7/3.66 mm GS wire)

Type of tower	A	B	C	D	E	F	G	H	I
DA (0 to 2°)	46030	4660	8100	8300	24970	4140	11838	8840+150	4730
DB (2 to 15°)	45069	8241	8000	8000	20828	0	11838	8840+150	6000
DC (15 to 30°)	45069	8241	8000	8000	20828	0	11833	8840+150	6740
DD (30 to 60°)	45578	8250	8200	8300	20828	0	11838	8840+150	5450

Type of tower	J	K	M	N	O	P	Q	R	S	L
DA (0 to 2°)	1800	4730	4750	2245	4750	5400	2700	5400	10200	6200
DB (2 to 15°)	2500	5230	5975	3750	5225	6500	5000	5720	14500	8960
DC (15 to 30°)	2600	5330	6740	4100	5380	7200	5600	5820	14607	10648
DD (30 to 60°)	2800	5450	5454	4192	5454	6400	5600	6400	15700	8500

Note: - 1. Dimensions are for standard towers only   2. "E" length of suspension string including hanger rope
3. "G" max. sag of conductor at 85° C   4. " H" minimum ground clearance including sag error
These points are to be considered for tower dimension.

DATA ON TRANSMISSION LINE CONDUCTOR

DATA FOR SOME ACSR CONDUCTORS

Code name	Eq. area (mm²)		Size/diameter (No./mm)		OD (mm)	Wt. (kg/km)	Breakdown strength (kg)	Rdc_{20° (Ohm/km)	A
	Cu	Al	Al	Steel					
Parrot	25	41.87	6/3.0	1/3.0	9	171	1503	0.6795	115
Mink	40	62.32	6/3.66	1/3.66	11	255	2207	0.4565	165
Beaver	45	74.07	6/3.99	1/3.99	12	303	2613	0.3841	176
Otter	50	82.85	6/4.22	1/ 4.22	12.6	339	2923	0.3434	185
Lion	140	232.5	30/3.18	7/3.18	22.3	1097	10210	0.1223	405
Deer	260	419.3	30/4.27	1/ 4.27	29.9	1977	16230	0.06756	590
Camel	300	464.5	54/3.35	7/3.35	30.2	1804	14750	0.06125	610
Lark	–	–	30/2.924	7/2.924	20.47	923	9080	–	–
Bear	–	–	30/3.353	7/3.353	23.5	1219	11300	–	–
Kundah	–	–	42/3.595	7/1.96	26.88	1218	9050	–	–
Canary	–	–	54/3.280	7/13.28	29.51	1721	14650	–	–
Dove	–	–	26/3.72	7/2.89	23.55	1137	10180	–	–
Redwing	–	–	30/3.92	19/2.35	27.46	1648	15690	–	–
Bersimis	–	–	42/4.57	7/2.34	35.1	2185	15734	–	–
Curlew	–	–	54/3.51	7/3.51	31.62	1976	16850	–	–
Duck	–	–	54/2.69	7/2.69	24.18	1158	10210	–	–
Gopher	–	–	6/2.36	1/2.36	7.09	106	952	1.098	85

COMMONLY USED ACSR LINE CONDUCTORS AND EARTH CONDUCTORS (GALVANIZED)

Code name	Size/diameter (No./mm)		Voltage (kV)	OD (mm)	Area (mm^2)	Wt. kg/km	Breakdown strength (kg)	$Rdc_{20°}$ (ohm/km)	A	Max. allow temp. (°C)
	Al	Steel								
Dog	6/4.72	7/1.57	66	14.15	118.5	394	3290	0.281	218	75
Panther	30/3	7/3	110, 132	21	261.5	974	9242	0.14	350.7	75
Zebra	54/3.18	7/3.18	220	28.62	484.5	1621	13466	0.06915	510.8	75
Twin Moose	54/3.53	7/3.53	400	31.77	597	2004	16594	0.05552	571.6	75
Commonly used earth conductors (galvanized)										
Steel	7/3.15 Steel		66–220	9.54	54.55	428	5913	3.375	–	53
Two steel	7/3..66 Steel		400	10.98	73.65	583	7848	2.5	–	53

CHARACTERISTICS OF ACSR CONDUCTOR AS PER IS 398 PART II 1976

	Area (mmsq.)	OD (mm)	Wt (kg)	Rdc20° (ohm/km)	Breakdown strength kg	Rac67 (ohm/km)	Rac75 (ohm/km)	Amp75°	Amp90°
ACSR Mole 6+1/1.50	10.6	4.5	42.8	2.779845	379.1	3.79784	3.9	56.3	66.9
ASCR 6+1/1.96	18.1	5.88	73.2	1.617859	647.2	2.21034	2.27	77.7	92.7
ASCR Squirrel 6+1/2.11	21	6.33	84.8	1.393964	750	1.90446	1.956	84.8	101
ASCR Weasel 6+1/2.59	31.6	7.77	128	0.929104	1130	1.26938	1.304	107.9	130

(Contd.)

	Area (mm sq.)	OD (mm)	Wt (kg)	Rdc20° (Ohm/km)	Breakdown (kg)	Rac 67 (Ohm/km)	Rac 75 (Ohm/km)	Amp 75°	Amp 90°
ASCR Rabbit 6+1/3.35	**52.9**	**10.05**	**214**	**0.555761**	**1891**	**0.75935**	**0.78**	**146**	**177**
ASCR Racon 6+1/4.09	78.8	12.27	319	0.371222	2818	0.50727	0.521	184.7	225
ASCR Horse 12+7/2.79	73.4	13.95	536	0.399665	6214	0.54605	0.561	181.8	225
ASCR Cat 6+1/4.50	95.4	13.5	386	0.3007	3412	0.41094	0.422	208.5	255
ASCR Dog 6 /4.72+7/1.57	**105**	**14.15**	**394**	**0.279197**	**3290**	**0.38158**	**0.392**	**218**	**267**
ASCR Leopard 6/5.28+7/1.75	13.14	15.81	452	0.223462	4102	0.30547	0.314	247.9	305
ASCR Tiger 30+7/2.36	131.2	16.52	602	0.225908	5719	0.26931	0.277	265.8	328
ASCR Coyote 26/2.54+7/1.90	131.7	15.86	519	0.224771	4462	0.26799	0.275	264.8	326
ASCR Wolf 30+7/2.59	158.1	18.13	725	0.187139	6888	0.22314	0.229	296.1	366
ASCR Lynx30+7/2.79	183.4	19.53	842	0.161	7993	0.19202	0.197	222.7	401
ASCR Panther 30+7/3.00	**212.1**	**21**	**973**	**0.139037**	**9242**	**0.16588**	**0.17**	**350.7**	**437**
ASCR Goat 30+7/3.71	324.3	25.97	1488	0.091057	14134	0.10882	0.112	445.3	562
ASCR Sheep 30+7/3.99	375.1	27.93	1721	0.078605	16348	0.09403	0.965	483.2	613
ASCR Kundah 42.3.53+7/1.96	404.1	26.88	1280	0.073106	9015	0.09017	0.093	491.2	622
ASCR Zebra 54+7/3.18	**428.9**	**28.62**	**1620**	**0.068691**	**13466**	**0.08463**	**0.087**	**510.8**	**649**
ASCR Moose 54+7/3.53	**528.5**	**31.77**	**1996**	**0.055959**	**16594**	**0.06919**	**0.071**	**571.6**	**732**
ASCR Drake 26/4.44+7/3.45	402.6	28.11	1625	0.07094	14197	0.08503	0.087	508.5	645
ASCR Morkulla 42/4.13+7/2	30563	31.68	1780	0.052576	12515	0.06533	0.067	588.1	753

DETAILS OF GI WIRES AND STAY WIRES

Description	Size	Weight per km (kg)	Diameter of wire (mm)	Max. breaking load in kg			
				40 ton	45 ton	60 ton	70 ton
Stay wire	7/8 SWG	729	4.06	5730	6446	8270	10029
Stay wire	7/10 SWG	456	3.25	3669	4135	5505	6418
Stay wire	7/12 SWG	306	2.64	2421	2724	3634	4238
Stay wire	7/14 SWG	209	2.03	1432	1613	2151	2507
G.I. wire	4 SWG	214	5.89	1721	–	2582	–
G.I. wire	6 SWG	146	4.87	1179	–	1768	–
G.I. wire	8 SWG	103	4.06	819	–	1228	–
G.I. wire	10 SWG	65	3.25	524	–	746	–

CHARACTERISTICS OF AAAC CONDUCTOR AS PER IS 398 PART IV 1979

Size	Area (mm²)	OD (mm)	Wt (kg)	Rdc$_{20°}$ (ohm/km)	Breakdown strength (kg)	Rac67 (Ohm/km)	Rac 75 (Ohm/Km)	Rac 90 (Ohm/km)	Amp 67°	Amp 75°	Amp 90°
10 (7/1.49)	12.2	4.47	33.4	2.697011	347.9	3.15342	3.231	3.376735	61.7	73.4	90.4
20 (7/2.09)	24	6.27	65.7	1.370785	684.4	1.60277	1.642	1.716271	92.3	111	137
30 (7/2.56)	36	7.68	98.5	0.913655	1027	1.06832	1.095	1.143968	117	141	176
50 (7/3 .31)	60.2	9.39	165	0.54652	1717	0.63912	0.655	0.068436	159	192	241
80 (19/2.46)	903	12.3	248	0.366307	2574	0.42848	0.439	0.458795	201	245	309
100 (19/2.79)	116.2	13.95	319	0.284778	3311	0.3321	0.341	0.356773	233	285	360
150 (37/2.49)	160.2	17.43	496	0.184006	5135	0.21552	0.221	0.230732	300	371	471
200 (37/2.88)	241	20.16	664	0.137545	6869	0.16124	0.165	0.172683	354	441	537
400 (37/3.92)	446.5	27.44	1230	0.074243	12727	0.08775	0.09	0.093831	499	633	821
420 (61/3.19)	487.5	28.71	1346	0.68153	13163	0.08071	0.083	0.086283	523	666	865
520 (61/3.55)	603.8	31.95	1666	0.055031	16302	0.0556	0.067	0.070077	587	753	984
560 (61/3.68)	648.8	33.12	1791	0.061212	17518	0.06123	0.063	0.06538	610	785	1027
7/2.00	22	6	60.1	1.496932	626.7	1.75026	1.793	1.874203	87.6	105	130
7/2.50	34.4	7.5	94	0.958036	979.3	1.12021	1.148	1.19953	114	137	171
7/3.15	54.6	9.45	149	0.603449	1555	0.70567	0.072	0.755628	150	181	227
7 /3.81	79.8	11.43	218	0.412489	2275	0.48245	0.494	0.516595	187	228	286
7/4.26	99.8	12.78	273	0.329946	2844	0.38599	0.395	0.413291	213	260	328
19/2.89	124.6	14.45	343	0.465411	3552	0.31058	0.318	0.332542	242	298	376

(Contd.)

19/3.15	148.1	15.75	407	0.223406	4220	0.26152	0.268	0.279998	268	330	418
19/3.40	172.5	17	474	0.19176	4916	0.22457	0.23	0.240425	292	631	459
19/3.66	199.9	18.3	549	0.165483	5697	0.49391	0.499	0.207581	318	394	403
19/3.94	231.7	19.7	634	0.142798	6602	0.16745	0.172	0.179242	346	430	550
37/3.15	288.3	22.05	794	0.114977	8218	0.13504	0.138	0.14452	391	490	629
37/3.45	345.9	24.15	953	0.9585	9858	0.1128	0.116	0.120686	433	545	703
37/3.71	400	25.97	1102	0.082886	11400	0.09775	0.1	0.104561	470	594	768
37/4.00	465.1	28	1281	0.71303	13251	0.08435	0.086	0.090186	510	448	841
61/3.31	524.9	29.79	1449	0.063301	14172	0.07511	0.077	0.08028	545	695	904
61/345	570.2	31.05	1574	0.058268	15397	0.06932	0.071	0.074065	570	729	950
61/3.55	603.8	31.95	1666	0.055031	16302	0.0656	0.061	0.070077	587	753	984
61/3.66	641.8	32.94	1771	0.051773	17328	0.06187	0.063	0.66069	607	780	1020
61/3.81	695.5	34.29	1919	0.047777	18777	0.0573	0.059	0.061167	633	816	1070
61/4.00	766.5	36	2116	0.043346	20697	0.05226	0.053	0.055751	666	862	1134

ALL ALUMINIUM CONDUCTORS (AAC)

Code name	Eq. area (mm^2)		Size/diameter (No./mm)	OD (mm)	Weight (kg/ km)	Break down strength (KN)	Rdc$_{20°}$ (Ohm/ km)
	Cu	Al					
Rose	13	20	7/1.96	5.88	58	2.6	2.156
Gnat	16	25	7/2.21	6.63	74	4.52	1.093
Pancy	25	40	7/2.78	8.34	116	7.55	0.6756
Ant	30	50	7/3.1	9.3	145	8.25	0.5561
Fly	40	60	7/3.4	10.2	174	11.64	0.4244
Slutrottle	45	70	7/3.66	10.95	201	12.6	0.3915
Clap	60	90	7/4017	12.61	261	14.05	0.3369
Wasp	65	100	7/4.39	13.17	290	15.96	0.277
Crafter	130	200	19/3.78	18.9	586	34.05	0.1393
Spider	104	240	19/3.55	19.95	654	35.74	0.1244
Butterfly	185	300	19/4.65	23.25	888	48.74	0.0917
Moth	225	370	19/5.00	25	1025	52.5	0.0837

SOME COMMONLY USED AAAC CONDUCTORS

Description	400 kV, AAAC 61/3.45	220 kV, AAAC 61/3.45	132 kV, AAAC 37/3.45
Profile	DC single AAAC conductor per phase in vertical position		
Span in metres			
i. Design span	400	350	220
ii. Max. span	1100	1100	1000
iii. Min. span	100	100	100
Tensile load in kg f per conductor for wind 50 m/s (Zone 5)			
A) For temp. 5° C (still air)	3265	2919	1791
B) For temp. 5° C (36 % wind speed)	4646	4090	2735
C) For temp. 32° C (36 % wind speed)	7805	6551	4469
Armorous rod used	Standard preformed armour rods/AGS		
Max. permissible dynamic strain	+150 microstrains		

DETAILS OF EARTH WIRES

Standing no./diameter (mm)	Weight per m (kg)	Overall diameter (mm)	Total sectional area (mm²)	Ultimate tensile strength 700 kg N/(mm²)	Ultimate tensile strength 1100 kg N/(mm²)	Ultimate tensile strength 1570 kg N/(mm²)	Modulus of elasticity kg/cm²	Coefficient of linear expansion (per°C)
7/3.15	0.429	9.45	54.552	3899	5913	8297	1.933×10^6	11.50×10^{-6}
7/3.50	0.523	10.50	67.348	4587	7177	10243	–	–
7/3.66	0.583	10.98	73.648	4994	7848	11201	–	–
7/4.00	0.690	12.00	87.965	5985	9374	13379	–	–
19/3.15	1.163	15.75	148.069	10041	15778	–	–	–
19/3.50	1.436	17.50	182.801	12396	19479	–	–	–
19/3.66	1.570	18.30	199.897	13555	–	–	–	–
19/4.00	1.875	20.00	238.761	16191	–	–	–	–
Optical fibre glass								
1/5+8/3.2	0.458	11.40	71.41	8000	–	–	1.520×10^6	013.40×10^6

DATA ON ELECTRICAL CLEARANCES

DIFFERENT CLEARANCES UP TO 66 kV (AC)

Category of clearances	Unit	LV	MV	11 kV	33 kV	66 kV	Remark
Clearances of overhead electrical lines across the street	m	5.8	5.8	6.1	6.1	6.4	Rule 77 IE 1956
Along the street	m	5.5	5.5	5.8	6.1	6.4	Rule 77 IE 1956
Elsewhere	m	4.6	4.6	5.2	5.5	5.8	Do
Minimum clearance between conductor and tree	m	–	–	2.6	2.8	3.4	IS 5613
Clearance to ground at crossing with tram-way/trolley	m	1.2	1.2	2.44	2.44	3.05	Rule 78 IE 1956
Clearance to ground at crossing with telecom lines	m	–	–	–	–	2.44	IS 5613 (Part II/ Sec-I) 1985
Clearance to ground at railway crossing	m						
1. Inside station		–	–	13.28	13.28	13.59	IS 5613 (Part II/ Sec-I) 1985
2. Outside station		–	–	11.28	11.59	12.20	
Clearance to ground at electrified railway crossing	m						
1. Inside station		–	–	15.28	15.28	15.59	IS 5613 (Part II/Sec-I) 1985
2. Outside station		–	–	13.28	13.28	13.59	
Vertical clearance of overhead lines from any building roof	m	2.5	2.5	3.7	3.7	4.0	Rule 77 &79 IE 1956
Horizontal clearance of overhead lines from any building roof	m	1.2	1.2	1.2	2.0	2.0	Rule 79 & 80 IE 1956
Ground clearance in outdoor Sub station	m	–	–	2.75	3.7	4.0	Rule 64-2 (a) (ii) IE 1956

(Contd.)

Category of clearances	Unit	LV	MV	11 KV	33 KV	66 KV	Remark
Sectional clearance in outdoor sub station	m	–	–	2.6	2.8	3.0	”
Clearance from power conductor to earthed metal parts	m	–	–	–	0.33	0.915	IS–4513 (Pt-2/ Sec-1) 1985
Clearance live conductor to earthed metal parts of switchgear	m	–	–	0.12	0.32	0.63	IS 10118 (Pt-III 1982
Phase to phase clearance of overhead lines	m	–	–	0.4	0.4	0.75 to 1.6	IS -5613
Phase to phase clearance inside Sub station	m	–	–	0.4	0.4	0.75	”
Phase to phase clearance inside switchyard	m	–	–	0.12	0.32	0.63	”
Clearance of right way width	m	–	–	7	15	18	IS–5613 (Pt-2/ Sec-1) 1985
Clearance between line crossing each other	m	–	–	2.44	2.44	2.44	Do
Clearance line conductor to earth wire of transmission line (midspan)	m	–	–	–	1.5	3.0	Do- (Pt-2/Sec-2)
Minimum ground clearance of transmission lines	m	–	–	–	–	5.5	Rule 77-4 IE 1956
Minimum clearance above highest flood level of transmission lines	m	–	–	–	–	3.65	IE 1956
Clearance line conductor to earth wire of at tower (angle)	θ	–	–	–	–	30	IE 1956
Minimum clearance between phases (cable box)	mm	50.8	50.8	50.8	127	–	
Minimum clearance to earth (cable box)	mm	50.8	50.8	50.8	101.6	–	

DIFFERENT CLEARANCES FROM 132 kV TO 400 kV (AC) AND 500 kV (DC)

Category of clearances	Unit	132 kV	220 kV	400 kV	800 kV	± 500 kV	Remarks
Clearances of overhead electrical lines across the street	m	7.01	7.83	9.5	–	–	Rule 77 IE 1956
Along the street	m	7.01	7.83	9.5	–	–	Rule 77 IE 1956
Elsewhere	m	6.41	7.23	8.9	–	–	do
Minimum clearance between conductor and tree	m	4.0	4.6	5.5	–	–	IS 5613
Clearance to ground at crossing with tram-way/ trolley	m	3.57	3.85	–	–	–	Rule 78 IE 1956
Clearance to ground at crossing with relecom lines	m	3.05	4.58	5.49	7.94	–	IS 5613 (Part II/ Sec-I) 1985
Clearance to ground at railway crossing	m						IS 5613 (Part II/ Sec-I) 1985
i. Inside station		14.2	15.11	–	–	–	
ii. Outside station		13.11	–	–	–	–	
Clearance to ground at electrified railway crossing	m						IS 5613 (Part II/ Sec-I) 1985
i. Inside station		16.2	18.66	–	–	–	
ii. Outside station		14.2	5.11	–	–	–	
Vertical clearance of overhead lines from any building roof	m	4.6	5.4	7.04			Rule 77 and 79 IE 1956

(Contd.)

Category of clearances	Unit	132 kV	220 kV	400 kV	800 kV	±500 kV	Remarks
Horizontal clearance of overhead lines from any building roof	m	2.3	2.9	3.7	5.34	–	Rule 79 & 80 IE 1956
Ground clearance in outdoor substation	m	4.6	5.5	8.8	–	–	Rule 64-2 (a) (ii) IE 1956
Sectional clearance in outdoor substation	m	3.6	4.3	5.57	–	–	”
Clearance from power conductor to earthed metal parts	m	1.53	2.13	–	–	–	IS –4513 (Pt-2/ Sec-1) 1985
Clearance live conductor to earthed metal parts of switchgear	m	0.9	1.3	–	–	–	IS 10118 (Pt-III) 1982
Phase to phase clearance of overhead lines	m	1.35 to 2.7	2.3 to 5.2	3.4 to 6.1	–	–	IS -5613
Phase to phase clearance inside substation	m	1.35	2.3	3.4	–	–	”
Phase to phase clearance inside switchyard	m	1.3	1.5	2.4	–	–	”
Clearance of right way width	m	27	35	52	–	–	IS –5613 (Pt-2/ Sec-1) 1985

(Contd.)

Clearance between line crossing each other	m	3.05	4.58	5.49	7.94	–	,,
Clearance line conductor to earth wire of transmission line (midspan)	m	6.1	8.5	9.0	12.4	8.5	" (Pt-2/ Sec-2)
Minimum ground clearance of transmission lines	m	6.1	7.0	8.8	12.4	12.5	Rule 77-4 IE 1956
Min. clearance above highest flood level of transmission lines	m	4.3	5.1	6.4	9.4	6.75	IE 1956
Clearance line conductor to earth wire of at tower (angle)	θ^0	30	30	20	20	10	IE 1956
Minimum clearance between phases (cable box)	mm	–	–	–	–	–	"
Minimum clearance to earth (cable box)	mm	–	–	–	–	–	"

DATA on ELECTRICAL EARTHING SYSTEM

TYPICAL VALUE OF EARTH RESISTIVITY IN OHM-METER

Particulars	Value
General average	100
Swampy ground	10–100
Pure slate	10^7
Sea water	0.01–1.0
Dry earth	1000
Sandstone	10^8

TYPE OF ELECTRODE USED FOR EARTHING

Type	Rod	Pipe	Strip	Round conductor	Plate
Diameter less than	16 mm 12.5 mm	38 mm Steel 100 mm GI	–	–	–
Length / depth of burial not less than	2.5 mm Ideal 3 to 3.5 mm	2.5 mm	0.5 m	15 m	1.5 m
Size	–	–	25 × 1.6 mm Cu 25 × 4 mm Steel	3 mm² Cu 6 mm² steel	60 × 60 cm
Thickness					6.3 mm steel 3.15 mm Cu

TESTING OF SOIL RESISTIVITY

Wenner's 4 point method

$$= 2\ A\,R,$$

where,

= Soil resistivity

A = Distance between the electrodes in m

R = Earth resistance (meter reading)

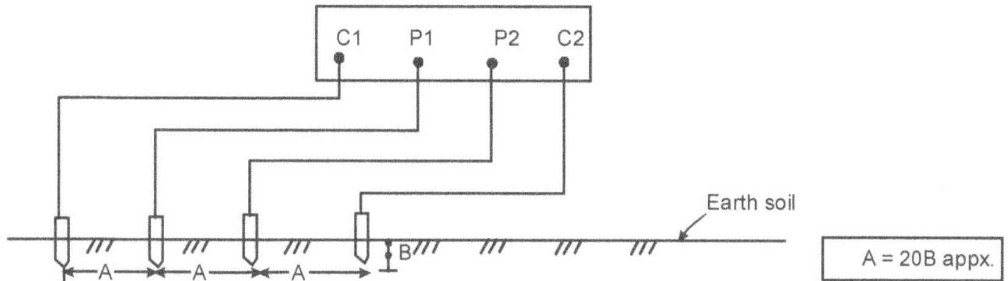

Points to be noted during measurement

1. 4 spikes to be kept at straight line at equal distance,
2. Link (terminals) of instrument P_1–C_1 and P_2–C_2 to be connected individually,
3. Reading of meter to be taken at 150 RPM.

MEASUREMENT OF EARTH ELECTRODE RESISTANCE

Ground Electrode (ROD)

Fall of potential method or three point method

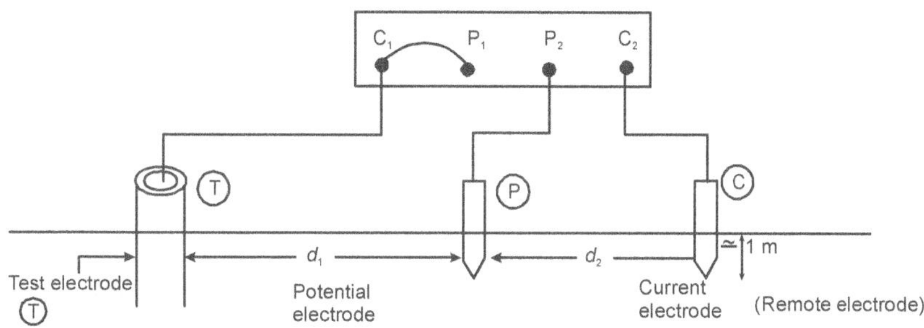

Points to be noted during measurement

1. $(d_1 + d_2)$ 30 m
2. Depth of current electrode (C) and potential electrode should be approximately 1 m or as per the dimension of spike.
3. 3 (d_1 / d_2) 1,
4. Reading of meter at 150 RPM

Note   Take $(d_1 + d_2) = 50$ m, depth of electrode = 1 m and $(d_1/d_2) = 1.5$ for better result

Ground Mat

Method of measurement same as ground rod.

FORMULA FOR DESIGN OF GROUNDING IN POWER SYSTEM

Safe Body Current

1. $I = 0.165/\sqrt{t}$ (As per AIEE 80/1961)
2. $I_{50} = 0.116/\sqrt{t}$ for person of 50 kg wt. (As per IEEE 80/1976)
3. $I_{70} = 0.157/\sqrt{t}$ for person of 70 kg wt. (As per IEEE 80/1986)
4. $I_{50} = 0.155/\sqrt{t}$ (As per IEEE 80/1976)
   where,
   > I = RMS current through human body
   >
   > t = Duration of current in second

Step and Touch Potential

1. Tolerable step potential (E step) = $(1000 + 6\, C_s \text{ps})\, 0.155/\sqrt{t}$
2. Tolerable touch potential (E touch) = $(1000 + 1.5 C_s \text{ps})\, 0.155/\sqrt{t}$
3. Attainable step potential (E step max) = $(K_s K_{Ip}\, I)/L$,
   where, $K_s = 1/\ [(1/2h) + 1/(D + h) + 1/(2D) + ... + n$ terms]
4. Attainable touch potential (E touch max) = $(K_m K_{Ip}\, I)/L$,
   where,

   $K_m = 1/2\ [\ln (D/16\ HD)] + 1/\ \times \ln 0.25$

   $K_{Ip} = 0.655 + 0.172\, n$

   C_s = Co-relation factor (As per IEEE 80/1985), its value is obtained according to thickness of crushed rock layer (*hs*) and

   Reflection factor $K = (p - ps)/(p + ps)$

   P = Earth resistivity in ohm-m

   ps = Surface layer resistivity in ohm-m

   t = Duration of allowable current in sec.

   I = Maximum current to the grounding system

   n = Number of parallel conductor in the grid

   h = Depth of burial in m

   L = Length of buried conductors in m

   D = Spacing of conductors in grid in m

   d = Diameter of conductor (equivalent to flat section).

Resistance of Earthing System (MAT)

Laurent Formula

$R = \{(p/4r) + (p/L)\}$ ohm

where,

p = Soil resistivity in Ohm-m

r = Radius of a circle in m having same area as that covered by earth mat

L = Length of buried conductors in m.

Size of Earth MAT Conductor

1. A (steel) = $0.0243\,I\sqrt{t}$ in mm² for welded joint as per CBIP.
2. A (steel) = $0.0292\,I\sqrt{t}$ in mm² for bolted joint as per CBIP.
   where,

   I = Fault current in A

   t = Duration of allowable current in sec

Assumptions Made for Design of Earth Mat

Terms	Value	Terms	Value
Surface resistivity for crushed rock	3000 ohm-m	Ambient temp.	50°C
Resistivity of steel	15 micro ohm-m	Density of material	7.86 g/cc
Specific heat of material	0.114 cal/gm/°C	Max. temp. of bolted joint	500°C
Depth of burial mat	0.6 m	Soil resistivity	100 ohm-m
Duration of fault current	1 sec.	Type of electrode	MS Flat (50 × 6 mm)
Human body resistance	1000	Allowable corrosion	1% per annum for the 1st 12 yrs and 0.5 % for the next 12 yrs as per ISS 3043/ 1987

Earth Resistance of Stations

Voltage class	Earth resistance
33 kV	5 and below
110 to 400 kV	Less than 1
66 kV	2 and below
Generating station	0.5 and below

OTHER DATA (EARTHING)

1. Size of earth conductor for domestic system

    = 2.5 mm² min. and 120 mm² max. for aluminum

    = 1.5 mm² min. and 70 mm² max. for copper

2. Resistance of earth electrode for domestic = 5   and below

3. Human body allowable limit of current

    Perception = 1 mA, let go current = 1 to 6 mA, muscular contraction = 9 to 25 mA

    Ventricular fibrillation = 100 mA

4. Resistance of human body = (500–3000)

DATA ON DISTRIBUTION SYSTEM
(Up to 33 kV System)

STANDARD WIRE GAUGE AND TINNED COPPER FUSE WIRES

SWG	Diameter (mm)	Ampere rating	Approx. fuse (A)	SWG	Diameter (mm)	Ampere rating	Approx. fuse (A)
40	0.122	1.5	3	16	1.63	73	166
39	0.132	2.5	4	15	1.83	78	197
38	0.152	3	5	14	2.03	102	230
37	0.173	3.5	6	13	2.34	130	295
36	0.193	4.5	7	12	2.64	DNT	DNT
35	0.213	5	8	11	2.95	DNT	DNT
34	0.234	5.5	9	10	3.25	DNT	DNT
33	0.254	6	10	9	3.66	DNT	DNT
32	0.274	7	11	8	4.06	DNT	DNT
31	0.295	8	13	7	4.47	DNT	DNT
30	0.315	8.5	13	6	4.88	DNT	DNT
29	0.345	10	16	5	5.38	DNT	DNT
28	0.376	12	18	4	5.89	DNT	DNT
27	0.417	13	23	3	6.40	DNT	DNT
26	0.457	14	28	2	7.01	DNT	DNT
25	0.508	15	30	1	7.62	DNT	DNT
24	0.559	17	33	1/0	8.23	DNT	DNT
23	0.610	20	38	2/0	8.84	DNT	DNT
22	0.711	24	48	3/0	9.45	DNT	DNT
21	0.813	29	58	4/0	10.20	DNT	DNT
20	0.914	34	70	5/0	11.00	DNT	DNT
19	1.02	38	81	6/0	11.80	DNT	DNT
18	1.22	45	106	7/0	12.70	DNT	DNT
17	1.42	65	135				

DNT—Data not available

SAG AND TENSION TABLE FOR WEASEL AND RABBIT CONDUCTOR

Span	ACSR 7/2.59 mm weasel		ACSR 7/3.35 mm rabbit	
	Sag (m)	Tension (kg)	Sag (m)	Tension (kg)
60	0.401	143	0.428	225
80	0.681	150	0.701	244
100	0.997	160	1.02	261
120	1.35	170	1.38	277
140	1.75	178	1.79	292
160	2.189	187	2.24	305

VOLTAGE REGULATION FOR THREE PHASE PER 100 kVA/km

Voltage (kV)	Code name of conductor	Equ. spacing in mm "D_e"	% Regulation for lag. pf		
			0.8	0.9	1
0.4	Gnat–AAC (7/2.21)	450	41	41.6	37
0.4	Ant–AAC (7/3.10)	450	35	34.6	33
11	Weasel–ACSR	1070	0.0897	0.0868	0.0840
11	Rabbit–ACSR	1070	0.0590	0.0589	0.0502
11	Raccoon–ACSR	1070	0.0450	0.0435	0.0337
33	Weasel–ACSR	1368	0.0100	0.0098	0.0097
33	Rabbit–ACSR	1368	0.0068	0.0067	0.0055
33	Raccoon–ACSR	1368	0.0053	0.0050	0.0037
33	Dog–ACSR	1368	0.0044	0.0041	0.0028
33	Coyote–ACSR	1368	0.0038	0.0037	0.0022
33	Weasel–ACSR	1520	0.00986	0.0101	0.0092
33	Rabbit–ACSR	1520	0.0069	0.00675	0.0055
33	Raccoon–ACSR	1520	0.00531	0.00504	0.0037
33	Dog–ACSR	1520	0.0045	0.00412	0.0028
33	Coyote–ACSR	1520	0.0037	0.0034	0.0022

1. Regulation is taken at temperature of 45°C
2. % regulation for 100 kVA/km
    = (Voltage drop in line × 100)/per phase voltage
    D_e = Equivalent spacing = $(D_1 \times D_2 \times D_3)^{1/3}$

SAG AND TENSION TABLE FOR RACCOON AND DOG CONDUCTOR

Span	ACSR 7/4.09 mm Raccoon		ACSR 6/4.72 + 7/1.57 mm, Dog	
	Sag (m)	Tension (kg)	Sag (m)	Tension (kg)
60	0.433	331	0.487	364
80	0.708	360	0.777	406
100	1.03	386	1.11	443
120	1.395	411	1.49	476
140	1.80	433	1.91	506
160	3.25	453	2.37	532

kW-km LOADING OF 3 PHASE LINES

Voltage	Conductor	Equation Spacing (mm)	kW-km loading for 1% voltage drop		
			0.8 lg	0.9 lg	1
400 V	AAC 16 mm^2	450	1.4	1.21	1.38
	AAC 30 mm^2	450	1.95	2.16	2.70
11 kV	ACSR 7/2.59 mm	1216	899	983	1190
	ACSR 7/3.35 mm	1216	1347	1520	1992
	ACSR 7/4.09 mm	1216	1758	2054	2967
33 kV	ACSR 7/2.59 mm	1216	8113	8910	10870
	ACSR 7/3.35 mm	1520	11594	13333	18182
	ACSR 7/4.09 mm	1520	15066	17857	27027
	ACSR 6/4.72 mm + 7/1.57	1520	17778	21845	35714
	ACSR 26/2.54 mm + 7/1.91	1520	21622	26470	45455

HORN GAP FUSE PROTECTION
FOR DISTRIBUTION TRANSFORMERS

33 kV Side with horn gap of 15 inch or 385 mm				
Transformer capacity (kVA)	Rated current	Tinned Cu fuse SWG	Fuse rating	Approx. fusing current
100	1.75	33	6	10
200	3.5	30	8.5	13
250	4.37	28	12	18
300	5.25	28	12	18
500	8.75	26	14	28
750	13.12	23	20	38
1000	17.5	22	24	48
1500	26.24	21	29	58
2000	34.99	21	29	58
11 kV side with horn gap of 8 inch or 205 mm				
16	0.876	37	3.5	6
25	1.31	35	5	8
50	2.62	35	5	8
63	3.31	35	5	8
100	5.25	30	8.5	13
163	8.15	25	15	29
250	13.12	23	20	38
500	26.24	20	34	70
400 V side fuse size				
16	21.33	20	34	70
25	33.33	17	65	130
50	66.5	14	102	230
63	84	13	130	295
100	133.33			
163	217			
250	333			
500	666			

WEATHERPROOF ALUMINIUM WIRE CONNECTION FOR SERVICE CONNECTION (As Per IS 694–Part II)

Type of connection	Wires		
	Type	**Phase**	**Neutral**
1 Ph, 5 A, 240 V up to 2 kW	Twin core	$(1/1.4 \text{ mm})/1.5 \text{ mm}^2$	$(1/1.4 \text{ mm})/1.5 \text{ mm}^2$
1 Ph, 10 A, 240 V up to 4 kW	Twin core	$(1/1.8 \text{ mm})/2.5 \text{ mm}^2$	$(1/1.8 \text{ mm})/2.5 \text{ mm}^2$
1 Ph, 20 A, 240 V up to 10 kW	Twin core	$(1/3.55 \text{ m})/10 \text{ mm}^2$	$(1/3.55 \text{ mm})$
3 Ph, 10 A, 440 V 3 HP (2.24 kW)	Single core	$(1/1.8 \text{ mm})/2.5 \text{ mm}^2$	$(1/1.8 \text{ mm})$
3 Ph, 15 A, 440 V 6 HP (4.5 kW)	Single core	$(1/1.24 \text{ mm})/4 \text{ mm}^2$	$(1/1.24 \text{ mm})$
3 Ph, 25 A, 440 V 6 to 15 HP	Single core	$(1/3.55 \text{ mm})/10 \text{ mm}^2$	$(1/3.55 \text{ mm})$
3 Ph, 50 A, 440 V 15 to 35 HP	Single core	7/3 mm	7/3 mm
3 Ph, 100 A, 440 V 35 to 70 HP	Single core	19/2.44 mm	7/3 mm

TRANSFORMER CABLES

Type/size in mm^2	Rating of transformer
16 mm^2 four core, (unarmoured) or 10 mm^2 , 4-single core unarmoured IS 1554 (Pt-2)	25 kVA
70 mm^2 four core, (unarmoured), with reduced neutral or 50 mm^2, 4-single core (unarmoured), IS 1554 (Pt-1) or 70 mm^2, 3 core solid sector shaped aluminum armoured IS 4288	63 kVA
120 mm^2 four core, (unarmoured), with reduced neutral or 95 mm^2, 4-single core (unarmoured), IS 1554 (Pt-1) or 70 mm^2, 3 core solid sector-shaped aluminum armoured IS 4288	100 kVA

11 kV VOLTAGE CLASS DISTRIBUTION TRANSFORMER

TRF rating (kVA)	Iron loss	Copper loss at full load	Copper loss at 75% full load
10	0.0365	0.35	0.197
15	0.085	0.40	0.213
25	0.130	0.65	0.366
50	0.195	0.92	0.52
75	0.28	1.23	0.68
100	0.31	1.7	0.965
150	0.50	2.1	1.18
200	0.615	2.6	1.46
250	0.720	3.4	1.90
300	0.850	3.9	2.19
500	1.28	5.8	3.25

PVC FLEXIBLE HOUSE WIRING, WIRES WITH COPPER CONDUCTOR SINGLE CORE 1.1 kV GRADE AS PER IS : 694/1990

Nominal area (mm^2)	No. and size (mm)	Current capacity
1.0	14/0.3	11
1.5	22/0.3	14
2.5	36/0.3	19
4	56/0.3	26
6	85/0.3	33

CORE CABLES WITH SINGLE-STRAND SOLID COPPER, 1100 V GRADE, AS PER IS : 694/1990

Nominal area (mm^2)	No. and size (mm)	Current capacity
1.5	1/1.4	15
2.5	1/1.8	20
4	1/2.24	27
6	1/2.80	34

RELAY COORDINATION CHART/
PROCEDURE FOR O/C AND E/F RELAY SETTING

1. Data required
   - 3 phase fault MVA for O/C relay setting
   - Single line to ground fault MVA for E/F relay
   - CTR used
   - Relay characteristics
2. Standard data to be used
   - Operating time can be selected as follows.

System voltage (kV)	Operating time (seconds)
11	0.25
33	0.5
132	1.0
220	1.25

Note   The selection and choice of the operating time of the relays depend upon the type of protection scheme used in the system and type of radial system and type of relay used in the system.

- PSM of O/C relay

   It is to be selected according to the load required by the system and thermal characteristics of the line conductor.

   Standard allowable over loading factor is 1.25 to 1.5.

- PSM of E/F relay

   It is to be selected according to the residual current and allowable unbalanced current in the a system.

Procedure of Calculation with a Typical Example of an O/C Relay

Example 1

- Suppose 3 phase fault MVA = 1000 MVA
- Single line to ground fault MVA = 750 MVA
- CTR used = 500/1
- 3 sec. of IDMT characteristics is in use

$$OT \quad \frac{3 \;\; TDS}{\log_{10} CM}$$

where,

OT = Operating time in sec.

CM = Current multiplier = (secondary current/PSM)

PSM = Plug setting multiplier

TDS = Time dial setting = Time setting multiplier (TSM)

⊙ System voltage = 33 kV
⊙ Suppose operating time = 0.5 sec.
⊙ PSM of the O/C relay = 0.75 (75%)

Step 1 (Calculation of CM)

Secondary fault current = (Primary fault current/CTR)

Primary fault current (O/C) = (3 phase fault current $\times 10^3$)/($\sqrt{3}$ × system voltage in kV)

$$= 1000 \times 10^3 / \sqrt{3} \times 33 = 17495.5 \text{ A}$$

So, secondary fault current = 17495.5/(500/1) = 34.99

$$CM = 34.99/0.75 = 46.65$$

Step 2 (Calculation of TDS)

$$OT = \frac{3 \ TDS}{\log_{10} CM} \quad 0.5 = \frac{3 \ TDS}{\log_{10} 46.65}$$

$TDS = 0.278 \approx 0.3$

So, final PSM = 0.75, TSM= 0.3

Note For E/F relay setting SLG fault MVA should be considered.

Example 2

Consider a radial system feeders from a 33/11 kV substation as shown below.

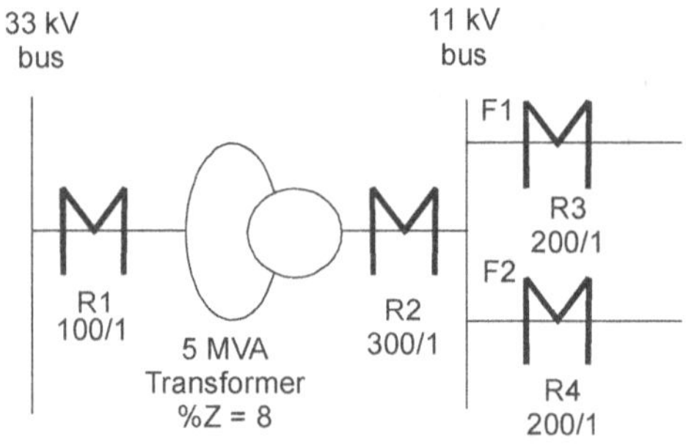

Suppose

- ⊙ Operating time for relays at R_3 and R_4 = 0.25 sec.
- ⊙ PSM of the relays = 1 (100%)
- ⊙ 3 secs. of IDMT characteristics is in use

Calculation for relays (R_3 and R_4)

Calculation on the basis of transformer rating to be taken.

So, fault MVA = (Transformer rating/pu reactance) with assumption of zero source impedance.

Fault MVA = 5/8 % = (5 × 100)/8 = 62.5

1. CM = 16.401(refer step 1)
2. TSM = 0.101 (refer step 2)
   So, PSM = 100 %, TSM = 0.1 (nearest value).

Calculation for relay R_2

Here the concept of Discrimination Time has to be selected.

Assume 1. Discrimination Time (DT) between primary and backup = 0.4 sec[#] and 2. Discrimination Time (DT) for the same feeder = 0.05[@] sec.

So, operating time of the relay (R_2) = (0.25 + 0.4[#]) = 0.65 sec.

suppose the PSM = 0.75

1. CM = 14.58 (refer step 1), 2. TSM = 0.252 (refer step 2) so, PSM = 75%, TSM = 0.25 (nearest value)

Calculation for relay R_1

Suppose PSM = 100%, OT of the relay (R_1) = (0.65 + 0.05[@]) = 0.70 sec.

1. CM = 10.935 (refer step 1), 2. TSM = 0.242 (refer step 2)
   So, PSM = 75% , TSM = 0.25 (nearest value)

[#]   Main protection and back-up protection relay are to be operated with time gap. This gap is selected as 0.4 sec.

[@]   Simillarly for same feeder case it is chosen as 0.05 sec.

ESTIMATED CURRENT RATING FOR COPPER AND ALUMINIUM CONDUCTORS

(VIR, PVC or Polythene Insulated Cables) for Single, Twin, Three and Four Cores(as Per IS 694/1990 AND BS 2004 / 1961)

Standard copper conductor			Continuous current rating in A (at 30°C)					Standard aluminium conductor	
Area (sq. mm)	Strand (mm)	SWG code	2 SC cable in conduit through or casing	Bunched in free air or open trench 2 SC cable	3 or 4 SC cable	One twin core	One 3/4 core for 3 phase	Area (sq.mm)	Strand (mm)
1.00	1/1.12	1/18	5	5	5	5	5	–	–
1.25	3/0.737	3/22	10	10	10/9	10	8/7	1.5	1/1.40
2.00	3/0.914	3/20	15	15	13/12	15	11/10	2.5	1/1.80
3.00	7/0.737	7/22	20	20	15	20	15	–	–
–	–		20	20	17	20	14	4.0	1 /2.24
–	–		27	27	24	27	19	5.0	1 /2.80
4.5	7/0.914	7/20	28	28	25	28	20	–	–
–	–		34	34	31	34	24	6.00	1/3.55
6.75	7/1.12	7/18	36	36	32	36	25	–	–
10.00	7/1.35	7/17	43	43	39/38	43	30	16	7/1.70
14.00	7/1.65	7/16	53	53	48	53	37	–	–
–	–		59	59	54	59	42	25	7/2.24
18.00	19/1.15	19/18	62	62	46	62	43	–	–
–	–		69	69	62	69	48	35	7/2.50
25.00	19/1.32	19/17	74	74	67	74	52	–	–
–	–		91	91	82	91	62	50	7/3.00, 19/1.80

(*Contd.*)

Standard copper conductor			Continuous current rating in A (at 30° c)					Standard aluminium conductor	
Area (sq. mm)	Strand (mm)	SWG code	2 SC cable in conduit through or casing	Bunched in free air or open trench		One twin core	One 3/4 core for 3 phase	Area (sq. mm)	Strand (mm)
				2 SC cable	3 or 4 SC cable				
40.00	19/1.65	19/16	–	97	88	97	68	–	–
50.00	19/1.83	19/15	–	125	107	115	78	–	–
–	–		–	134	118	118	82	70	19/2.24
–	–		–	143	138	135	94	95	19/2.50
65.00	19/2.11	19/14	–	160	140	140	98	–	–
70.00	37/1.65	37/16	–	177	158	158	109	–	–
–	–		–	184	170	162	114	120	37/2.06
95.00	37/1.85	37/15	–	205	185	180	126	–	–
–	–		–	210	185	181	127	150	37/2.23
–	–		–	246	216	209	146	185	37/2.50
130.00	37/2.15	37/14	–	250	220	218	153	–	–
–	–		–	290	248	240	169	225	37/2.80
160.00	37/2.36	37/13	–	293	260	252	178	–	–
200.00	37/2.62	37/12	–	335	295	284	199	–	–
–	–		–	364	302	281	202	300	61/2.50
260.00	61/2.36	61/13	–	425	360	342	240	–	–
–	–		–	435	372	–	–	400	61/3.33
325.00	61/2.62	61/12	–	480	410	–	–	–	–
–	–		–	480	411	–	–	500	91/2.65
–	–		–	565	484	–	–	625	61/3.00
500.00	91/2.62	91/12	–	610	620	–	–	–	–
670.00	127/2.62	127/12	–	704	630	–	–	–	–

DATA ON SUB-STATION

BREAKING CURRENT AND FAULT MVA CAPACITY

Voltage (kV)	Breaking current in kA	Fault MVA
132	25/31	150
220	31.5/40	320
400	40	1000
765	40	2500

NORMAL BUS BAR SIZE, BAY LENGTH AND BAY WIDTH

Voltage	Bus bar size		Bay width (mm)	Bay length (mm)
	Strung bus	Rigid bus		
33 kV	50 mm² copper equivalent	40 mm Al pipe	5550	28350
132 kV	61/3.18 mm conductor	80 mm Al pipe	10400	44300
220 kV	61/3.18 mm conductor duplex	80 mm Al pipe duplex	17000	63400

TUBULAR BUS AND STRINGING BUS ARRANGEMENT

	Tubular bus		Strain bus	
Voltage (kV)	Nominal diameter in mm		Voltage (kV)	Bus conductor
	External	Internal		
72.5, 145	42	35	72.5	Lynx ACSR
	60	52		Kundah ACSR
	60	49.25		19/3.53 AAC
245	89	78	145	Panther ACSR
	89	74		Kundah ACSR
	101.6	90.1		19/4.22 AAC
	101.6	85.4	245	Moose ACSR
420	114.3	102.3		Kundah ACSR
	114.3	97.2		19/5.36 AAC
	114.3	102.3		37/5.23 AAC
	114.3	97.2	420	Moose ACSR
	127	114.5		
	127	109		

CONDUCTOR TENSION IN kgf/CONDUCTOR TO BE TAKEN FOR BUS ARRANGEMENT

Arrangements	110/132 kV	220 kV	400 kV
Line termination	1000	1000	2000
Main bus/sub-bus	800	900	1000
Interconnection between yards	800	900	1000
Earth wires	600	600	800

VARIOUS FORMULAE USED FOR THE DESIGN OF BUS ARRANGEMENT

Bus Capacity Load Current

$$I \quad K \quad A\sqrt{1/t} \quad 10^4$$

where,

I = Symmetrical RMS current in A
A = Cross-sectional area in inches
t = Time in secs.
K = Coefficient of alloys at maximum temperatures

Max. temp (°C)	Value of K for different aluminium alloys
200	5.50 to 5.71
250	6.28 to 6.52
300	6.94 to 7.18

Note   For general practice the temperature of rigid aluminium bus to 100° C and for emergency rating 250° C for short-circuit duty.

Vibration in Bus

$f = (K^2/24L^2) \times (E_i m/M)^{1/2}$
where,

f = Natural frequency of span in Hz
L = Span length in feet
E_i = Modulus of elasticity
m = Moment of inertia
M = Mass per unit length

Short-Circuit Force in Bus

For line to line fault

$$F_{SC} = \frac{37.4}{10^7} \frac{I^2_{sc}}{D}$$

where,

F_{sc} = SC force in lb/in^2

I_{sc} = Symmetrical RMS SC current (A)

D = Conductor spacing centre to centre (inch)

For 3 phase fault

$$F_{SC} = \frac{43.2}{10^7} \frac{I^2_{sc}}{D}$$

EHV SUB-STATION SYSTEM TECHNICAL PARTICULARS

Description of technical particulars	Unit	System voltages			
Normal system voltages	kV RMS	33	132	220	400
Max. system voltages	kV RMS	36	145	245	420
Power frequency withstand voltage	kV RMS	70	275	460	520
					630
Switching surge withstand voltage (for 250/2500 ms)					
1. Line to earth	kV$_p$	Not	Not	Not	1050
2. Across isolating gap		applicable	applicable	applicable	900 + 345 kV RMS
Lightning impulse withstand voltage					
1. Line to earth	kV$_p$	170	650	1050	1425
2. Across isolating gap	1.2/50 S)	195	750	1200	1425 + 240 RMS
Power frequency withstand (1 min.)					
Dry	kV RMS	70	275	460	520
Wet		80	315	530	610
Frequency	Hz			50	
Vibration frequency	%			2.5	
Corona extinction Voltage	kV	–	84	156	320
Radiation interface voltage		–	1000 V at 93 kV	1000 V at 167 kV	1000 V at 266 kV

Description of technical particulars	Unit	System voltages			
System neutral rating		Solidly earthed			
Continuous current rating	A	600	800	1600	1600–2000
Sym. Short-circuit current	kA	25	31.5	40	40
Duration of short-circuit current	sec.	3	1	1	1
Dynamic short-circuit current	kAp	62.5	79	100	100
Conductor spacing for AIS layout					
Phase to ground	m	1.5	3	4.5	6.5
Phase to phase	m	1.5	3	4.5	7.0
Design temp.	°C			50	
Pollution level as per IEC 815 & 71				III	
Creepage distance	mm	900	3625	6125	10500
Max. fault clearing time	msec.	150	100	100	100
Bay width	m	5.5	10.4–12	16.4–18	27
Height of bus equip. connection from ground	m	4	5	5.5	8
Height of strung bus bar	m	5.5	8	10	>15

LIGHTNING AT DIFFERENT PLACES FOR SUB-STATION

Sl	Places	Illumination level (lux)
1.	Switchyard	25
2.	Control room	300–500
3.	Carrier room, LT panels, charger room, offices, conference room, rest room, workshop, repair bay, etc.	300
4.	Battery room, corridor, PLCC room, toilets, store room, cloak room, stairs, etc.	100

ELECTRICAL STANDARDS

GENERAL INDIAN STANDARDS USED

Particulars	Indian standards	Particulars	Indian standards
Battery	1651/1979	Disconnecting switch	1818/1972
Insulation coordination	2165 (Pt–1 & 2)	AC circuit breakers	2516 (Pt 1 to 5)
Voltage transformers	3156/1978 (Pt 1 to 4)	Lightning arresters	3070/1974 (Pt 1 & 2)
Porcelain insulators	731/1971	Bushings	2099/1973
Transformers	2026/1977 (Pt-1to 4)	Current transformer	2705/1981 (Pt 1 to 4)
Shunt reactors	5553/1970		

VOLTAGE STANDARD

Statutory Limits as per Section of IE Rules 1956 and as per IEGC

	As per IE rule 1956			As per IEGC		
Voltage	Ranges	Limits		Voltage	max.	min.
Low voltage	Voltage < 250 V	6%		132 kV	145 kV (+10%)	120 kV (–10%)
Medium voltage	650 V > Voltage > 250 V	6%		220 kV	245 kV (+10%)	200 kV (–10%)
High voltage	33 kV > Voltage > 650 V	6% to 9%		400 kV	420 kV (+5 %)	360 kV (–10%)
EHV	Voltage > 33 kV	+ 10% to –12.5%				

OTHER STANDARDS

Indian standards	Title	International standards
A) Transmission lines		
IS 226-1975	Structural steel (standard quality)	ISO/R/630-1967 CAN/CSA G40.2 BSEN 10025
IS 269-1976	Ordinary rapid hardening and low heat Portland cement	ISO/R/597-1967
IS 383-1970	Course and fine aggregates from natural sources for concrete	CSA A23.1/A23.2
a) IS 398-1982 Part-I	Specification for aluminum conductors for overhead transmission purposes	ICE 1089-1991 BS 215-1970
b) IS 398-1982 Part-II	Aluminum conductor galvanized steel reinforced	BS 215-1970 ICE 1089-1991
c) IS 398-1982 Part-IV	Aluminum alloy standard conductor	BS 3242-1970 ICE 1089-1991 ASTMB399M86
d) IS 398-1982 Part-V	Aluminum conductor galvanized steel reinforced for extra high voltage (400 kV and above)	BS 215-1970 IEC 1089-1991
IS 278-1978	Specification for barbed wire	ASTM A 121
IS 456-1978	Code of practice f or plain and reinforced concrete	ISO 3893-977
IS 731-1971	Porcelain insulators for overhead power lines with nominal voltage greater than 1000 volts	BS 137-1982 (Part-I&II) IEC 383-1993 (Part I & II)
a. IS 802-1995 (Part-I/Sec.I) (Part –I/ Sec.II)-1992	Code of practice for use of structural steel in overhead transmission line: materials, loads and permissible stresses	IEC 826 ANSI/ASCE 10-90 (1991) BS 8100
b. IS 802-1978 (Part-II)	Code of practice for use of structural steel in overhead trans. line fabrication, galvanizing, inspection and packing	ANSI/ASCE 10-90 (1991)

(Contd.)

Indian standards	Title	International standards
c. IS 802-1978 (Part-III)	Code of practice for use of strut. steel in overhead trans. line towers: testing	ANSI/ASCE 10-90 (1991) IEC 652
IS 1778-1980	Reels and drums for bare conductors	BS 1559-1949
IS 1893-1965	Criteria of earthquake-resistant design of structures	IEEE 693
IS 2016-1967	Plain washers	
IS 2121	Specification for conductor and earth wire accessories for overhead power lines	ISO/R 887-1968 ANSI B 18.22.1
a) Part-I 1981	Armour rods, binding wires and tapes for conductors	
b) Part-II 1981	Mid-span joints and repair sleeve for conductors	
c) Part-III 1992	Accessories for earth wire	
d) Part-IV 1991	Non-extension joints	
IS 2486	Specification for insulator fittings for overhead power lines with nominal voltage greater than 1000 V	
Part-I	General requirements and tests	BS 3288 IEC 1284
Part-II	Dimensional requirements	IEC 120- 1984
Part-III	Locking devices	IEC 372-1984
IS 1367-1967	Technical supply conditions for threaded fasteners	
IS 1521-1972	Method of tensile testing of steel wires	ISO 6892-1984
IS 3043-1972	Code of practice for earthing (with amendment no. 1 and 2)	

(Contd.)

IS 3188-1965	Dimensions for disc insulators	IEC 305-1978
IS 4091-1967	Code of practice for design and construction of foundation for transmission line towers and poles	ANSI/ASCE 10-90 (1991)
IS 5613 (Part-II/Sec-1)-1985 (Part-III /Sec1)-1989	Code of practice for design, install. and maintain of overhead power lines (section 1 : design)	
IS 5613	Code of practice for design, installation and maintenance of overhead power lines	
(Part-II/Sec-2)-1985	(Section 2 installation and maintenance)	
1985(Part-III/Sec.2)-1989		
IS 6639-1972	Hexagonal bolts for steel structure	ISO/R 272-1968 ASTM A 394 CSA B 33.4
IS 6745-1972	Methods for determination of wt. of Zn. Coating of zinc-coated iron and steel articles	ASTM A90 ISO 1460
IS 1363	Hexagonal head bolts, screws and nuts of product grade C	
IS 2633-1972	Testing uniformity of coating of zinc coated	ASTM A 123 CAN/CSA
IS 1367	Tech. spec. condns. for threaded steel fasteners	

(Contd.)

Indian standards	Title	International standards
Part III	Mechanical properties and test methods for bolts, screws and studs with full load	ISO 898-1
Part VI	Mechanical properties and test methods for nuts with full load ability	ISO/DIS 898/II
	Indian electricity rules 1956	
	Indian electricity act 1910	
IS 1888-1982	Method of load test on soils	
IS 2911-1979 (Part-I)	Code of practice for design and construction of pile foundations.	
IS 2629-1966	Recommended practice for hot dip galvanizing of iron and steel.	ASTM A 123 CAN/CSA G164 BS729
IS 14000-1994	Quality management and quality assurance standards	ISO 9000-1994
IS 2131 1967	Method of standard penetration test for soils ASTM D 1883	
	B) Power transformers	
IS 1885 (Part 38)1977	Electro technical vocabulary of transformer	
IS 2026 (Part 1)1977	Power transformer – general	
IS 2026 (Part 2)1977	Power transformer – temperature rise	
IS 2026 (Part 3)1981	Power transformer – insulation level and dielectric test	
IS 2026 (Part 4)1977	Power transformer – terminal marking, tapping and connection	
IS 2026 (Part 5)1994	Transformer/reactor bushings min. clearance in air specification	

(Contd.)

IS 6600 /1972	Guide for loading of oil immersed transformer
IS 10561/1983	Application guide for power transformers
IS 11171/1985	Dry type transformer
IS 10028 (Part 1) 1985	Code of practice for insulation and maintenance of transformer
IS 10028 (Part 2) 1985	Installation
IS 10028 (Part 3) 1985	Maintenance

C) Instrument transformer

IS 1885(Part 28) 1993	Electrotechnical vocabulary of instrument transformer
IS 2705 (Part 1)1992	Current transformer – general requirement
IS 2705 (Part 2)1992	Current transformer – measuring CT
IS 2705 (Part 3)1992	Current transformer – protective CT
IS 2705 (Part 4)1992	Current transformer – protective CT special purpose
IS 3156 (Part 1) 1992	Voltage transformer – general requirement
IS 3156 (Part 2)1992	Voltage transformer – measuring VT
IS 3156 (Part 3) 1992	Voltage transformer – protection VT
IS 3156 (Part 4) 1992	Voltage transformer – capacitive VT
IS 4146/1983	Application guide for VT
IS 4201/1983	Application guide for CT
IS 5547 /1983	Application guide for capacitive voltage transformers

(Contd.)

Indian standards	Title
D) Circuit breakers	
IEC 56 :1987/IS 13118:1991	High voltage AC circuit breaker
IEC- 376:1971/ IS 13072 :1992	Specification and acceptance of new SF_6
IEC –427 : 1973/ IS 13516 :1993	Report on synthetic testing of high voltage AC circuit breaker
IEC- 694 :1980	Common clauses for HV control gears
IS 1885 (Pt 17)/1979	Electrotechnical vocabulary of switchgear and control gear
IS 12032 (Pt –7)/987	Graphical symbol for diagrams in the field of electro technology. Pt-7 Switchgear, control gears and protective devices
IS 2516 (Pt-1/Sec1) Pt 1 & 2/sec 1) :1985	Circuit breakers Pt 1 & 2 requirements and tests. section 1 voltages not exceeding 1000 V AC or 1200 V DC
IS 12729/1988	General requirements for switchgear and control gear exceeding 1000 V AC
IS 7567: 1982	Specification for automatic reclosing circuit breakers for AC distribution
IS 9135: 1979	Guide for testing of circuit breaker with respect to our phase switching
IS 10118 (Pt 1 to 4 :1982)	Code of practice for installation and maintenance of switchgear and control gear
E) Transformer oil and oil testing	
IS 335 : 993	New insulating oil
IS 1448 (Pt–10) 1970	Test method for petroleum products by cloud point and pour point
IS 1448 (Pt–16) 1970	Density of crude petroleum, liquid petroleum products by hydrometer method
IS 1448 (Pt–21) 1970	Flash point by Pensky–Martens apparatus

(Contd.)

IS 1448 (Pt – 25) 1970	Determination of kinematics and dynamic viscosity
IS 1886 :1983	Code of practice for maintenance and supervision of oil in service
IS 6103:1971	Test method of sp. resistance
IS 12463 :1988	Inhibited mineral insulating oil
IS 2362 : 1973	Determination of water content by Karl Fischer method
IS 6104 / 1971	Test method of IFT of oil against water by ring method
IS 6262 :1971	Test method for power factor and dielectric constant of insulating materials
IS 6792:1992	Method of determination of electric strength
IS 6855 :1973	Method of sampling for liquid dielectric
IS 9434 :1992	Guide for sampling and analysis of free and dissolved gases and oil from oil-filled equipment
IS 10593 :1992	Method of evaluating analysis of free and dissolved gases and oil from oil-filled equipment.
IS 13567 :1992	Methods of test for determination of water in insulating liquids and in oil-impregnated paper and press board by automatic coulomatric Karl Fischer titration method
IS 13631 :1992	Method of test for determination of antioxidant additives in insulating oil
IS 12177: 1987	Method of test for oxidative ageing of electrical insulation of petroleum oils by open-beaker method

(Contd.)

Indian standards	Title
	F) Isolators
IEC 129	AC current isolators (disconnectors and earth switches)
IS : 3921 Pt –1/1981	AC current isolators (disconnectors and earth switches for voltages above 1kV (general and definitions)
IS : 3921 Pt –2/1982	-do- (rating)
IS : 3921 Pt –3/1982	AC current isolators (disconnectors and earth switches for voltages above 1kV (general and definitions) construction
IS : 3921 Pt –4/1985	”(type test and routine test)
IS : 3921 Pt –5/1985	” Information for tender order and inquires
	G) Surge arresters
IEC: 99-4	Specification for surge arresters without gap for AC system
IEC: 99-1	Non-linear resistor type gapped arresters for AC systems
IEC 270	Partial discharge measurement
IS : 3070 (Pt-3)/1993	LA for AC systems specification for metal-oxide LA without gap

EXTRACTS FROM INDIAN ELECTRICITY RULES 1956

Rules	Title	Rules	Title
2 (av)	Definition of voltage	76	Maximum stresses and factor of safety in overhead lines
50	Supply and use of energy	77	Clearance above ground of the lowest conductor
54	Declared voltage of supply to consumer	79	Clearances from buildings of low and medium voltage lines and service line
55	Declared frequency of supply to consumer	80	Clearances from buildings for HT, EHT lines
56	Sealing of meters and cut-outs	83	General clearances
57	Meters, maximum demand indicators and other apparatus on consumers' premises	84	Routes, proximity to aerodrome
58	Point of commencement of supply	85	Maximum interval between supports
74	Materials and strength in overhead lines	87	Line crossing or approaching each other
75	Joints in overhead lines		

DATA ON POWER TELECOMMUNICATION

PLCC Technical Data (Nominal Value)

Data	Value
DC supply required for carrier set	48 V
AC current required for single carrier	1 A
Characteristic impedance of co-axial cable	75/125 ohm
Capacitance of CC/CVT	4400 pF (220 kV) 5575/6000 pF (132 kV)
Inductance of wave trap	0.5 mH (132 kV) 1 mH (220 kkV)

(Contd.)

Data	Value
Line current capacity of wave trap	630/800/1200 A
Nominal Tx at co-axial termination (By SLM)*	+ 4 to +12 dBM
Nominal Tx at co-axial termination (By SLM)*	–26 to 0 dBM

* Measurement procedure—Tune SLM to Tx/Rx frequency + pilot frequency of that carrier type¥

Local loop Using dummy load in place of hybrid print, local Tx/Rx can be looped back for ensuring healthiness of local carrier set

¥Different Types of Carrier and Pilot Frequency

Carrier type	Frequency (Hz)
ETI	3600
ETL	3780
BPL	3570
PUNCOM	3923

TELEPROTECTION BASIC SYSTEM

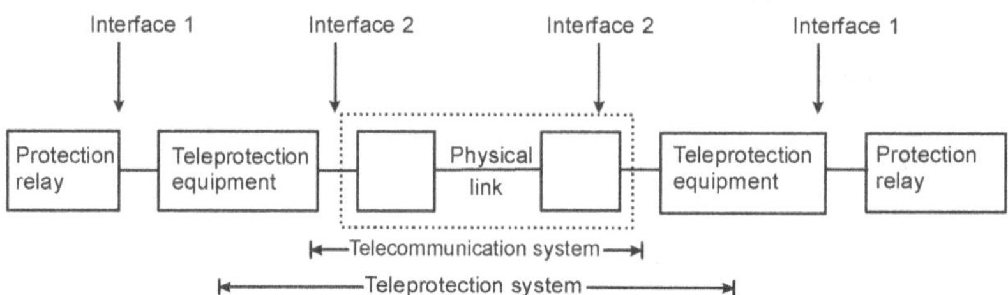

Interface 1: Wetted contact/relay

Interface 2: Voice frequency => "analog channel"

Digital data   => "digital channel"

POWER ALLOCATION TO SPEECH AND VFT CHANNEL

AF signal	Signal levels		Absolute voltage levels at the test sockets TXAF (–10dBr)
	dBmO	Weighting	dBu
Speech test tone 800 Hz Internal test tone 1000 Hz	0	1.0	–10
Speech with safety margin	+3	1.41	–7
VFT channels and modems			
50 Bd	–14	0.2	–24
100 Bd	–11	0.28	–21
200 Bd	–8	0.4	–18
600 Bd	–3	0.71	–13
1200 Bd,V.23	0	1.0	–10
1200 Bd+speech	–3	0.71	–13
2400Bd	0	1.0	–10
Maximum permissible load with speech plus superimposed channels	+10.8	3.48	+0.8
Pilot tone	–6	0.5	–16

BASIC TERMS TO PLCC

Terms	Formula	Unit	Reference
Absolute power level(L)	10 log (Px)/1 mW	dBm	$P_0 = 1$ mW
Absolute voltage level (L_u)	20 log (Ux)/775 mV	dBu	$U_0 = 775$ mV
Relative level (L_{rel})	Magnitude of difference with respect to virtual reference point (0 dBr)	dBr	Reference = 0 dBr
Absolute signal level (L_0)	Magnitude of difference with respect to virtual reference point (0dBr)	dBmO	Reference = 0 dBr
Conversion to another system impedance L_u	$*L_u = $ L (dBm) – 10 log (600 ohm/Z ohm)		

* **Example** The power level at 75   RF O/P of a PLC equipment is given as 40 dBm. Then the voltage level L_u would be L_u= 40 (dBm) –10 log (600 ohm/75   ) = 31 dBu

TYPES OF DIFFERENT CHANNEL

Channels	Data
PLCC	1. 50–450 kHz frequency for power sector 2. 50–150 kHz freely available to power sector 3. Rest with permission of DOT (Dept. of telecomm) and WPC (wireless planning coordination) 4. 500 ± 5 kHz used for international distress calling
Microwave/VHF radio link	1. Power system network with PCM (pulse code modulation), multi channel digital circuits based on multiple of 30 channels PCM multiplex operating at 2048 kpbs 2. The modulation method is either of two-level frequency shift keying (FSK) or four-level 3. WPC has assigned 2.30 to 2.50 GHz and 8.3 to 8.5 GHz bands to power sectors for narrow band network. Other frequency band is 7.11 to 7.125 GHz, 7.725 to 7.8 GHz and 10.5 to 10.68 GHz
Satellite links	SCPC technique is used with 64 kbps PCM or 32/16 kbps delta modulation or 9.6/16 K voice coding
Fiber optics	1. Bandwith of 100 GHz per km with repeater span of 50 km is used with mono mode fiber 4 to 24 cores as per CCITT Rec. G652 having 1300 nm wave length with attenuation figure of 0.5 db per km 30 channels (primary multiplex) – 2 Mbps 120 channels (2nd order multiplex) – 8 Mbps 480 channels (3rd order multiplex) – 34 Mbps 1920 channels (4th order multiplex) – 140Mbps

OTHER DATA RELATED TO TELECOMMUNICATION

Data	Value
Communication channel width	0–4 kHz
Unused (vacant) width	0–300 Hz (To avoid 6th harmonics)
Speech channel	300–2200 Hz
Pilot frequency dialing purpose	3600 30 Hz
Centre frequency for healthiness test	3600 Hz
For VFT band	2200–3570 Hz
Frequency shift keying (FSK)	50, 100, 200 baud as per CCITT recommendation
Data transmission	i. 1200 baud rate FSK to CCITT Rec.V23 on a basic 4-wire, 4 kHz PLC ii. 2400 baud phase modulated 4-wire leased telephone type circuit as CCITT Rec.V29 iii. PCM 64 kbps digital data channel to CCITT Rec.C702 iv. For FAX, CCITT Rec.T3 used with vestigial side band modulation (BW 800–2600 Hz)
Leased circuits	i. 2- wire audio cables (up to 20 km) ii. CCITT M1 020 and M1 040 types of circuit (similar to PLC and microwave system) iii. Wide spectrum signals on either group (60 to 108 kHz) or super group (312 to 520 kHz) as per CCITT Rec. M900 and M910 iv. Satellite leased circuit such as SCPC, FM modulated, analogous
Minimum clearances power and telecom line	i. LT lines (230/400 V)—1.22 m ii. 11 kV lines—1.83 m iii. 33/66 kV lines—2.44 m iv. 132 kV lines—3.05 m v. 220 kV lines—4.58 m vi. 400 kV lines—5.49 m vii. 800 kV lines—7.94 m

CHANNEL SPECIFIC DATA WITH VARIABLE CENTRE FREQUENCIES (MODEM)

Channel (Bd)	50	100	200	200	300
Maximum baud rate (Bits/s)	75	150	225	200	300
Nominal baud rate (Bits/s)	50	100	200	200	300
ITU-T channel	R.35	R.37	R.38B	R.38A	
Channel bandwidth/spacing Hz	120	240	360	480	480
Lowest centre frequency Hz	420	480	540	600	600
Highest centre frequency Hz	3900	3840	3780	3840	3840
Frequency shift Hz	30	60	90	120	120
Maximum isochronous distortion:					
a) At nominal baud rate %	5	5	7	5	8
b) At maximum baud rate %	8	8	8	5	8
c) With receive level margin of dB	5	5	5	5	5
Regenerator on					
d) Receive distortion %<25%	3	3	3	3	3
e) <50%	5	5	5	5	5
Channel delay times:					
a) Propagation delay ms	30.5	18.3	13.5	11.4	10.5
b) Max RTS on to DCD on delay (ms)	67.0	36.0	24.0	19.0	16.0
c) Max RTS off to DCD off delay (ms)	54.0	45.0	31.0	24.0	22.0
d) RTS on to CTS on channel	60.0	30.0	15.0	15.0	10.0
dependent delay (ms)					
e) TX channel turn-off time (ms)	20.0	12.5	7.5	5.6	4.4
(for half-duplex operation)					
Maximum 2-wire attenuation:					
a) No frequency gap	20	30	25	30	25
b) 120 Hz frequency gap	35	38	35	38	38
with own channel & <15% dist.					
Maximum 4-wire attenuation: dB	40	40	40	40	40
Signal to noise ratio:					
a) For bit error rate of dB 10^{-3}	12.5	12.5	11.5	12.5	11.0
10^{-4}	14.5	14.5	13.0	14.5	13.0
10^{-6}	16.5	16.5	15.5	16.5	15.5
b) Measured with ITU-T V.52					
Noise bandwidth Hz	50	100	200	200	300

MEANINGS OF INDICATIONS FOR DIFFERENT DISTANCE RELAYS

Relay indications	Meanings	Relay indications	Meanings
English electric MR3V & RR3V			
30 (A–B)	Phase A – B shorted	30 G	Zone 1 fault
30 (B–C)	Phase B –C shorted	30 H	Zone 2 fault
30 (C–A)	Phase C– A shorted	30 J	Zone 3 fault
30 A	Phase A with earth fault	86 Y	Distance relay operated
30 B	Phase B with earth fault	186A,186B	CB lockout
30 C	Phase C with earth fault	VARM	Auto reclose operated
English electric SSM3V			
R	Phase (R–Y) shorted	Zone II	Zone 2 fault
B	Phase (Y–B) shorted	Zone III	Zone 3 fault
Zone I	Zone 1 fault	86 X	Aux. relay trip
English electric MM3V			
Same as MR3V & RR3V relay with following indications			
Y3AXX	Aux. relay for Mho start	Y3CXX	Aux. relay for Mho start
Y3BXX	Aux. relay for Mho start	85X2	Carrier receive
English electric pyts (Static)			
LED		LED	
A, B, C	Phase faults	Zone III	Zone 3 fault
Zone I	Zone 1 fault	SOTF	Switch on to fault
Zone II	Zone 2 fault	PSB	Power swing block
English electric (quadramho) SHPM			
LED		LED	
A, B, C	Phase faults	Aided trip	Carrier inter trip
Zone 2	Zone 2 fault	SOTF	Switch on to fault
Zone 3	Zone 3 fault	V ~ fail	P.T. fuse fail
English electric SR3V			
30 A	Phase A with earth fault	30 H	Zone 2 fault
30 B	Phase B with earth fault	30 J	Zone 3 fault
30 C	Phase C with earth fault	86 AX	Distance relay operated
30 D	Short to earth	86 BX	Relay operated in remote scheme
30 G	Zone 1 fault		

Relay indications	Meanings	Relay indications	Meanings
English electric SSR3V			
ABC	Phase fault	Zone II	Zone 2 fault
AB	Phase A–B shorted	Zone III	Zone 3 fault
BC	Phase B–C shorted	186A, 186B	CB lockout
CA	Phase C–A shorted	86 X1	Aux. relay with zone 1(carr. receive)
Zone I	Zone 1 fault	85 X1	Switch on to fault
ASEA RYZOD, RYZOE, RYZFB			
UD	Distance relay operated	2	Zone 2 fault
UA	Auto reclose OP	3	Zone 3 fault
RST	Phase faults	4	Zone 4 fault
English electric M3V			
30 (A-B), 30 (B-C), 30 (C-A), 30G, 30H 30 J same AS MR3V & RR3V relay, 30 E–earth fault			
English electric MM3T			
AB	Phase A–B shorted	86A	A phase trip
BC	Phase B–C shorted	86B	B phase trip
CA	Phase C–A shorted	86C	C phase trip
Zone 2	Zone 2 fault	86T	Trip relay for phase fault
Zone 3	Zone 3 fault	97X	VT fuse fail
85S	Carrier end operated	2	Zone 2 timer-operated
AN	Phase A with earth fault	59X	Over-voltage
BN	Phase B with earth fault	64	Backup E/F trip operated
CN	Phase C with earth fault	2L	LBB relay
85X2	Carrier receive		
ASEA RAZOG			
Same as RYZOD relay with following extra indications			
CR	Carrier receive	CS	Carrier send
ABB RAZFE			
U	General trip	3Ph	3 Ph. fault
RN	R ph. to earth fault	TK2	Zone 2 timer
SN	R ph. to earth fault	TK3	Zone 3 timer
TN	R ph. to earth fault	P	Power swing block
2Ph	Ph to Ph fault	=	DC fail
ABB RAZOA			
R	R phase fault	2	Zone 2 fault
S	S phase fault	3	Zone 3 fault
T	T phase fault	4	Zone 4 fault
N	E/F fault		

Relay indications	Meanings	Relay indications	Meanings
ABB LZ96			
D	General trip	BCL	Auto reclosing blocked
DH	Trip with signal from remote station	E	Ground fault
RST	General start	EM	Own measure
RST	Starting, phase selective	RI	Direction
T, T2...T4	Time step other than basic time	DCE	Power supply failed
BD	Blocking of trip	DEF	Relay disturbed
PSB	Ground fault	XX	Relay under test
CL	Auto reclosing	AUE	Pilot wire supervision
		FBL	Blocking protection
ABB RELZ 100			
ZM1	Zone 1 measuring	TRZ1	Trip zone 1
ZM2	Zone 2 measuring	TRZ2	Trip zone 2
ZM3	Zone 3 measuring	TRZ3	Trip zone 3
ZM3R	Zone 3 reverse	TRZ3R	Trip zone 3 reverse
PSR	Phase selection R	TRC	Carrier aided trip
PSS	Phase selection S	TRWEI	Trip weak end infeed
PST	Phase selection T	TRSOTF	Trip switch on to fault
CRZ	Carrier receive	TRSTUB	Trip stub protection
CSZ	Carrier send	TROC	Trip on O/C
ECHO	Carrier ECHO from week infeed logic	TRULOW	Trip loss of potential
PSB	Power swing block	TRVTF	Trip VT fuse fail
SUP	System supervision	TRIP3PH	Trip 3 phase
VTS	Fuse fail supervision	TRBL	Trip dir. E/F blocking
START	Start from protection	TRIPZ	Trip from distance protection
VTF	Fuse fail detection		
Brown boveri L3WYAS, L3WYS			
PAR	R Phase fault	PTrH	Carrier started
PAS	S Phase fault	PD3	Tripping initiated from internal relay
PAT	T Phase fault	ZA	Impedance start relay
PD	Distance relay operated	CM	Rotating field relay
PE	Ground fault	RLv	O/C relay for E/F
PtaE	Persistent E/F	PSA	Distance step contactor for overlap

Relay indications	Meanings	Relay indications	Meanings
PTA	3 Phase fault	U	Auxillary current transformer
PS II	Zone 2 fault	V	Balancing transformer
PS III	Zone 3 fault	W	Auto reclose switch
PtaW	Breaker trip with recloser	Wa	Manual switch for overlap
PSW	Switch over to lockout		

ABB REL 316			
Trip RST	General trip of phase	Delay 2	Zone 2 time delay
Trip CB 1P	Single-phase trip	Delay 3	Zone 3 time delay
Trip CB 3P	Three-phase trip sig.	Delay 4	Zone 4 time delay
Trip CB R	R Phase fault	Start R + S + T	General start of phase R.S.T
Trip CB S	S Phase fault	Start R	General start of phase R
Trip CB T	T Phase fault	Start S	General start of phase S
Trip CB	Gen trip of phase R.S.T	Start T	General start of phase T
Comm. Fail	PLCC Channel failure	Start E	General start of phase E/F
Com. Rx	Signal received	Start OC	General start O/C
Com Tx	Signal transmitted by PLCC	3 Ph. Trip	Always 3 phase trip
SOTF	Tr switch on to fault	VT Sup	VT supervision
Delay 1	Zone 1 time delay		

Alstom PD 521			
LED CON. ADDRESS		**LED CON. ADDRES**	
5701	LED func assgn. H1	5713	LED func assgn. H7
5703	LED func. assgn. H2	5715	LED func assgn. H8
5705	LED func assgn. H3	5717	LED func assgn. H9
5707	LED func. assgn. H4	5719	LED assgn. H10
5709	LED func assgn. H5	5721	LED assgn. H11
5711	LED func assgn. H6	5723	LED assgn. H12
Fault signals		**Fault signals**	
3600	General start	3618	Dist. fault forward
3601	Starting A	3619	Dist. fault backward
3602	Starting B	3635	Carrier send
3603	Starting C	3626	Carrier receive
3604	Starting ground fault	3627	Dist t1 elapsed
3605	General trip signal	3628	Dist t2 elapsed
3609	Dist. trip signal	3629	Dist t3 elapsed

Relay indications	Meanings	Relay indications	Meanings
3613	Backup O/C start	3670	Warning
3614	Backup O/C trip	3671	General trip
Easun reyrolle, THR 3PE–1			
P.O	Protection operated	Z2	Zone 2 fault
R, Y, B	Phase faults	Z3	Zone 3 fault
E	Earth faults	LOR	Lock out relay
Easun reyrolle, THR 10			
Ic	O/C operated	1, 2, 3	Zone 1, 2, 3
PO	Protection operated	ry, yb, br	Phase faults
Pt	Pt supply	r, y, b	Earth faults
Brown boveri LIZ 613			
R, S, T	Phase fault	T2	Zone 2 fault
E	Earth fault	T3	Zone 3 fault
T	Breaker trip	T4	Non-directional
Easun reyrolle, SEL 321, SEL321-1			
LED		**LED**	
EN	Relay in normal cond.	4	Zone 4 or level 4
INSTANT	Instantaneous tripping	A	A Phase involved
TIME	Time delay tripping	B	B Phase involved
COMM	Comm. aided tripping	C	C Phase involved
SOTF	Trip switch on to fault	G	Ground involved
1	Zone 1 or level 1	Q	Negative sequence current
2	Zone 2 or level 2	51	Time O /C
3	Zone 3 or level 3	50	O/C high set
DISPLAY		Display	
AG, BG, CG	Phase to ground fault	M1PT, M2PT, M3PT	Phase dist. time delay
AB	AB phase fault	LOP	Loss of potential
BC	BC phase fault	67Q1, 67Q2, 67Q	Dirr. neg. seq. O/C inst
CA	CA phase fault	67N1, 67N2, 67N3	Dir. E/F time delay
ABG	Two phase to ground	67Q2T, 67Q3T, 67Q4	Dirr. neg. seq. time delay
BCG	Same	3P27	3 Phase U/V
CAG	Same	50H	High set O/C, inst
ABCT	3 phase fault	50Q1	Neg. seq Non–dirr
Z1G, Z2G, Z3G	Zone Mho or Quadr.	50N1	High set E/F non-dir
M1P, M2P, M3P	Zone phase dist. inst	3PT	3 Pole trip
Z1GT, Z2GT, Z3GT	Zone dist. time delay	3P59	3 Phase O /V

Relay indications	Meanings	Relay indications	Meanings
General electric GLY			
21L (R–Y)	R–Y shorted	21GB	Phase B fault
21L (Y–B)	Y–B shorted	21GX	Timer for 21G relay
21L (B–R)	B–R shorted	85MX	Carrier tripping for phase fault
21GR	Phase R fault	85GX	Carrier tripping for ground fault
21GY	Phase Y fault	21 LX	Timer for 21L relay
ALSTOM, EPAC 3100/3500			
Dec A	Phase A tripping	\Z1	Zone 1
Dec B	Phase B tripping	Z2	Zone 2
Dec C	Phase C tripping	Z3	Zone 3
Dec M	One phase tripping	Mro	Start up
Dec f	Fuse failure tripping	Poly	Multi phase fault
Sel A	Phase A selection	Mono	One phase fault
Sel B	Phase B selection	Tele	Carrier send
Sel C	Phase C selection	Ffus	Fuse failure
Aval	Forward direction	Varc	Auto reclose blocking
Amo	Reverse direction	Dpom	Power swing detection

DIAGNOSTIC SYSTEM/MAINTENANCE SCHEDULE OF ELECTRICAL EQUIPMENT

POWER TRANSFORMER

Periodicity	Checking/ testing	Notes on checking	Actions to be taken
Pre-commissioning	1. Components checking	1. All components (OLTC, PRV, bushing others)	1. Compare with factory value
	2. Electrical testing	2. Routine test	2. ”
Hourly/ continuous	1. Temperature monitoring	1. Reading of OTI, WTI	1. Follow-up action 2. Follow-up action
	2. Electrical parameters	2. Voltage, current, loads	3. Follow-up action
	3. Oil flow (OFAF)	3. Ensure about the flow	

Periodicity	Checking/testing	Notes on checking	Actions to be taken
Daily	1. Oil levels in transformer	1. Check up to Level	1. Top up oil if required
	2. Cooling system	2. Check contacts, bearings, etc.	2. Change or replace the faulty parts
	3. Breather reagent	3. Check colour of reagent	3. Change reagents
Quarterly	1. Bushings, OTI, WTI, other equipments	1. Clean the tank, bushing and trial operation of components	1. Suitable action
	2. Oil condition in transformer	2. Dielectric strength of oil	2. Filter the oil if required.
Half-yearly	1. DGA of oil	1. Check sample of main tank oil and OLTC	Compare with initial result and follow-up action
Yearly	1. Electrical testing	1. Routine test	Compare with initial result and follow-up action
	2. PI, tan value	2. Record values	
	3. DGA	3. DGA analysis	
	4. Detailed checking of clamps, joints, etc.	4. Check	
	5. OLTC	5. OLTC checking	

CURRENT TRANSFORMER

Periodicity	Checking/testing	Actions to be taken
Daily	1. Oil leakage	Follow-up action
	2. Abnormal noise	
	3. Other visual check	
Monthly	1. Oil level	1. Fill the oil
	2. Terminal checking of marshal box	2. Tight the terminals
	3. Current reading of secondary circuits	3. Record the readings
Yearly	1. IR value	Compare with initial results and follow-up action
	2. All connections	
	3. Earth resistance	
	4. tan value	
	5. Cleaning of insulators	

CVT and PT

Periodicity	Checking/testing	Actions to be taken
Daily	1. Oil leakage 2. Abnormal noise 3. Other visual check	Follow-up action
Monthly	1. Oil level 2. Terminal checking of marshal box 3. Voltage reading of secondary circuits 4. Earthing of PLCC link 5. HF Bushing	1. Fill the oil 2. Tight the terminals 3. Record the readings 4. Check its correctness 5. Check its correctness
Yearly	1. Capacitance measurement 2. All connections 3. Earth resistance 4. Cleaning of insulator	Compare with initial results and follow-up action

BATTERY MAINTENANCE

Periodicity	Items to be checked	Periodicity	Items to be checked
Daily	1. Measure and record the pilot cell voltage, sp. gravity and electrolyte temperature 2. Battery voltage by switching off charger 3. Hourly reading of DC voltage, charger output current and trickle charge current	Monthly	1. Specific gravity, voltage of each cell and electrolyte temperature 2. Give equalizing charge 3. Switch off charger and test tripping/closing of any one feeder from battery source 4. Check all connection of battery and charger
Weekly	1. Cleaning of terminals, topping up distilled water if required 2. Check pilot cell reading and adjust the trickle charge current if required	Yearly	1. Allow condition charging

BATTERY SYMPTOMS AND REMEDIES

Symptoms	Remedies
1. Corrosion	1. Remove electrolyte 2. Pour distilled water 3. Refill with new electrolyte
2. Sulphation Under-charging Over-discharge Too strong/weak acid Too rapid discharge Short-circuit Plate exposed to air	1. Initial stage, repeat the low-rate charging and discharging 2. Later stage, special treatment should be provided
3. Shedding of active material Over-charging of plates Charging done at high rate Defective material Improper application of material on plate	1. Maintain the proper charging and discharging limits (2.4 V for charging and 1.85 V for discharge) 1200 specific gravity – (charging) 1190 specific gravity – (discharging)
4. Over-charging Excessive gassing Deterioration of active material from +ve plate Temperature rise Bending of plates	1. Reduce the charging rate and add distilled water to attain the sp. gravity of 1200.
5. Under-charging Low specific gravity Change of the colour of plate Deterioration of active material from +ve plate Bending of plates	1. Increase the charging rate to attain the specific gravity of 1200.
6. Abnormal reduction of capacity Abnormal drop of voltage Sulphur formation Development of pores in lead sponge Loss of active material Loss of electrolyte	1. Allow proper charging and discharging method 2. Contact manufacturer for replacement

(Contd.)

Symptoms	Remedies
7. Buckling or bending of plates a) Discharge at rapid rate. b) Unequal distribution of current c) Defective plates d) Direct discharge e) Expansion of plates	1. Allow proper charging and discharging method 2. Contact manufacturer for replacement
8. Reversal of negative plate a) It results due to connection of weak cell in the set, its discharge is ended before the other cells and causes the reversal of negative plate.	Use correct electrolyte and charge the cell or replace with a good one.
9. Internal discharger local action a) Gassing of the cells even under idle condition b) Abnormal gassing in charge condition c) Reduction of battery capacity	Use correct electrolyte and charge the cell or replace with a good one.
10. Hardening of – ve in air a) Oxidation and heating in air or exposed plates	Pour distilled water and proper electrolyte and cover the plates in proper manner

SURGE ARRESTER

Periodicity	Checking/testing	Actions to be taken
Daily	1. Surge counter reading 2. Leakage current reading 3. Other visual check	1. Record purpose 2. Within the green zone 3. Like crack, damage
Monthly	1. Earth resistance 2. Leakage current analysis	1. Within the limit 2. Compare results
Yearly	1. Capacitance measurement 2. All connections 3. Cleaning of insulator 4. Ammeter calibration	Compare with initial results and follow-up action

INDEX